THE HIGH LUMINOSITY
LARGE HADRON COLLIDER

The New Machine for Illuminating the Mysteries of Universe

ADVANCED SERIES ON DIRECTIONS IN HIGH ENERGY PHYSICS

ISSN: 1793-1339

Published

Vol. 1   – High Energy Electron–Positron Physics (*eds. A. Ali and P. Söding*)

Vol. 2   – Hadronic Multiparticle Production (*ed. P. Carruthers*)

Vol. 3   – CP Violation (*ed. C. Jarlskog*)

Vol. 4   – Proton–Antiproton Collider Physics (*eds. G. Altarelli and L. Di Lella*)

Vol. 5   – Perturbative QCD (*ed. A. Mueller*)

Vol. 6   – Quark–Gluon Plasma (*ed. R. C. Hwa*)

Vol. 7   – Quantum Electrodynamics (*ed. T. Kinoshita*)

Vol. 9   – Instrumentation in High Energy Physics (*ed. F. Sauli*)

Vol. 10 – Heavy Flavours (*eds. A. J. Buras and M. Lindner*)

Vol. 11 – Quantum Fields on the Computer (*ed. M. Creutz*)

Vol. 12 – Advances of Accelerator Physics and Technologies (*ed. H. Schopper*)

Vol. 13 – Perspectives on Higgs Physics (*ed. G. L. Kane*)

Vol. 14 – Precision Tests of the Standard Electroweak Model (*ed. P. Langacker*)

Vol. 15 – Heavy Flavours II (*eds. A. J. Buras and M. Lindner*)

Vol. 16 – Electroweak Symmetry Breaking and New Physics at the TeV Scale
          (*eds. T. L. Barklow, S. Dawson, H. E. Haber and J. L. Siegrist*)

Vol. 17 – Perspectives on Higgs Physics II (*ed. G. L. Kane*)

Vol. 18 – Perspectives on Supersymmetry (*ed. G. L. Kane*)

Vol. 19 – Linear Collider Physics in the New Millennium (*eds. K. Fujii, D. J. Miller and A. Soni*)

Vol. 20 – Lepton Dipole Moments (*eds. B. Lee Roberts and William J. Marciano*)

Vol. 21 – Perspectives on Supersymmetry II (*ed. G. L. Kane*)

Vol. 22 – Perspectives on String Phenomenology (*eds. B. Acharya, G. L. Kane and P. Kumar*)

Vol. 23 – 60 Years of CERN Experiments and Discoveries (*eds. H. Schopper and L. Di Lella*)

Vol. 24 – The High Luminosity Large Hadron Collider:
          The New Machine for Illuminating the Mysteries of Universe
          (*eds. O. Brüning and L. Rossi*)

Forthcoming

Vol. 8   – Standard Model, Hadron Phenomenology and Weak Decays on
          the Lattice (*ed. G. Martinelli*)

Advanced Series on
Directions in High Energy Physics — Vol. 24

THE HIGH LUMINOSITY LARGE HADRON COLLIDER

The New Machine for Illuminating the Mysteries of Universe

Editors

Oliver Brüning and **Lucio Rossi**

CERN

W⊖ World Scientific

NEW JERSEY · LONDON · SINGAPORE · BEIJING · SHANGHAI · HONG KONG · TAIPEI · CHENNAI · TOKYO

Published by

World Scientific Publishing Co. Pte. Ltd.

5 Toh Tuck Link, Singapore 596224

USA office: 27 Warren Street, Suite 401-402, Hackensack, NJ 07601

UK office: 57 Shelton Street, Covent Garden, London WC2H 9HE

Library of Congress Cataloging-in-Publication Data
The High Luminosity Large Hadron Collider : the new machine for illuminating the mysteries of universe /
edited by Oliver Brüning (CERN) and Lucio Rossi (CERN).
pages cm -- (Advanced series on directions in high energy physics ; vol. 24)
Includes bibliographical references and index.
ISBN 978-9814675468 (hardcover : alk. paper) -- ISBN 978-9814678148 (pbk. : alk. paper)
1. Large Hadron Collider (France and Switzerland). I. Brüning, O., editor. II. Rossi, Lucio, 1955–   editor.
III. Series: Advanced series on directions in high energy physics ; v. 24.
QC787.P73H54 2015
539.7'36--dc23

2015022489

British Library Cataloguing-in-Publication Data
A catalogue record for this book is available from the British Library.

Foreword

CERN management usually changes once every five years. In January 2009, the new management of CERN had three high priority tasks.

1. To repair the LHC following the damage resulting from a serious technical accident in September 2008 (this work had already been started in collaboration with the outgoing management in October 2008). It was also clear that this repair should also ensure that, in the longer term, repetition of such an accident could never occur.
2. To commission the collider and begin operation with relatively high luminosity as soon as possible to provide the possibility for the experiments to make new discoveries in particle physics.
3. To prepare the upgrades of the LHC for the longer term future.

Priority 1 was reviewed during "Chamonix" 2009 and a clear plan was proposed for the repair and subsequent operation for the period 2010–2013.

During 2009, in parallel with the repair and redesign of the defected parts of the LHC, discussions started on the future upgrades of the collider (priority 3). During 2009, several reviews were made of the existing proposals, which involved building new injectors (a new Proton Synchrotron, PS2, and a Superconducting Proton Linac, SPL), upgrading the Super Proton Synchrotron, as well as phased new insertions for the LHC interaction points.

In January 2010, the results of these discussions were globally reviewed at the "Chamonix" meeting. Following this meeting, a new injector upgrade program for the LHC was born which did not necessitate the construction of new injectors. This new proposal involved upgrading of the **existing** injectors: increasing the energy of the PS booster and faster pulsing of the PS. The new scheme was considered to be less demanding on CERN manpower and financial resources as well as being much better matched to the requirements on "useful integrated luminosity" in the detectors.

In addition, based on experience on upgrades from previous colliders (where the beam down time needed for the implementation of the upgrade was barely compensated in integrated luminosity in the years following the upgrade), it was decided not to perform the upgrade of the insertions in two phases but in a single project using the newer superconducting technologies. The development of these newer technologies also paves the way for higher energy in a future collider such as the Future Circular Collider (FCC).

For optimization of the "useful integrated luminosity", it was realized that luminosity "leveling" was needed since the luminosity lifetime was comparable with the "physics to physics" turn-round time. A new scheme emerged using crab cavities with a crossing angle at the interaction points. The luminosity could be maintained constant for several hours either by changing the crossing angle or by reducing the beam size by increasing the focusing. This new scheme allowed a very high "constant" luminosity and therefore a constant "event pile-up" and lower radiation in the detectors.

Following Chamonix 2010, two new CERN major projects were created: LHC Injectors Upgrade (LIU), and High Luminosity LHC (HL-LHC). Since then, HL-LHC has become a multi-continent project (inclusion of the USA and Japan), involving research and development of many new cutting edge technologies such as very high field superconducting magnets, crab cavities, superconducting power links, new materials for collimators, etc. These technologies are not only necessary for the HL-LHC but are critical for future High Energy Physics projects.

The successful completion of the HL-LHC is CERN's flagship project and will allow the LHC to continue data taking at fantastic and constant collision rates until the middle of the 2030s. It will also serve as a critical test bed for future HEP projects such as the CERN Future Circular Colliders (FCC) project which was first proposed in 2012 in preparation for the European HEP strategy update.[a]

Steve Myers
Former CERN Director for Accelerator & Technology

---

[a]O. Brüning, B. Goddard, M. Mangano, S. Myers, L. Rossi, E. Todesco and F. Zimmerman, "High Energy LHC", CERN-ATS-237.

Preface

The High Luminosity LHC (HL-LHC) Project was set up in 2010 by the then CERN Director for Accelerator and Technology, Dr. Steve Myers, following a change of the CERN plan resulting in a down scoping of the previous LHC injector upgrade plan and a merging of the LHC upgrade Phase I and Phase II into one unique project. To this end CERN in consortium with 15 European Institutions applied in November 2010 to the call for European funding under the 7th Framework Programme, Design Study category: the application was approved with full budget in 2011 with the name FP7 High Luminosity Large Hadron Collider Design Study (nicknamed as HiLumi LHC, Grant No. 284404).

The book gives a fairly detailed description of the project, covering both project structure, governance, physics goals, accelerator layouts and key technologies like: new generation superconducting magnets, beam deflecting superconducting cavities, new materials for collimators, new high current superconducting links, etc. The book is structured in such a way that each of the 22 contributions can be read as an independent paper, although cross-references among papers are numerous. The emphasis is more on highlighting the accelerator challenges, the machine layout and required technology advances, rather than giving complete and detailed technical descriptions. The most conventional parts on infrastructure and civil engineering are not treated as a consequence of this approach, while the interplay with the companion LIU (LHC Injector Upgrade) project and review of operation modes and challenges are discussed.

This book reflects the work done in the first part of the HiLumi LHC Design Study, from November 2010 up to the end of 2014. Meanwhile in the first part of 2015, a few fine tunings have been carried out. For instant, a small change of the operating parameters of the inner triplet quadrupole magnets and a revision of the number of collimators in the Dispersion Suppressor associated with the 11 T dipole magnets have been implemented. However, these and other minor changes, which will likely continue until the end of 2016, do not affect the substance of the layout or of the importance of the technological advances needed for the ambitious HL-LHC project that, with its 1.2 km replacement of the existing LHC infrastructure is one of the most ambitious "new accelerator" projects approved for construction during the next ten years.

Editors:
Oliver Brüning (HL-LHC Deputy Project Leader)
Lucio Rossi (HL-LHC Project Leader)

Contents

List of Authors

C. Adorisio (CERN, TE Department, Genève 23, CH-1211, Switzerland)

G. Ambrosio (Fermi National Accelerator Laboratory, Batavia, IL 60510, USA)

R. B. Appleby (University of Manchester and the Cockcroft Institute, UK)

G. Arduini (CERN, BE Department, Genève 23, CH-1211, Switzerland)

T. Baer (CERN, TE Department, Genève 23, CH-1211, Switzerland)

A. Ballarino (CERN, TE Department, Genève 23, CH-1211, Switzerland)

M. Barnes (CERN, TE Department, Genève 23, CH-1211, Switzerland)

I. Bejar (CERN, Accelerator and Technology Sector, Genève 23, CH-1211, Switzerland)

A. Bertarelli (CERN, EN Department, Genève 23, CH-1211, Switzerland)

E. Bravin (CERN, BE Department, Genève 23, CH-1211, Switzerland)

K. Brodzinski (CERN, TE Department, Genève 23, CH-1211, Switzerland)

R. Bruce (CERN, BE Department, Genève 23, CH-1211, Switzerland)

M. Brugger (CERN, TE Department, Genève 23, CH-1211, Switzerland)

O. Brüning (CERN, Accelerator and Technology Sector, Genève 23, CH-1211, Switzerland)

H. Burkhardt (CERN, BE Department, Genève 23, CH-1211, Switzerland)

J. P. Burnet (CERN, TE Department, Genève 23, CH-1211, Switzerland)

G. Burt (University of Lancaster, STFC, Lancaster, UK)

R. Calaga (CERN, BE Department, Genève 23, CH-1211, Switzerland)

F. Caspers (CERN, BE Department, Genève 23, CH-1211, Switzerland)

F. Cerutti (CERN, TE Department, Genève 23, CH-1211, Switzerland)

S. Claudet (CERN, TE Department, Genève 23, CH-1211, Switzerland)

B. Dehning (CERN, BE Department, Genève 23, CH-1211, Switzerland)

L. S. Esposito (CERN, TE Department, Genève 23, CH-1211, Switzerland)

S. Fartoukh (CERN, BE Department, Genève 23, CH-1211, Switzerland)

G. Ferlin (CERN, TE Department, Genève 23, CH-1211, Switzerland)

P. Ferracin (CERN, TE Department, Genève 23, CH-1211, Switzerland)

P. Fessia (CERN, TE Department, Genève 23, CH-1211, Switzerland)

R. Garoby (CERN, BE Department, Genève 23, CH-1211, Switzerland)

B. Goddard (CERN, TE Department, Genève 23, CH-1211, Switzerland)

E. Jensen (CERN, BE Department, Genève 23, CH-1211, Switzerland)

R. Jones (CERN, BE Department, Genève 23, CH-1211, Switzerland)

J. M. Jowett (CERN, BE Department, Genève 23, CH-1211, Switzerland)

M. Lamont (CERN, BE Department, Genève 23, CH-1211, Switzerland)

A. Lechner (CERN, EN Department, Genève 23, CH-1211, Switzerland)

T. Lefevre (CERN, BE Department, Genève 23, CH-1211, Switzerland)

J. Lettry (CERN, BE Department, Genève 23, CH-1211, Switzerland)

R. Losito (CERN, EN Department, Genève 23, CH-1211, Switzerland)

M. Mangano (CERN, PH Department, Genève 23, CH-1211, Switzerland)

D. Manglunki (CERN, TE Department, Genève 23, CH-1211, Switzerland)

V. Mertens (CERN, TE Department, Genève 23, CH-1211, Switzerland)

E. Métral (CERN, BE Department, Genève 23, CH-1211, Switzerland)

N. Mounet (CERN, BE Department, Genève 23, CH-1211, Switzerland)

T. Nakamoto (KEK, 1-1 Oho, Tsukuba, Ibaraki 305-0801, Japan)

T. Pieloni (CERN, BE Department, Genève 23, CH-1211, Switzerland)

A. Ratti (Lawrence Berkeley National Laboratory, Berkeley, CA 94720, USA)

S. Redaelli (CERN, BE Department, Genève 23, CH-1211, Switzerland)

J. M. Rifflet (CERN, TE Department, Genève 23, CH-1211, Switzerland)

S. Roesler (CERN, TE Department, Genève 23, CH-1211, Switzerland)

L. Rossi (CERN, Accelerator and Technology Sector, Genève 23, CH-1211, Switzerland)

G. Rumolo (CERN, BE Department, Genève 23, CH-1211, Switzerland)

G. L. Sabbi (Lawrence Berkeley National Laboratory, Berkeley, CA 94720, USA)

B. Salvant (CERN, BE Department, Genève 23, CH-1211, Switzerland)

M. Schaumann (CERN and RWTH Aachen University, D-52056 Aachen, Germany)

H. Schmickler (CERN, BE Department, Genève 23, CH-1211, Switzerland)

R. Schmidt (CERN, TE Department, Genève 23, CH-1211, Switzerland)

M. Segreti (CEA, Saclay, 91400, France)

L. Tavian (CERN, TE Department, Genève 23, CH-1211, Switzerland)

E. Todesco (CERN, TE Department, Genève 23, CH-1211, Switzerland)

R. Tomas (CERN, BE Department, Genève 23, CH-1211, Switzerland)

C. Urscheler (Bundesamt fuer Gesundheit, Direktionsbereich Verbraucherschutz, Zchwarzenburgstrasse 165, 3003 Bern, Switzerland)

J. Uythoven (CERN, TE Department, Genève 23, CH-1211, Switzerland)

R. van Weelderen (CERN, TE Department, Genève 23, CH-1211, Switzerland)

R. Versteegen (CERN, BE Department, Genève 23, CH-1211, Switzerland)

H. Vincke (CERN, TE Department, Genève 23, CH-1211, Switzerland)

M. Vretenar (CERN, Accelerator and Technology Sector, Genève 23, CH-1211, Switzerland)

U. Wagner (CERN, TE Department, Genève 23, CH-1211, Switzerland)

J. Wenninger (CERN, BE Department, Genève 23, CH-1211, Switzerland)

D. Wollmann (CERN, TE Department, Genève 23, CH-1211, Switzerland)

Q. Xu (KEK, 1-1 Oho, Tsukuba, Ibaraki 305-0801, Japan)

M. Zerlauth (CERN, TE Department, Genève 23, CH-1211, Switzerland)

F. Zimmermann (CERN, BE Department, Genève 23, CH-1211, Switzerland)

Chapter 1

Introduction to the HL-LHC Project[*]

L. Rossi and O. Brüning

CERN, Accelerator and Technology Sector, Genève 23, CH-1211, Switzerland

The Large Hadron Collider (LHC) is one of largest scientific instruments ever built. It has been exploring the new energy frontier since 2010, gathering a global user community of 7,000 scientists. To extend its discovery potential, the LHC will need a major upgrade in the 2020s to increase its luminosity (rate of collisions) by a factor of five beyond its design value and the integrated luminosity by a factor of ten. As a highly complex and optimized machine, such an upgrade of the LHC must be carefully studied and requires about ten years to implement. The novel machine configuration, called High Luminosity LHC (HL-LHC), will rely on a number of key innovative technologies, representing exceptional technological challenges, such as cutting-edge 11–12 tesla superconducting magnets, very compact superconducting cavities for beam rotation with ultra-precise phase control, new technology for beam collimation and 300-meter-long high-power superconducting links with negligible energy dissipation.

HL-LHC federates efforts and R&D of a large community in Europe, in the US and in Japan, which will facilitate the implementation of the construction phase as a global project.

1. Context and Objectives

The Large Hadron Collider (LHC) was successfully commissioned in March 2010 for proton–proton collisions with a 7 TeV center-of-mass energy and has delivered 8 TeV center-of-mass proton collisions since April 2012. The LHC is pushing the limits of human knowledge, enabling physicists to go beyond the Standard Model: the enigmatic Higgs boson, mysterious dark matter and the world of super-symmetry are just three of the long-awaited mysteries that the LHC might unveil. The announcement given by CERN on 4 July 2012 about the discovery of new

---

[*]The project is partially supported by the EC as *FP7 HiLumi LHC Design Study* under grant no. 284404. In addition to the FP7-Hilumi LHC consortium, the Project relies on the special contributions by: USA (LARP), Japan (KEK), France (CEA), Italy (INFN-Milano and Genova) and Spain (CIEMAT).

boson at about 125 GeV, the long awaited Higgs particle, is hopefully the first fundamental discovery of a series that LHC can deliver. Thanks to the LHC, Europe has decisively regained world leadership in High Energy Physics, a key sector of knowledge and technology. The LHC can act as catalyst for a global effort unrivalled by other branches of science: out of the 10,000 CERN users, more than 7,000 are scientists and engineers using the LHC, half of which are from countries outside the EU.

The LHC baseline programme till 2025 is schematically shown in Fig. 1. After entering in the near-to-nominal energy regime of 13 TeV center-of-mass energy in 2015, (hoping to reach the 14 TeV in the subsequent year) it is expected that the LHC will reach the design peak **luminosity**[1] of 10^{34} cm^{-2}s^{-1} and a total integrated luminosity over a one year of about 40 fb^{-1}. Then in the period 2015–2022 LHC will hopefully increase the peak luminosity: indeed margins have been taken in the design to allow, in principle, to reach about two times the nominal design performance. The baseline programme for the next ten years is depicted in Fig. 1, while Fig. 2 shows the graphs of the possible evolution of peak and integrated luminosity.

After 2020 the statistical gain in running the accelerator without an additional considerable luminosity increase beyond its design value will become marginal. The running time necessary to half the statistical error in the measurements will be more than ten years after 2020. Therefore to maintain scientific progress and to explore its full capacity, the LHC will need to have a decisive increase of its luminosity. That is why, when the CERN Council adopted the European Strategy for Particle Physics in 2006 [1], its first priority was agreed to be: "*to fully exploit the physics potential of the LHC. A subsequent major luminosity upgrade, motivated by physics results and operation experience, will be enabled by focused R&D*". The European Strategy for Particle Physics has been integrated into the ESFRI Roadmap of 2006 and its update of 2008 [2]. The priority to fully exploit the potential of the LHC has been recently confirmed as *first priority* among the "High priority large-scale scientific activities" in the new European Strategy for Particle Physics – Update 2013 [3], approved in Brussels on 30 May 2013 with the following wording: "*Europe's top priority should be the exploitation of the full potential of the LHC, including the high-luminosity upgrade of the machine and detectors with a view to collecting ten times more data than in the initial design, by around 2030.*"

---

[1]**Luminosity** is the number of collisions per square centimeter and per second, cm^{-2}s^{-1}.

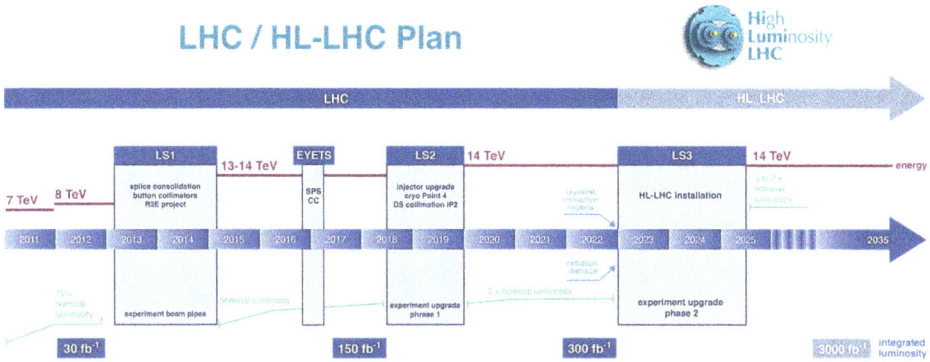

Fig. 1. LHC baseline plan for the next decade and beyond. In terms of energy of the collisions (upper line) and of luminosity (lower lines). The first long shutdown (LS1) 2013–14 is to allow design parameters of beam energy and luminosity. The second one, LS2 in 2018–19, is for securing luminosity and reliability as well as to upgrade the LHC Injectors. After LS3, in 2025 the machine should have the High Luminosity configuration (HL-LHC).

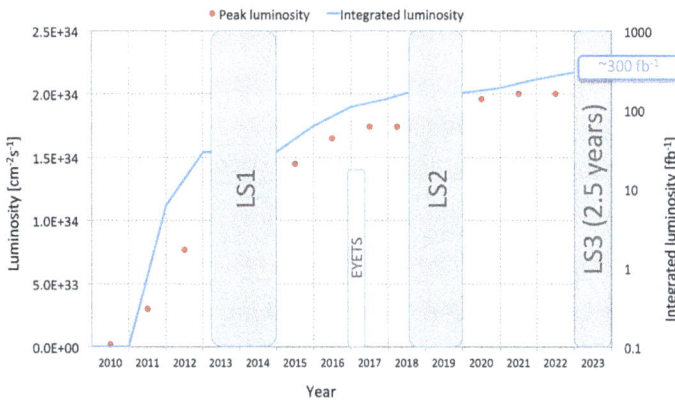

Fig. 2. Possible peak luminosity evolution (till the so-called "ultimate" limit) with consequent best forecast for integrated luminosity for the first decade of operation of LHC. Superimposed are the three long shutdowns (LS1, LS2, LS3) and the Extended Year End Technical Stop, as proposed in RLIUP and approved by CERN management and endorsed by CERN Council of December 2013. Also indicated the integrated luminosity goal of the LHC baseline program: 300 fb^{-1}.

The importance of the LHC upgrade in luminosity for the future of High Energy Physics has been also recently re-affirmed by the May 2014 resolution of the so-called P5 panel in the USA [4], a critical step in updating the USA strategy for HEP, with the following wording: "*Recommendation 10: ... The LHC upgrades constitute our highest-priority near-term large project.*"

In this context, CERN has put in place, at the end of 2010, the High Luminosity LHC (HL-LHC) project [5, 6]. Started as a Design Study, HL-LHC has become CERN's major construction project for the next decade after the approval of CERN

Council of 30 May 2013 and the insertion of the budget in the CERN Medium Term Plan approved by Council in the June 2014.

The main objective of High Luminosity LHC is to determine a set of beam parameters and the hardware configuration that will enable the LHC to reach the following targets:

(1) A peak luminosity of 5×10^{34} cm^{-2}s^{-1} with leveling, allowing:
(2) An integrated luminosity of 250 fb^{-1} per year, enabling the goal of 3000 fb^{-1} in about a dozen years after the upgrade. This luminosity is about ten times the luminosity reach of the first twelve years of the LHC lifetime.

The time horizon foresees the installation of the main hardware for HL-LHC during LS3 and commissioning the new machine configuration in the period 2023–2025.

All hadron colliders in the world have so far produced a total combined integrated luminosity of about 10 fb^{-1}; LHC has delivered nearly 30 fb^{-1} at the end of 2012 and should reach 300 fb^{-1} in its first 10–12 years of life. The High Luminosity LHC is a major and extremely challenging upgrade. For its successful realization, a number of key novel technologies have to be developed, validated and integrated. The work is initiated with the FP7 Design Study HiLumi LHC which, approved by EC in the Seventh Framework Programme (FP7-INFRA) in 2011 with the highest mark [7], is instrumental in initiating a new global collaboration for the LHC that matches the spirit of the worldwide user community of the LHC experiments.

The High Luminosity LHC project is working in close connection with the companion ATLAS and CMS upgrade projects of 2018–2023 and the upgrade foreseen in 2018 for both LHCb and Alice, as discussed in [8]. Furthermore, the performance of the high luminosity machine will depend on the performance of the injector chain, which is also being upgraded by a companion program, the LHC Injector Upgrade (LIU) program [9].

2. Approach for the Upgrade

The (instantaneous) luminosity L can be expressed as:

$$L = \gamma \frac{n_b N^2 f_{rev}}{4\pi \beta^* \varepsilon_n} R; \; R = 1 \Big/ \sqrt{1 + \frac{\theta_c \sigma_z}{2\sigma}}$$

where:
γ is the proton beam energy in unit of rest mass;
n_b is the number of bunches in the machine: 1380 for 50 ns spacing and 2808 for 25 ns;

N is the bunch population. $N_{\text{nominal 25 ns}}$: 1.15×10^{11} p ($\Rightarrow 0.58$ A of beam current at 2808 bunches);

f_{rev} is the revolution frequency (11.2 kHz);

β^* is the beam beta function (focal length) at the collision point (nominal design 0.55 m);

ε_n is the transverse normalized emittance (nominal design: 3.75 μm);

R is a luminosity geometrical reduction factor (0.85 at 0.55 m of β^*, down to 0.5 at 0.25 m);

θ_c is the full crossing angle between colliding beam (285 μrad as nominal design);

σ, σ_z are the transverse and longitudinal r.m.s. size, respectively (16.7 μm and 7.55 cm).

2.1. *Present luminosity limitations and hardware constraints*

There are various expected limitations to a continuous increase in luminosity, either in beam characteristics (injector chain, beam impedance and beam-beam interactions in the LHC) or in technical systems. Mitigation of potential performance limitations arising from the LHC injector complex are addressed by the LIU project, which should be completed in 2019 (LS2). Any potential limitations coming from the LHC injector complex put aside, it is expected that the LHC will reach a performance limitation from the beam current, from cleaning efficiency at 350 MJ beam stored energy and from the acceptable pile-up level. The ultimate value of bunch population with nominal LHC beam parameters should enable to reach $L = 2 \times 10^{34}$ cm^{-2}s^{-1}. Any further performance increase of the LHC will require significant hardware and beam parameter modifications with respect to the designed LHC configurations.

Before discussing the new configuration, it is useful to recall the systems that need to be changed, and possibly improved, just because they become more vulnerable to breakdown and accelerated wear out. This goes well beyond the on-going basic consolidation.

(1) *Inner Triplet Magnets*: At about 300 fb^{-1} some components of the low-beta triplet quadrupoles and their corrector magnets, we will have received a dose of 30 MGy, entering in the region of radiation damage. The quadrupoles may withstand 400–700 fb^{-1} but some corrector magnets of nested type are likely to wear out are already above 300 fb^{-1}. Damage must be anticipated because the most likely way of failing is through sudden electric breakdown, entailing serious and long repairs. That is why replacement of the triplet must be envisaged before damage. Replacement of the low-beta triplet is a long intervention, requiring one to two years shutdown and must be coupled with a major detector upgrade.

(2) *Cryogenics*: To increase flexibility of intervention and then availability (i.e. integrated luminosity) we plan to install a new cryo-plant in P4 for a full

separation between SCRF and Magnets cooling. In the long term, the cooling of the inner triplets and matching section magnets must be separate from the magnets of arc, to avoid that an intervention in the triplet region requires warm up of the entire arc (an operation of three months, not without risk).

(3) *Collimation*: The collimation system has been designed for the first phase of LHC life, but will certainly need a renovation plan mainly concerning the momentum and betatron cleaning in P3 and P7, as well as the tertiary colli-mators protecting the triplets. Any small gain in triplet aperture and perfor-mance must be accompanied by an adequate consolidation or modification of the collimation system. A second area that will require a special attention to the collimation system is the Dispersion Suppressor (DS), where a leakage of off-momentum particle into the first and second main superconducting dipole, has been already identified as a possible LHC performance limitation. The most promising concept is to substitute an LHC main dipole with a dipole of equal bending strength ($121 \, T \cdot m$) obtained by a higher field ($11 \, T$) and shorter length ($11 \, m$) than those of the LHC dipoles ($8.3 \, T$ and $14.2 \, m$). The room gained is sufficient for placing special collimators. A further improvement of the collimation system will be the use of new material for the jaws, in order to reduce the impedance (half of the LHC impedance is attribute to collimators). A molybdenum-graphite composite, coated with molybdenum, seems the best solution, capable to reduce the impedance of factor five to ten, keeping the robustness of the present design.

(4) *R2E and SC links for remote cold powering*: A considerable effort is under way to study how to replace the radiation sensible electronic boards with rad-hard cards. A complementary solution is also pursued for special zones: removal of the power supplies and associated DFBs (electrical feed-boxes, delicate equipment today in line with the continuous cryostat) out of the tunnel, possibly on the surface. LHC availability will be improved. In particular for Point 7 where a set of 600 A power converters are placed in front of the betatron cleaning collimators, removal will be done in a lateral tunnel since here ground surface is not accessible. Displacement of power converter to far away distance or surface is possible only thanks to a novel technology, not yet developed at the LHC design and construction: Superconducting links (SCLs) made out of HTS (YBCO or Bi-2223) or MgB_2 superconductors.

(5) *QPS, machine protection and remote manipulation*: Other systems will become a bottleneck along with aging of the machine and higher performance of 40 to 60 fb^{-1} per year:

(a) *Quench Protection System* (QPS) of the superconducting magnets, which is based on a design of almost twenty years ago.

   (b) *Machine protection*: improving vulnerability to mis-injected beams, to kickers sparks and asynchronous dumps. The kicker system is, with collimation and TDI, the main barrier against severe beam induced damage. Not only the kicker system, but also the interlock system needs renovation after 2020.

   (c) *Remote manipulation*: the level of activation from 2020, and even earlier, requires a carefully study and development of special equipment to allow replacing collimators, magnets, vacuum components, etc., according to ALARA principle. While full robotics is difficult to implement, given the real conditions, remote manipulation, enhanced reality and supervision is the key to minimize the radiation dose to operators.

2.2. *Upgraded systems for the high luminosity*

2.2.1. *Luminosity leveling and availability*

Both consideration of energy deposition by collision debris in the interaction region magnets, and the necessity to limit the peak pile up in the experimental detector, impose "*a priori*" a limitation of the peak luminosity. The consequence is that the HL-LHC operation will have to rely on luminosity leveling. As shown in Fig. 3 (left), the luminosity profile without leveling quickly decreases from the initial peak value, due to "proton burning" (protons consumed in collisions). By designing the collider to operate with a constant luminosity, i.e. "leveling" it and suppressing its decay for a good part of the fill, the average luminosity is almost the same as the one of a run without leveling, see Fig. 3 (right), however with the advantage that the maximum peak luminosity is smaller.

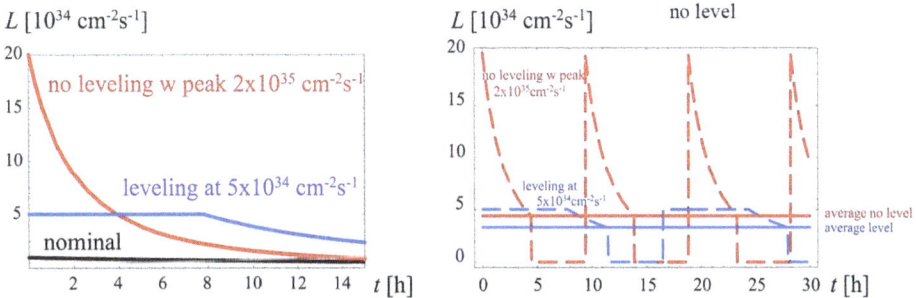

Fig. 3.   Left: Luminosity profile for a single long run starting at nominal peak luminosity (black line), with upgrade no leveling (red line) with leveling (dotted line). Right: Luminosity profile with optimized run time, without and with leveling (blue and red dashed lines), and average luminosity in both cases (solid lines).

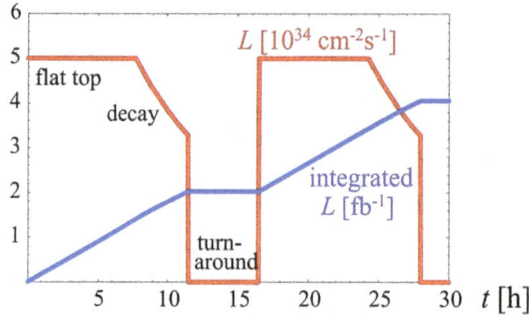

Fig. 4. Luminosity cycle for HL-LHC with leveling and a short decay (optimized for integrated luminosity). The set of parameters generating cycle are the 25 ns column of Table 1, standard.

The fact that the maximum leveled luminosity is limited, means that to maximize the integrated value one needs to maximize the run length, which can be obtained by filling the maximum number of protons, i.e. by maximizing the beam current: $I_{beam} = n_b \times N$. Other key factors for maximizing the integrated luminosity and obtaining the challenging goal of 3 fb^{-1}/day, see Fig. 4, are a short average machine turnaround time and a good overall machine "efficiency" defined as the ratio between actual time spent in physics production and the physics time of the ideal cycle. Clearly, for maximizing the integrated luminosity the efficiency counts almost as much as the virtual peak performance.

HL-LHC with 150 days of physics needs an efficiency of ca. 40%. During the 2011 run the efficiency varied, without luminosity leveling and the added system complexity of the HL-LHC (e.g. Crab Cavity operation), between 20% and 40%. Requiring an efficiency much higher than the one of the present LHC, with a (leveled) luminosity five times the nominal one and additional technically challenging hardware, will be a real challenge. The project must foresee a vigorous consolidation for the high intensity and high luminosity regime: the High Luminosity LHC must also be a High Availability LHC.

2.2.2. *Upgrade parameters*

Table 1 lists the main parameters foreseen for the high luminosity operation. Although the 25 ns bunch spacing remains the baseline, given the experience of the first years of operation, 50 ns is kept as a viable alternative, in case the e-cloud or other unforeseen effects undermine the 25 ns performance. For similar reasons, a slightly different parameter set with very small emittance beams (BCMS) is also maintained in case the LHC operation at with high beam intensities reveals unexpected sources for emittance blow-up during the beam injection and acceleration.

Table 1. High Luminosity LHC parameters (LHC nominal ones for comparison).

Parameter	Nominal LHC (design report)	HL-LHC 25ns (standard)	HL-LHC 25ns (BCMS)	HL-LHC 50ns
Beam energy in collision [TeV]	7	7	7	7
N_b	1.15E+11	2.2E+11	2.2E+11	3.5E+11
n_b	2808	2748	2604	1404
Number of collisions in IP1 and IP5	2808	2736 [1]	2592	1404
N_{tot}	3.2E+14	6.0E+14	5.7E+14	4.9E+14
beam current [A]	0.58	1.09	1.03	0.89
x-ing angle [µrad]	285	590	590	590
beam separation [σ]	9.4	12.5	12.5	11.4
β^* [m]	0.55	0.15	0.15	0.15
ε_n [µm]	3.75	2.50	2.50	3
ε_L [eVs]	2.50	2.50	2.50	2.50
r.m.s. energy spread	1.13E-04	1.13E-04	1.13E-04	1.13E-04
r.m.s. bunch length [m]	7.55E-02	7.55E-02	7.55E-02	7.55E-02
IBS horizontal [h]	80 -> 106	18.5	18.5	17.2
IBS longitudinal [h]	61 -> 60	20.4	20.4	16.1
Piwinski parameter	0.65	3.14	3.14	2.87
Geometric loss factor R0 without crab-cavity	0.836	0.305	0.305	0.331
Geometric loss factor R1 with crab-cavity	(0.981)	0.829	0.829	0.838
beam-beam / IP without Crab Cavity	3.1E-03	3.3E-03	3.3E-03	4.7E-03
beam-beam / IP with Crab cavity	3.8E-03	1.1E-02	1.1E-02	1.4E-02
Peak Luminosity without crab-cavity [cm^{-2} s^{-1}]	1.00E+34	7.18E+34	6.80E+34	8.44E+34
Virtual Luminosity with crab-cavity: Lpeak*R1/R0 [cm^{-2} s^{-1}]	(1.18E+34)	19.54E+34	18.52E+34	21.38E+34
Events / crossing without levelling and without crab-cavity	27	198	198	454
Leveled Luminosity [cm^{-2} s^{-1}]	-	5.00E+34 [5]	5.00E+34	2.50E+34
Events / crossing (with leveling and crab-cavities for HL-LHC)	27	138	146	135
Peak line density of pile up event [event/mm] (max over stable beams)	0.21	1.25	1.31	1.20
Leveling time [h] (assuming no emittance growth)	-	8.3	7.6	18.0
Number of collisions in IP2/IP8	2808	2452/2524 [7]	2288/2396	0 [4]/1404
N_b at SPS extraction [2]	1.20E+11	2.30E+11	2.30E+11	3.68E+11
n_b / injection	288	288	288	144
N_{tot} / injection	3.46E+13	6.62E+13	6.62E+13	5.30E+13
ε_n at SPS extraction [µm] [3]	3.40	2.00	< 2.00 [6]	2.30

[1] Assuming one less batch from the PS for machine protection (pilot injection, TL steering with 12 nominal bunches) and non-colliding bunches for experiments (background studies…). Note that due to RF beam loading the abort gap length must not exceed the 3 µs design value.

[2] An intensity loss of 5% distributed along the cycle is assumed from SPS extraction to collisions in the LHC.

[3] A transverse emittance blow-up of 10% to 15% on the average H/V emittance in addition to the 15% to 20% expected from intra-beam scattering (IBS) is assumed (to reach the 2.5 µm/3.0 µm of emitance in collision for 25 ns/50 ns operation).

[4] As of 2012 ALICE collided main bunches against low intensity. Satellite bunches (few per-mill of main bunch) produced during the generation of the 50 ns beam in the injectors rather than two main bunches, hence the number of collisions is given as zero.

[5] For the design of the HL-LHC systems (collimators, triplet magnets,…), a design margin of 50% on the stated peak luminosity was agreed upon.

[6] For the BCMS scheme emittances well below 2.0 µm have already been achieved at LHC injection.

[7] The lower number of collisions in IR2/8 wrt to the general purpose detectors is a result of the agreed filling scheme, aiming as much as possible at a democratic sharing of collisions between the experiments.

An upgrade should provide the possibility of performance increase over a wide range of parameters, such that the machine experience and experiments can eventually find the practical best set of parameters in actual operations.

Beam current and brightness: The total beam current may be a hard limit in the LHC since many systems are affected by this parameter. RF power system and RF cavity, Collimation, Cryogenics, Kickers, Vacuum, beam diagnostics, QPS, various controllers, etc. Radiation effects put aside, all systems have been designed in principle for $I_{beam} = 0.86$ A, the so-called "ultimate" beam current. However this is still to be experimentally proven and for the goal of HL-LHC we need to go beyond the ultimate value by 30% with 25 ns bunch spacing.

For HL-LHC it is needed to increase the beam brightness, which is a property that must be maximized at beginning of the beam generation and then preserved throughout the entire injector chain and LHC itself, i.e. it is a global property. The LIU project has as primary objective to increase the brightness at the LHC injection, basically increasing the number of protons per bunch by a factor two above what we have today while keeping the emittance at the present low value.

β^* *and canceling the reduction factor R*: A classical route to the luminosity upgrade is to reduce β^*, the optical function at the Interaction Points (IPs), by means of stronger and larger aperture low-β triplet quadrupoles. However a reduction in β^* value implies an increase of beam sizes inside the low-β triplet quadrupoles and a wider crossing angle, which both require in turn larger aperture low-β triplet quadrupole magnets, a larger D1 (first separation/recombination dipole) and a few modifications in the matching section, too. Stronger chromatic aberrations coming from the larger β-functions inside the triplet magnets may exceed the strength of the existing correction circuits. The peak beta-function inside the triplet magnets is also limited by the possibility to match the optics to the regular beta functions of the neighboring arcs. A previous study has shown that a practical limit in LHC is $\beta^* = 30$–40 cm, compared to the 55 cm foreseen in nominal operation. However a novel scheme called Achromatic Telescopic Squeeze (ATS) uses the adjacent arcs as enhanced matching sections and the increase of the beta-functions in those arcs to boost at constant strength the efficiency of the lattice sextupoles. In this way a β^* value of 15 cm can be envisaged and a flat optics with a β^* as low as 5 cm in the plane perpendicular to the crossing plane is enabled. For the β^* reduction the quadrupole magnets need to double the aperture, with a peak field 50% above the present LHC, requiring a new more advanced superconducting technology based on Nb₃Sn.

The drawback of very small β^* is that it requires larger crossing angle, which entails a reduction of the geometrical luminosity reduction factor 'R', see luminosity expression. In Fig. 5 the reduction factor is plotted for a constant normalized beam separation of 10σ vs. β^* values.

An efficient and elegant solution for compensating the geometric reduction factor is the use of special superconducting RF crab cavities, capable to generate transverse electric field to rotate each bunch by $\theta_c/2$, such as they collide effectively head on, overlapping perfectly at the collision point, see Fig. 6. Crab cavities make then accessible the full performance reach of the small β^* that the ATS scheme and the large low-beta triplet quadrupoles can generate: their primary function is boosting the virtual peak luminosity for attaining the full HL-LHC performance.

The lay-out and main hardware modifications required to meet the parameters listed in Table 1 are described in Chapter 3 of this book (The High Luminosity LHC Machine).

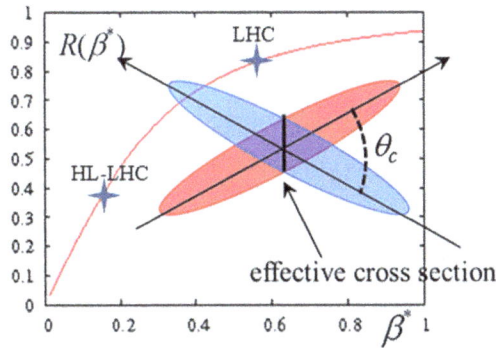

Fig. 5. Behavior of geometrical reduction factor of luminosity vs. β^* for constant normalized beam separation with indicated two operating points: Nominal LHC and HL-LHC. The sketch of bunch crossing shows the reduction mechanism.

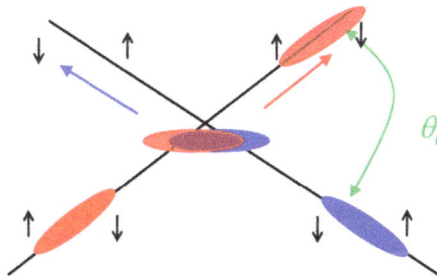

Fig. 6. Effect of the crab cavity on the beam (small arrows indicate the torque on the beam by transverse varying RF field).

2.3. *Project: performance, plan and cost*

The performance of the HL-LHC, both in terms of peak and integrated luminosity, is reported in the plot of Fig. 7.

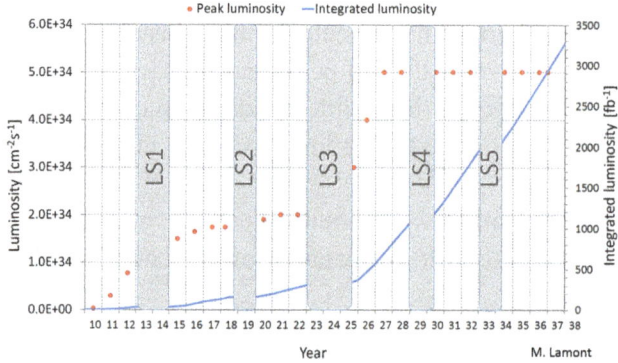

Fig. 7.   Peak luminosity (red dots) and integrated luminosity (blue line) vs time till 2035.

The plan is based on the following milestones:

2014: Preliminary Design Report (PDR)
2015: End of Design Phase, issue of the Technical Design Report (TDR)
2016: Proof on test bench of main hardware
2017: Test prototypes (including Crab Cavity test in SPS) and issue of TDR_v2
2017–2021: Construction and test of long lead hardware (Magnets, Crab Cavities, SC links, collimators)
2018–2019: LS2 – Installation of Cryo-plant P4, DS collimators (11T) in P2, SC link in P7
2021–2022: String test of Inner triplet
2023–2025: LS3 – Main installation and commissioning

The Cost-to-Completion of the full HL-LHC project, according to the initial evaluation of 2011, amounts to about 830 MCHF for Material (CERN accounting) and requires between 1000 and 1500 FTE-y.

In June 2014, CERN draft budget accounts for about 750 MCHF for the HL-LHC project till 2025, with certain guess of in-kind contributions both from the USA and Japan. The discrepancy is not critical at this stage, since modifications of certain equipment is not yet fully defined. LHC operation at full energy and intensity will give important indications, as well as the thorough investigation of the connection with LHC consolidation project and the various studies to make savings without compromising performance. Of course, additional in-kind contributions to the hardware baseline would be equivalent to a budget increase.

A further possibility is to stage the project by using also LS4, see Fig. 7. Indeed the performance "forecast" of Fig. 7 is somehow theoretical: there will certainly be a learning curve to pass from a luminosity of 2×10^{34} cm^{-2}s^{-1} to (leveled) 5×10^{34} cm^{-2}s^{-1}, favoring staging. However, only when installation of all equipment is completed, the 250 fb^{-1} annual integrated luminosity goal can be attained and, possibly, overcome.

2.4. *The international collaboration*

The LHC luminosity upgrade was born even more international than LHC, since USA laboratories started to work on it, with considerable resources, well before CERN. In 2002–2003 common work between US labs and CERN indicated the route for upgrade [10]. Right after the LARP (LHC Accelerator Research Program) was set up and approved by DOE [11] and become a ten-year program with a financing from 2008 of about 12 M$/year (in USA accounting). LARP heavily profited for the low-β triplet quadrupoles R&D of the DOE-Conductor Development Program, launched in 1998, which was instrumental for improving Nb$_3$SN to accelerator quality [12]. Meanwhile CERN was heavily engaged in the LHC construction and commissioning and could only participated to an EC-FP6 program in 2004–2008, called CARE that contained a modest program for the LHC upgrade. Then two EC-FP7 programmes helped to reinforce the Design and R&D for the LHC upgrade in Europe, although a modest level: SLHC-PP and EuCARD. KEK in Japan, in the framework of the permanent CERN-KEK collaboration, engaged in small activities for the LHC upgrade from 2008. LARP provided, until 2011, the largest part of the work for the LHC upgrade.

Finally with the approval of the EC-FP7 Design Study HiLumi LHC in 2011, and the maturing of all conditions illustrated in Section 1, the collaboration for HL-LHC took the present form. It is worth noticing that FP7 HiLumi Design Study covers only the design of a few components of the general lay-out, given the limited amount of funding in the program. However, it has allowed to form and structure a European participation to the upgrade at the very beginning of the project, something that was missing at the time of LHC. Since 2014, CEA (Saclay, FR), INFN (Milano and Genova, IT) and CIEMAT (Madrid, ES), have signed each a further collaboration agreement to carry out design, engineering and prototypal works for HL-LHC magnets in addition to the FP7-EC commitment. In all three cases the CERN funding is about 50%, the rest coming at charge of the collaborating Institutes. Figure 8 illustrates the various collaboration branches.

As stated above, the FP7-HiLumi LHC covers only a few WPs, which are the backbone of the upgrade. Work Packages are the basic structure of all FP7 projects: the WP structure, with task branching, is now the basic structure of the project.

LARP is a parallel structure, independently funded, but associated to FP7 with connections both at project management level as well as at WP/task level, to assure a maximum synergy. KEK is directly member of FP7-HiLumi. It is worth noticing that HiLumi LHC is the nickname to indicate the part of HL-LHC that is covered by FP7 funds, even if in practice has become a popular name to indicate the full project. Figure 9 shows the general governance of the project, while Fig. 10 illustrates the detailed structure in WP. Typically, each WPs is composed by 3 to 6 tasks.

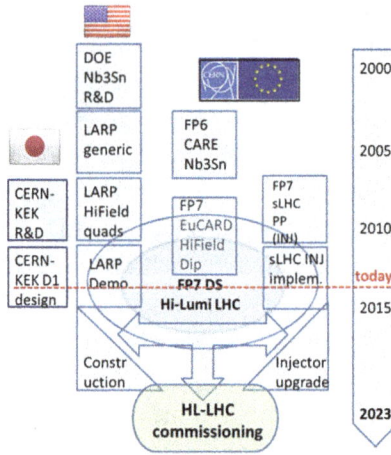

Fig. 8.   The International Collaboration and various paths toward the High Luminosity LHC.

High Luminosity LHC Project

Fig. 9.   Project structure and governance.

High Luminosity LHC Project

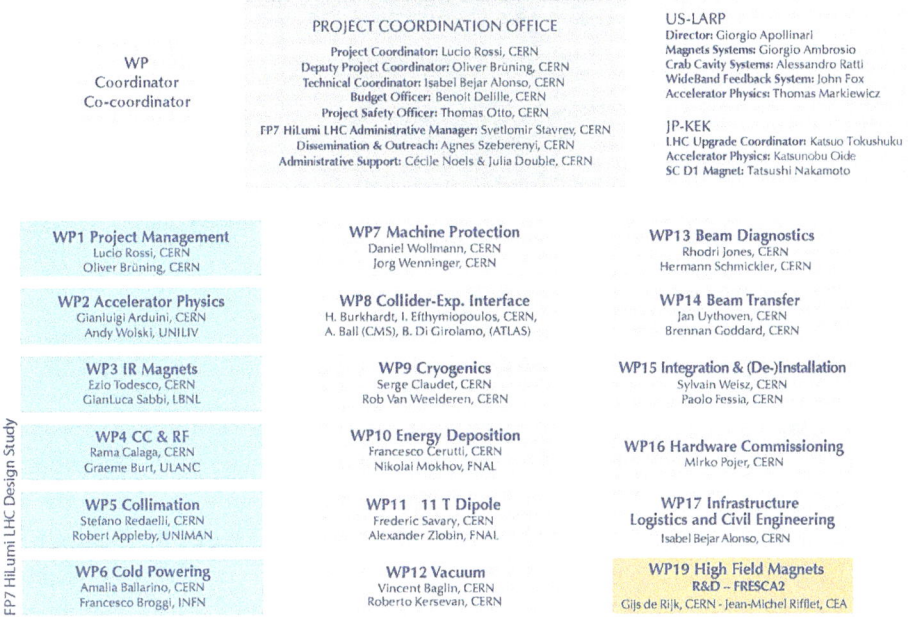

WP
Coordinator
Co-coordinator

PROJECT COORDINATION OFFICE

Project Coordinator: Lucio Rossi, CERN
Deputy Project Coordinator: Oliver Brüning, CERN
Technical Coordinator: Isabel Bejar Alonso, CERN
Budget Officer: Benoît Delille, CERN
Project Safety Officer: Thomas Otto, CERN
FP7 HiLumi LHC Administrative Manager: Svetlomir Stavrev, CERN
Dissemination & Outreach: Agnes Szeberenyi, CERN
Administrative Support: Cécile Noels & Julia Double, CERN

US-LARP
Director: Giorgio Apollinari
Magnets Systems: Giorgio Ambrosio
Crab Cavity Systems: Alessandro Ratti
WideBand Feedback System: John Fox
Accelerator Physics: Thomas Markiewicz

JP-KEK
LHC Upgrade Coordinator: Katsuo Tokushuku
Accelerator Physics: Katsunobu Oide
SC D1 Magnet: Tatsushi Nakamoto

WP1 Project Management
Lucio Rossi, CERN
Oliver Brüning, CERN

WP2 Accelerator Physics
Gianluigi Arduini, CERN
Andy Wolski, UNILIV

WP3 IR Magnets
Ezio Todesco, CERN
GianLuca Sabbi, LBNL

WP4 CC & RF
Rama Calaga, CERN
Graeme Burt, ULANC

WP5 Collimation
Stefano Redaelli, CERN
Robert Appleby, UNIMAN

WP6 Cold Powering
Amalia Ballarino, CERN
Francesco Broggi, INFN

WP7 Machine Protection
Daniel Wollmann, CERN
Jorg Wenninger, CERN

WP8 Collider-Exp. Interface
H. Burkhardt, I. Efthymiopoulos, CERN,
A. Ball (CMS), B. Di Girolamo, (ATLAS)

WP9 Cryogenics
Serge Claudet, CERN
Rob Van Weelderen, CERN

WP10 Energy Deposition
Francesco Cerutti, CERN
Nikolai Mokhov, FNAL

WP11 11 T Dipole
Frederic Savary, CERN
Alexander Zlobin, FNAL

WP12 Vacuum
Vincent Baglin, CERN
Roberto Kersevan, CERN

WP13 Beam Diagnostics
Rhodri Jones, CERN
Hermann Schmickler, CERN

WP14 Beam Transfer
Jan Uythoven, CERN
Brennan Goddard, CERN

WP15 Integration & (De-)Installation
Sylvain Weisz, CERN
Paolo Fessia, CERN

WP16 Hardware Commissioning
Mirko Pojer, CERN

**WP17 Infrastructure
Logistics and Civil Engineering**
Isabel Bejar Alonso, CERN

**WP19 High Field Magnets
R&D -- FRESCA2**
Gijs de Rijk, CERN - Jean-Michel Rifflet, CEA

FP7 HiLumi LHC Design Study

Fig. 10.　Simplified project governance and project work structure at July 2014. In dark green are evidenced the work packages which are co-financed by the European Commission under FP7. WP19 has a different color because it started as technological R&D program before the setting up of HL-LHC. The organigram reflect the status of summer 2014.

Table 2.　List of the Institutes that are members of the FP7-HiLumi LHC Consortium.

Short Name	Country	Logo	Short Name	Country	Logo
CERN	Geneva Switzerland		STFC*	Daresbury United Kingdom	Science & Technology Facilities Council
CEA	Saclay France		ULANC*	Lancaster United Kingdom	LANCASTER
DESY	Hamburg Germany		UNILIV*	Liverpool United Kingdom	UNIVERSITY OF LIVERPOOL
INFN	Frascati Italy		UNIMAN*	Manchester United Kingdom	MANCHESTER 1824 The University of Manchester
CSIC	Madrid Spain	CSIC	HUD	Huddersfield United Kingdom	University of HUDDERSFIELD
EPFL	Lausanne Switzerland	EPFL ECOLE POLYTECHNIQUE FEDERALE DE LAUSANNE	KEK	Tsukuba Japan	KEK-JAPAN
SOTON	Southampton United Kingdom	UNIVERSITY OF Southampton	BINP	Novosibirsk Russia	
RHUL	London United Kingdom	Royal Holloway University of London			

*Members of Cockcroft Institute

Table 3. List of the USA Institutions collaborating with the High Luminosity LHC Project.

Short Name	Country	Logo	
BNL	Upton, NY USA	BROOKHAVEN NATIONAL LABORATORY	
FNAL	Batavia, IL USA	✦ Fermilab	
LBNL	Berkeley, CA USA	BERKELEY LAB	*LARP*
SLAC	Menlo Park, CA USA	SLAC	
ODU	Norfolk, VA USA	OLD DOMINION UNIVERSITY	

The mechanism of FP7 funding is such that each of the thirteen European Institutions that are members of HiLumi LHC have to match the EC contribution with their internal funding: in case of HiLumi the matching funds equal the EC funds (except for CERN that receives from the EU only 17% of the total CERN cost for the design study, mainly for the management and coordination). Table 2 lists the 15 FP7-HiLumi Institutions and Table 3 the four USA-LARP institutions.

References

[1] European Strategy for Particle Physics, adopted by the CERN Council at a special session at ministerial level in Lisbon in 2006. http://cern.ch/council/en/EuropeanStrategy/ESParticlePhysics.html.

[2] European Strategy Forum for Research Infrastructures, ESFRI, http://ec.europa.eu/research/esfri.

[3] The European Strategy for Particle Physics Update 2013, CERN-Council-S/106, adopted at a special session in the Brussels on 30 May 2013. http://cern.ch/council/en/EuropeanStrategy/ESParticlePhysics.html.

[4] Building for Discovery: Strategic Plan for U.S. Particle Physics in the Global Context, in http://science.energy.gov/hep/hepap/reports/.

[5] L. Rossi, LHC Upgrade Plans: Options and Strategy, in *Proceedings of IPAC2011*, San Sebastián, Spain, 01/09/2011, pp. 908–912.

[6] L. Rossi and O. Brüning, High Luminosity Large Hadron Collider – A description for the European Strategy Preparatory Group, CERN-ATS-2012-236.

[7] European Commission – 7th Framework Programme for Research – Evaluation Summary Report: HiLumi LHC_284404.

[8] ECFA High Luminosity LHC Experiments Workshops – 2013 and 2014, https://indico.cern.ch/event/252045/ and https://indico.cern.ch/event/315626/.

[9] H. Damerau *et al.*, Upgrade Plans for the LHC Injector Complex, in *Proceedings of IPAC2012*, New Orleans, Louisiana, USA, pp. 1010–1014.

[10] J. Strait *et al.*, Towards a New LHC Interaction Region Design for a Luminosity Upgrade, in *Proceedings of the 2003 Particle Accelerator Conference (PAC2003)*, Portland, OR, USA, 12–16 May 2003, Vol. 1, pp. 42–44; also available as: CERN-LHC-Project Report 643.

[11] LARP Proposal (2003), http://www.uslarp.org/.

[12] R. M. Scanlan, *IEEE Trans. Appl. Supercond.* **11**, 2150 (2001).

Chapter 2

The Physics Landscape of the High Luminosity LHC

M. Mangano

CERN, PH Department, Genève 23, CH-1211, Switzerland

We review the status of HEP after the first run of the LHC and discuss the opportunities offered by the HL-LHC, in light of the needs for future progress that are emerging from the data. The HL-LHC will push to the systematic limit the precision of most measurements of the Higgs boson, and will be necessary to firmly establish some of the more rare decays foreseen by the Standard Model, such as the decays to dimuons and to a Z + photon pair. The HL-LHC luminosity will provide additional statistics required by the quantitative study of any discovery the LHC may achieve during the first 300 inverse femtobarn, and will further extend the discovery potential of the LHC, particularly for rare, elusive or low-sensitivity processes.

1. Introduction

The first run of the LHC, during the period 2009–2012, delivered three key messages: (i) the discovery [1, 2] of the Higgs boson [3–5], (ii) the lack of evidence for particles and interactions beyond those described by the Standard Model (SM) [6–8] and (iii) the excellent agreement between the data and the theoretical modeling of proton–proton (pp) collisions at a center-of-mass energy $\sqrt{S} = 7$ and 8 TeV. Each of these points contributes to sharpening the definition of the landscape for the future of the LHC programme, as we briefly summarize here, and elaborate in more detail in the rest of this Section.

The prospects for LHC physics with a dataset of 3000 fb^{-1} had been outlined already more than ten years ago, in a series of studies collected in [9]. Work has taken place within the experiments over the last couple of years [10, 11], to update these studies using the more realistic simulation tools certified by data, and to adapt them to the overall post-Run-1 physics scenario. The recent Snowmass process in the US has also contributed to further understand the physics case of future LHC operations [12, 13]. More progress, and further refinement of the goals and ambitions, will be possible after the analysis of the 14 TeV data.

To start with, the observation of a Higgs boson gives a compelling and concrete case to define, quantify and justify the goals of the long-term LHC exploration.

Table 1. SM branching ratios, in %, for various Higgs decay channels and at $m_H = 125.5$ GeV (see [14] for details).

$b\bar{b}$	$\tau^+\tau^-$	$\mu^+\mu^-$	$c\bar{c}$	gg	$\gamma\gamma$	$Z\gamma$	WW^*	ZZ^*
56.9	6.2	0.02	2.9	8.5	0.2	0.16	22.3	2.8

Within the SM, the value of the Higgs mass allows to predict uniquely its production and decay properties. A large number of decay final states is accessible for exploration at the LHC, each of them sensitive, in different ways, to the possible effects of physics beyond the SM (BSM). The relevant decay branching ratios, for a SM Higgs of mass $m_H = 125.5$ GeV, are collected in Table 1. One of the primary goals of the future LHC programme is therefore to greatly extend the range and precision of Higgs studies, improving the accuracy of the measurements, searching for yet unobserved decay modes, and probing in more detail the mechanism of electroweak symmetry breaking (EWSB). Precision targets in the range of few percent provide a concrete reference against which to benchmark the performance of the plans for future detector and accelerator improvements.

The expectation that the LHC should find evidence for BSM phenomena is justified by decades of theoretical work on the foundations of the SM and its conceptual shortcomings, and on the possible interpretations of experimental facts that cannot be explained within the SM, such as dark matter (DM), the baryon asymmetry of the universe and neutrino masses. The lack of BSM signals from the first run of the LHC does not spoil this expectation. It just constrains the set of suitable BSM models, possibly reducing the appeal of some frameworks, which would now require a finer tuning of their parameters to remain viable. The search for BSM signals remains therefore a top priority for the LHC. Two directions emerge: searching for particles of higher mass, and searching for final states that are harder to distinguish from the SM backgrounds. The increase in energy from 8 to 14 TeV will greatly extend the LHC search potential at high mass: within few months the discovery reach of Run 1 will be matched and surpassed. Higher statistics will however be crucial to push it towards the kinematic limit, and to pursue the more *stealthy* manifestations of new physics. High luminosity will also be needed in case of an early discovery: the constraints on new physics already established during the first LHC run are such that, if discovered, new physics will appear as a rare phenomenon even at 14 TeV. A detailed study of its properties, therefore, will demand extraordinary amounts of luminosity.

In addition to the Higgs discovery, and to the tighter constraints on the existence of BSM phenomena, the first years of LHC physics have proven two facts, which corroborate the reliability of the projections for the physics potential of the HL-LHC phase. On one side, the performance of the detectors matches, and often

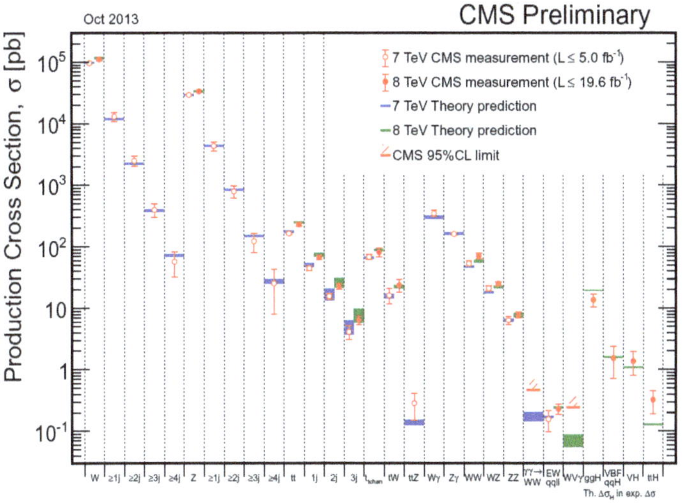

Fig. 1.   Collection of cross section measurements for various SM processes by the CMS experiment, compared against theoretical predictions [15].

surpasses, the expectations. This is particularly true of the ability to operate in a regime of very high pile-up, a critical test for effective data taking in the environment expected with the HL-LHC. On the other, the theoretical modeling of the properties of pp collisions at these energies has proven very accurate. Dedicated precise measurements of SM processes and cross sections have shown that data and theory agree over a broad dynamical range of phenomena, including the very complex final states with mixtures of gauge bosons, heavy quarks, and multijets, which characterize BSM processes. This is briefly summarized in Fig. 1. Where the theoretical predictions are limited in precision by the lack of higher-order calculations or by uncertainties in the knowledge of the quark and gluon content of the proton (the so-called parton distribution functions, PDFs), great progress is taking place to match the precision needs, with better calculations, and by using the LHC data themselves to validate the calculations and to refine the knowledge of PDFs. This progress will continue through Run 2 with more data, higher energy, and more powerful theoretical tools. For the specific case of PDFs, further progress would arise from a programme of ep collisions, such as the one studied by the LHeC project. Its results would surely fulfill, and even exceed, the precision goals of the HL-LHC.

1.1. *Status and prospects of Higgs studies*

The new particle discovered in July 2012 has been studied in greater detail since then, using the complete statistics of Run 1, and continuously refining the experimental analyses. There is no reasonable doubt by now that this particle is a scalar [17, 18], consistent with being an excitation of the Higgs field, responsible for the breaking of the $SU(2) \times U(1)$ electroweak (EW) symmetry, and for the masses of the W and Z bosons, as well as of the known quarks and leptons. While the simplest theoretical model describing the Higgs boson is what is built into the SM itself, it is well known that there are several alternative "Higgs mechanisms". A Higgs mechanism is defined by the spectrum of the Higgs states, and by the dynamics that leads the Higgs field to acquire a non-zero vacuum expectation value, resulting in the EWSB. In the SM version of the Higgs mechanism [7], there is a single complex $SU(2)$ doublet, corresponding to four real degrees of freedom. The symmetry breaking is driven by the minimization of the mexican-hat-shaped potential, described by two parameters associated to the mass of the Higgs boson and its expectation value. Three of the four degrees of freedom become the longitudinal modes of the W^+, W^- and Z^0 massive vector bosons, and the fourth is left as *the* SM Higgs boson.

Alternatives to the SM Higgs mechanism include theories with a more extended spectrum and/or with a different EWSB dynamics. For example, several theories, most notably supersymmetric models, have two doublets, instead of one. One doublet is used to give mass to the up-type quarks, the other doublet is used to give mass to the down-type quarks and to the charged leptons. In these models, there is a total of five Higgs particles (the 2×4 states of 2 complex $SU(2)$ doublets, minus the 3 states absorbed by $W^\pm$ and Z^0), three of them neutral (typically labeled h^0, H^0, A^0), the other 2 being the opposite-charge states of a charged $H^\pm$ Higgs.

Other SM extensions include the possibility of further doublets, or of different $SU(2)$ representations, including singlets, or vectors, in which case doubly-charged Higgs fields would also appear. In other scenarios, EWSB may arise from an underlying strong dynamics, the Higgs field emerging as a composite particle, a bound state of elementary fermions confined together (for a recent review see e.g. [28]). In some theories with extra space dimensions, the Higgs scalar could be a component of vector fields living in higher dimensions.

In the SM, the production and decay properties of the Higgs boson are completely determined by its mass and by the masses of the SM particles. In BSM theories such as those described above, those properties can change, whether because of a richer Higgs spectrum, or of a different EWSB dynamics, or because the other new BSM particles can influence the Higgs couplings: additional decay channels can be open, or new intermediate virtual states can modify the loop-mediated Higgs effective couplings, such as those to gluons and photons.

While the available experimental information is sufficient to claim with complete confidence that the discovered particle is a Higgs boson, the quantitative determination of its consistency with the SM predictions is limited by the statistical and systematics accuracy of the measurements. The Higgs has been searched, and detected, in a combination of different production and decay modes. The dominant production channels are: gluon fusion ($gg \rightarrow H$), vector-boson fusion ($qq \rightarrow qqH$, via intermediate $W^+W^- \rightarrow H$ or $ZZ \rightarrow H$), associated production with a gauge boson ($pp \rightarrow W/Z+H$) and with top quarks ($pp \rightarrow t\bar{t}H$). The decay modes that have been positively identified so far, including evidence from the Tevatron experiments [19], cover both decays to gauge boson pairs [20,21] ($\gamma\gamma$, $ZZ^* \rightarrow 4$ leptons, $WW^* \rightarrow 2$ leptons + 2 neutrinos) and to fermion pairs ($\tau^+\tau^-$ [22,23], and $b\bar{b}$ [25,26]).

For both ATLAS and CMS, the measurement of production rates times branching fractions, averaged over all channels, agrees with the SM expectations to within 15–20%, consistent with the current statistical and systematic uncertainty. All the individual channels are also consistent with the SM prediction, with varying degrees of significance. Higgs decays to the $\mu^+\mu^-$ and $Z\gamma$ final states have been searched, but no signal was found, as expected from the very small branching ratios predicted by the SM. Likewise, there are no signals of decays that are forbidden in the SM, such as $H \rightarrow \mu^\pm e^\mp$, or indications of BSM Higgs-like states, over a mass range extending up to several hundreds GeV. Therefore, today's LHC data support the case of a SM Higgs mechanism.

It must be said, however, that it would have been quite surprising to find deviations from the SM Higgs given the available statistics: precision EW measurements and direct BSM searches set indirect constraints on possible deviations of the Higgs properties from their SM expectations. Assuming new physics at a scale Λ to affect the Higgs properties, whether via modifications of Higgs dynamics due to the composite nature of the Higgs, or due to new particles affecting the Higgs interactions with SM particles, one expects the Higgs branching ratios to vary at the level of $\delta \times (1 \text{ TeV}/\Lambda)^2$, where $\delta \sim O$ (1–10%), depending on the model (for some explicit examples, see e.g. [12]).

The goal of the LHC is therefore to extend the measurements of the Higgs properties, and to reach a precision in the range of few percent, in order to match the sensitivity to direct manifestations of new physics appearing at the TeV scale. The prospects for such improvements, including estimates for the HL-LHC, have been studied by ATLAS [11] and CMS [10], and were summarized in the Snowmass study [12]. Those results are reproduced here in Table 2. The range in projections reflects two assumptions on the extrapolation of systematics uncertainties: the more conservative results assume uncertainties equal to those achieved today, while the more optimistic ones assume that experimental systematics scale with statistics like $1/\sqrt{\int \mathscr{L}}$ and theoretical systematics will be reduced by half.

Notice that only with the full statistics of the HL-LHC phase one can achieve a meaningful measurement of the important decay modes $H \rightarrow Z^0\gamma$ and $H \rightarrow \mu^+\mu^-$.

Table 2.   Projected accuracy per experiment, in %, in the extraction of Higgs couplings, for 300 and 3000 fb^{-1} of integrated luminosity at 14 TeV. The results are a fit to projected data of the parameters κ_X, representing the relative deviations from the SM of the Higgs couplings to the gauge bosons, $\kappa_{\gamma,g,W,Z}$, to the charm and top quarks, κ_u, to the bottom and strange quarks, κ_d, to the μ and τ leptons $\kappa_{\mu,\tau}$, and to the $Z\gamma$ final state, $\kappa_{Z\gamma}$. The last column gives the sensitivity to invisible Higgs decay modes. The ranges reflect different assumptions on the systematics uncertainties (see [12] for details).

L/fb^{-1}	κ_γ	κ_g	$\kappa_{W,Z}$	κ_u	κ_d	κ_τ	$\kappa_{Z\gamma}$	κ_μ	BR$_{\text{BSM}}$
300	5–7	6–8	4–6	14–15	10–13	6–8	41	23	< 14–18%
3000	2–5	3–5	2–5	7–10	4–7	2–5	10–12	8	< 7–11%

The former has a sensitivity complementary to the $H \rightarrow \gamma\gamma$ decay to virtual particles propagating in the internal loop; the latter provides the most direct probe of an anomalous flavor pattern in the Higgs couplings to leptons. The HL-LHC statistics will also permit to increase the number of combinations of SM production and decays channels, enabling e.g. precise measurements of $pp \rightarrow t\bar{t}H$ and $pp \rightarrow W/ZH$ with $H \rightarrow \gamma\gamma$ [10, 11]. Access to multiple production and decay channels is crucial to reduce the systematics in the extraction of the Higgs couplings to individual particles, and to disentangle the origin of possible anomalies induced by BSM physics. The HL-LHC will furthermore extend the sensitivity to exotic or forbidden decays, such as $H \rightarrow \gamma J/\psi$ [29], $H \rightarrow \mu^\pm\tau^\mp, \mu^\pm e^\mp, t \rightarrow Hc$ and others [30].

To further probe the mechanism underlying the breaking of electroweak symmetry (EWSB), to prove that the observed Higgs is indeed responsible for it, and to determine whether it is a fundamental particle (as predicted by the SM and other theories), or whether it is a composite object (as predicted by others), other measurements, in addition to the determination of decay rates, will prove useful. These include the study of Higgs pair production and the measurement of WW scattering at high energy [9]. The production rate of Higgs pairs is sensitive to the Higgs selfcoupling, and thus to the form of the Higgs field potential. Its measurement is made difficult by the small cross-section and by the large backgrounds that affect the dominant decay modes, such as $HH \rightarrow b\bar{b}b\bar{b}$. Viability studies, using various decay modes, are in progress [12], and indicate that, while the study of Higgs pair production is likely beyond the reach of 300 fb^{-1}, a sensitivity at the level of $\pm 50\%$ becomes possible at the HL-LHC.

The study of WW interactions at high invariant mass probes the unitarization of the scattering amplitude, to confirm the role of the Higgs boson, or to suggest the presence of a new strong dynamics underlying the EWSB mechanism and of a Higgs composite substructure. The nominal LHC luminosity is not sufficient to probe the relevant kinematical regions, where $m(WW) \gtrsim 1$ TeV. While the study of Higgs pair production and high-mass WW scattering would mostly benefit from an increase in the LHC beam energy, the HL-LHC luminosities will be necessary to perform the first tests ever of these sensitive probes of EWSB and of the true nature of the Higgs boson.

1.2. *Status and prospects of BSM searches*

Physicists have long anticipated the existence of new phenomena at the TeV scale, in order to address issues like the existence of DM and the hierarchy problem, namely the fine tuning of the Higgs self-coupling, to one part in $10^{\sim 30}$, necessary to justify the smallness of the Higgs mass with respect to the Planck scale. The LHC experiments have found no evidence, as yet, for such new phenomena, as exemplified in Fig. 2, which collects the exclusion limits obtained by ATLAS for a large number of (non-supersymmetric) BSM models. In most cases these limits already approach, or well exceed, the TeV scale. In the case of supersymmetry, partners of quarks and gluons are ruled out in the constrained minimal models, if their mass is below 1.5 TeV, and the partners of the top and bottom quarks have constrains reaching out to 500 GeV.

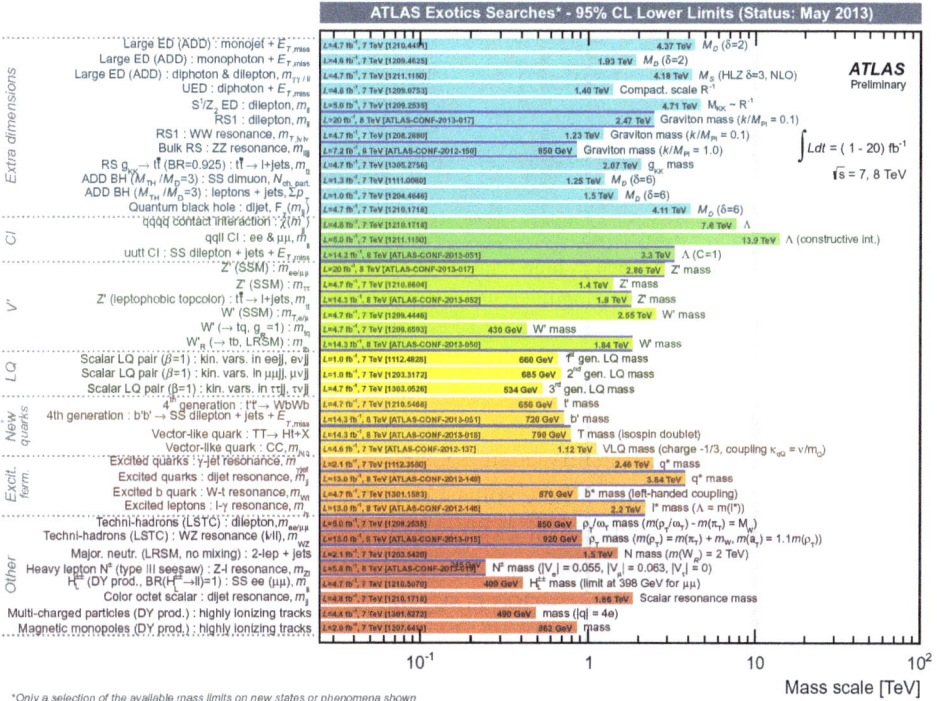

Fig. 2.   Collection of mass limits for new particles of BSM models, as obtained by the ATLAS experiment [16]. Results from the 8 TeV run are highlighted in blue (thick lines).

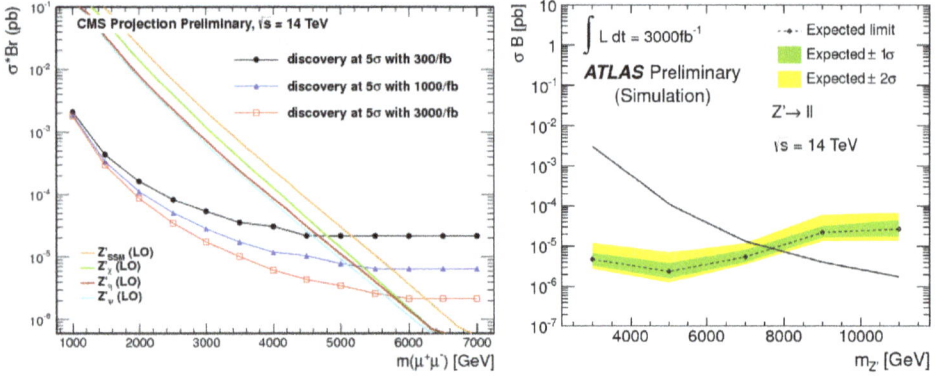

Fig. 3. Projections for the search of a new Z' gauge boson. Left: 5σ discovery reach, in the dimuon channel [10], for different coupling scenarios (Z'_{SSM} corresponds to SM-like couplings). Right: Exclusion limit, at 95%CL, in the case of SM-like couplings [11].

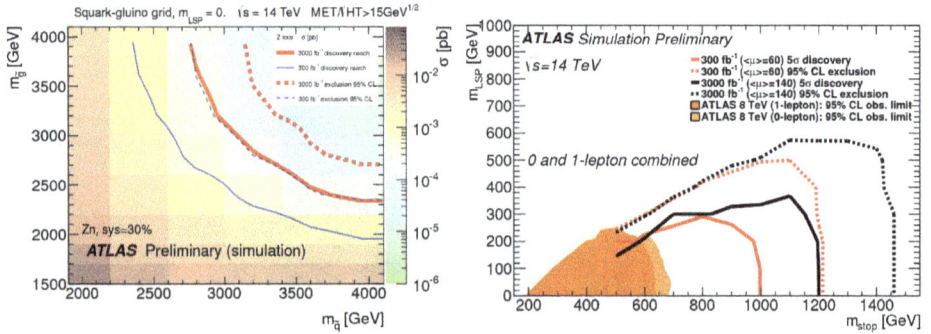

Fig. 4. Discovery/exclusion reach at high luminosity for squarks and gluinos (left panel) and for stop squarks (right panel) in some simplified supersymmetric scenarios [11, 31].

Needless to say, these results leave well open the possibility of discoveries at a higher mass, when the LHC reaches 14 TeV. Many examples have been studied in the past [9], and have recently been updated in view of the prospects for high luminosity [10, 11]. We collect here some indicative examples.

Figure 3 shows the increase in exploration reach of the HL-LHC for a new Z' gauge boson. The luminosity increase from 300 to 3000 fb^{-1} increases the discovery reach by 20%, up to $\sim$6300 GeV, and increases the 95%CL limit up to $\sim$7600 GeV. A similar reach improvement is observed for other searches, as shown in the case of squarks, gluinos and stops in Fig. 4.

In spite of the ample room for discoveries at 14 TeV, at these large masses the production rates will be small, and high-statistics precision measurements of their properties will call for a luminosity well beyond the nominal LHC phase. For example, while new Z' gauge bosons, signaling the existence of new weak interactions, can be discovered with 300 fb^{-1} up to $\sim$ 5 TeV, the full HL-LHC luminosity will be needed to determine their properties if their mass is above 2.5 TeV. Since the available LHC data already constrain the existence of Z' bosons below 2.5 TeV (see e.g. Fig. 2), we can state already today that the clarification of any such discovery will need the luminosity upgrade.

1.3. *Additional remarks*

We conclude this Section with few independent remarks.

There is no doubt that, however welcome an increase in the LHC luminosity, the most effective way to push the discovery reach at large mass is to ultimately increase the beam energy. But it must be pointed out that the efforts to increase the LHC luminosity, and thus to develop the relevant technologies and experimental skills, will remain crucial even should the avenue of the energy increase be undertaken. While the hadronic cross section for the production of states of a given mass increases with the beam energy ($E_{\text{beam}} = 0.5\sqrt{(S)}$), the cross sections to produce states whose mass M is a fixed fraction of E_{beam} decreases as $1/S$. This is shown in practice in Fig. 5, which gives the cross-sections for production of pairs of heavy quarks with mass m_Q, plotted for $E_{\text{beam}} = 7$ and 50 TeV as a function of m_Q. For example, the event rate of $1/(1000$ fb$^{-1})$ allows to reach $m_Q \sim 0.46E_{\text{beam}}$ for $E_{\text{beam}} = 7$ TeV, but only $m_Q \sim 0.31E_{\text{beam}}$ at 50 TeV, with a loss of 1/3 of the relative discovery reach.

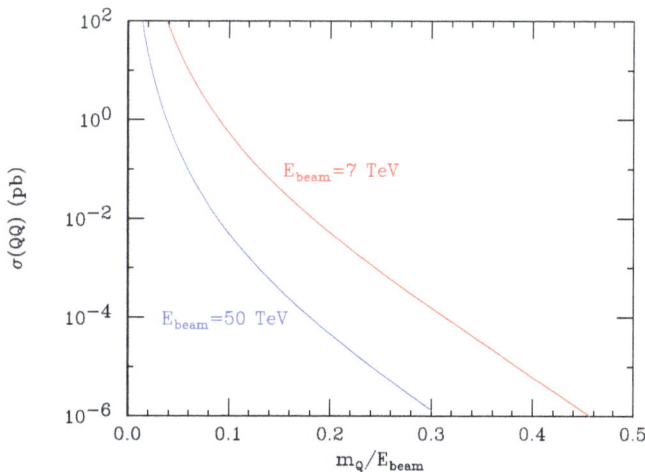

Fig. 5. Cross-sections for the production of heavy quark pairs, as a function of m_Q/E_{beam}, for pp collisions at $E_{\text{beam}} = 7$ and 50 GeV.

To fully exploit the exploration power of a high energy hadron collider at the highest mass scales, its luminosity must therefore scale like S. In addition to challenging the accelerator, such high luminosities would greatly challenge the ability of the experiments to cope with very high levels of pile-up. On the other hand, pushing the high-mass frontier implies dealing with highly energetic final states, where pile-up will likely be a minor issue. For example, at 100 TeV the hardest jets will have energies about 7 times larger than at 14 TeV. One can thus probably tolerate a pile-up contribution to the energy of these jets at a level 7 times higher than at the HL-LHC. Since the energy deposit of a generic minimum bias event grows only logarithmically with energy, and so does not change much between 14 and 100 TeV, this means that one could tolerate a pile-up rate 7 times larger. This naive argument may fail for observables more sophisticated than jet energies (such as tracking efficiency, lepton or photon isolation, tagging of bottom hadrons, etc.), but overall it is not unreasonable that the detectors for a 100 TeV collider, 30 years from now, will be able to address the physics of the highest mass objects with several hundreds or perhaps 1000 pile-up events.

With the discovery of the Higgs boson, a long era of more or less guaranteed discoveries is over. The W and Z bosons, the top quark, and the Higgs, are the necessary components of an extremely compelling theoretical framework, the SM, whose predictions have steadily grown in reliability over the years. The high-energy accelerators of the last 30 years were built to discover and study in detail those particles, whose existence and whose properties were anticipated with great confidence. While there is similar confidence in the existence of new physics beyond the SM, there is nowadays no certainty as to precisely what this new physics should be, and where or how it will appear in accelerators. There is no guarantee that it will be discovered during the Run 2 of the LHC, or in the subsequent phases of its upgrade. The programme of precision Higgs physics and continued exploration of EWSB (including WW scattering and Higgs-pair production) is therefore the most concrete and robust deliverable that the LHC upgrade can promise, and whose future returns can be anticipated today.

The achievement of the HL-LHC programme as discussed above will greatly enrich our knowledge of particle physics, even in absence of BSM discoveries, and by itself it motivates the upgrade effort. In parallel with this programme, a large set of ancillary measurements are necessary, in order to improve the precision of the theoretical predictions, to reduce the experimental systematics, and to improve the knowledge of the complete set of parameters or observables defining the SM. Among these, we mention m_{top}, m_W, $\sin^2 \theta_W$, as well as the flavor observables associated to the quark-mixing phenomena. Furthermore, while the flavor programme carried out by LHCb, and the programme of heavy ion collisions engaging ALICE and the other general purpose detectors, will not be exposed to the highest luminosities afforded by the HL-LHC, they will nevertheless greatly benefit from the extended operation of the LHC.

The longevity of a hadron collider is a great asset, as the example of the Tevatron Run 2 shows very well. If the Higgs boson had had a mass of ~ 160 GeV, the Run 2 luminosity would have guaranteed its discovery, and an extension of the run by few more years would have guaranteed a full discovery even at the actual mass of 125 GeV. It is also remarkable that some of the most impressive results from Run 2, like the oscillations of B_s mesons, the observation of single top production, the precision measurement of the W and of the top quark masses, among others, were achieved after 20 years since the first Tevatron collider run. Tantalizing hints of new phenomena, like the production asymmetry of top quarks, have also emerged only after a major fraction of the ultimate luminosity sample was collected. This demonstrates the scientific longevity of a hadron collider, and its potential to deliver surprises over a very long life span, provided a sufficiently rapid luminosity doubling time is attainable.

References

[1] G. Aad *et al.* [ATLAS Collaboration], *Phys. Lett. B* **716**, 1 (2012) [arXiv:1207.7214 [hep-ex]].

[2] S. Chatrchyan *et al.* [CMS Collaboration], *Phys. Lett. B* **716**, 30 (2012) [arXiv: 1207.7235 [hep-ex]].

[3] F. Englert and R. Brout, *Phys. Rev. Lett.* **13**, 321 (1964).

[4] P. W. Higgs, *Phys. Rev. Lett.* **13**, 508 (1964).

[5] G. S. Guralnik, C. R. Hagen and T. W. B. Kibble, *Phys. Rev. Lett.* **13**, 585 (1964).

[6] S. L. Glashow, *Nucl. Phys.* **22**, 579 (1961).

[7] S. Weinberg, *Phys. Rev. Lett.* **19**, 1264 (1967).

[8] A. Salam, Weak and electromagnetic interactions, in *Elementary Particle Physics: Relativistic Groups and Analyticity*, ed. N. Svartholm (Almqvist & Wiskell, 1968), p. 367 [Proceedings of the Eighth Nobel Symposium, *Conf. Proc.* **C680519**, 367 (1968)].

[9] F. Gianotti *et al.*, *Eur. Phys. J. C* **39**, 293 (2005) [arXiv:hep-ph/0204087].

[10] [CMS Collaboration], arXiv:1307.7135. See also: https://cern.ch/twiki/bin/view/ CMSPublic/PhysicsResultsFP.

[11] [ATLAS Collaboration], arXiv:1307.7292 [hep-ex]. See also: https://cern.ch/twiki/ bin/view/AtlasPublic/UpgradePhysicsStudies.

[12] S. Dawson *et al.*, arXiv:1310.8361 [hep-ex].

[13] R. Brock *et al.*, arXiv:1401.6081 [hep-ex].

[14] S. Heinemeyer *et al.* [LHC Higgs Cross Section Working Group Collaboration], arXiv:1307.1347 [hep-ph].

[15] CMS collaboration, see: https://twiki.cern.ch/twiki/bin/view/CMSPublic/Physics ResultsCombined.

[16] ATLAS collaboration, see: https://cern.ch/Atlas/GROUPS/PHYSICS/Combined SummaryPlots/EXOTICS/.

[17] S. Chatrchyan *et al.* [CMS Collaboration], *Phys. Rev. Lett.* **110**, 081803 (2013) [arXiv:1212.6639 [hep-ex]].

[18] G. Aad *et al.* [ATLAS Collaboration], *Phys. Lett. B* **726**, 120 (2013) [arXiv: 1307.1432 [hep-ex]].

[19] T. Aaltonen *et al.* [CDF and D0 Collaborations], *Phys. Rev. D* **88**, 052014 (2013) [arXiv:1303.6346 [hep-ex]].

[20] G. Aad *et al.* [ATLAS Collaboration], *Phys. Lett. B* **726**, 88 (2013) [arXiv:1307.1427 [hep-ex]].

[21] S. Chatrchyan *et al.* [CMS Collaboration], arXiv:1312.5353 [hep-ex].

[22] The ATLAS collaboration, ATLAS-CONF-2013-108.

[23] S. Chatrchyan *et al.* [CMS Collaboration], arXiv:1401.5041 [hep-ex].

[24] S. Chatrchyan *et al.* [CMS Collaboration], arXiv:1401.6527 [hep-ex].

[25] S. Chatrchyan *et al.* [CMS Collaboration], *Phys. Rev. D* **89**, 012003 (2014) [arXiv:1310.3687 [hep-ex]].

[26] The ATLAS collaboration, ATLAS-CONF-2013-079.

[27] M. Mangano, *Contemp. Phys.* **51**, 211 (2010).

[28] B. Bellazzini, C. Csáki and J. Serra, arXiv:1401.2457 [hep-ph].

[29] G. T. Bodwin, F. Petriello, S. Stoynev and M. Velasco, *Phys. Rev. D* **88**, 053003 (2013) [arXiv:1306.5770 [hep-ph]].

[30] D. Curtin *et al.*, arXiv:1312.4992 [hep-ph].

[31] ATLAS collaboration, https://cern.ch/Atlas/GROUPS/PHYSICS/PUBNOTES/ATL-PHYS-PUB-2013-011/.

Chapter 3

The HL-LHC Machine[*]

I. Bejar[1], O. Brüning[1], P. Fessia[2], L. Rossi[1], R. Tomas[3] and M. Zerlauth[2]

[1]*CERN, Accelerator and Technology Sector, Genève 23, CH-1211, Switzerland*
[2]*CERN, TE Department, Genève 23, CH-1211, Switzerland*
[3]*CERN, BE Department, Genève 23, CH-1211, Switzerland*

This chapter summarizes the baseline parameters and layout for the HL-LHC machine, discusses options for alternatives to the baseline configurations, comments on the integration issues and describes the overall planning for the HL-LHC upgrade.

1. HL-LHC Baseline Parameters

The performance of the HL-LHC machine is boxed in between the request for a high integrated luminosity (ca. 3000 fb^{-1} by the end of the HL-LHC exploitation over ca. 10 years of operation and translating to an annual integrated luminosity of ca. 250 fb^{-1} assuming scheduled 160 days for proton physics production per year and that the HL-LHC exploitation starts with an integrated luminosity of ca. 300 fb^{-1} at the end of the LHC Run III in 2022) and a maximum number of 140 events per bunch crossing. While the request for maximum integrated luminosity asks for the largest possible peak luminosity, the request for limited number of events per bunch crossing limits the peak luminosity to a maximum value of ca. $5 \cdot 10^{34}$ cm^{-2}s^{-1}. Operating the HL-LHC with the maximum number of bunches and utilizing luminosity leveling provides the best compromise for satisfying both requests. Table 1 shows the resulting baseline parameters approved by the HL-LHC Layout and Parameter Committee [1] for the standard 25 ns bunch spacing configuration together with the parameters for the nominal LHC configuration and two alternative scenarios which might become interesting in case the LHC operation during Run II reveals problems either related to the

[*] The project is partially supported by the EC as *FP7 HiLumi LHC Design Study* under grant no. 284404. In addition to the FP7-Hilumi LHC consortium, the Project relies on the special contributions by: USA (LARP), Japan (KEK), France (CEA), Italy (INFN-Milano and Genova) and Spain (CIEMAT).

emittance preservation along the LHC cycle for high intensity operation (the so-called BCMS filling scheme allows the preparation of small emittance beams at the price of a reduced number of bunches) or the electron-cloud effect. The fall back solution for the latter scenario is a 50 ns bunch separation scheme at which electron cloud effects are not expected to be an issue, but where the peak luminosity needs to be leveled at a lower value in order to keep the number of events per bunch crossing below 140. The luminosity leveling time is of the order of 8 hours and an efficient operation of the HL-LHC machine hence requires an average physics fill length that is slightly larger than the leveling time (e.g. ca. 10 hours). The required HL-LHC average fill length is approximately 50% larger than the average fill length of the LHC Run I period (ca. 6 hours).

The baseline parameters are based on a β^* value of 15 cm at the IP and the operation with Crab Cavities for compensating the geometric luminosity loss factor that becomes significant when operating with such small β^* values and a large crossing angle. These parameters coupled together imply larger aperture insertion magnets (triplet magnets, D1, D2 and Q4, Q5 magnets), lower operating temperatures for some of the insertion magnets (e.g. Q4 and Q5) and the exploitation of a novel optics matching scheme ATS [2] that utilizes the neighboring arcs for matching the insertion optics to the rest of the machine, requiring some upgrades in the non-experimental insertions (e.g. additional Q5 in IR6). The larger aperture triplet magnets of the HL-LHC insertion increases the peak fields at the coils for constant magnet gradients and implies for the HL-LHC the use of novel Nb_3Sn magnet technology and a reduction of the triplet magnet gradients with respect to the nominal LHC configuration. The use of lower quadrupole gradients implies in turn longer triplet magnets (the functional quantity is given by the integrated magnet gradients) and an increase in length of the common beam pipe region next to the IP. The use of superconducting recombination dipole magnets in IR1 and IR5 allows to a large extend a compensation of the length increase of the common vacuum beam pipe region and it limits the increase in unwanted parasitic collision points of the two beams at an acceptable level.

Figure 1 shows the HL-LHC baseline insertion layout (top) together with the layout of the present nominal LHC machine (bottom) [3].

The installation of additional collimators in the dispersion suppressors (DS) next to IR2, the insertion for the ion-physics detector ALICE, is also part of the HL-LHC baseline. In the DS of IR7 two collimators per side are foreseen to be installed to cope with diffractive proton losses, while additional collimators in the DS of IR1 and IR5 are considered as options for the HL-LHC upgrade pending further results from the LHC operation experience in Run II. Figure 2 shows the

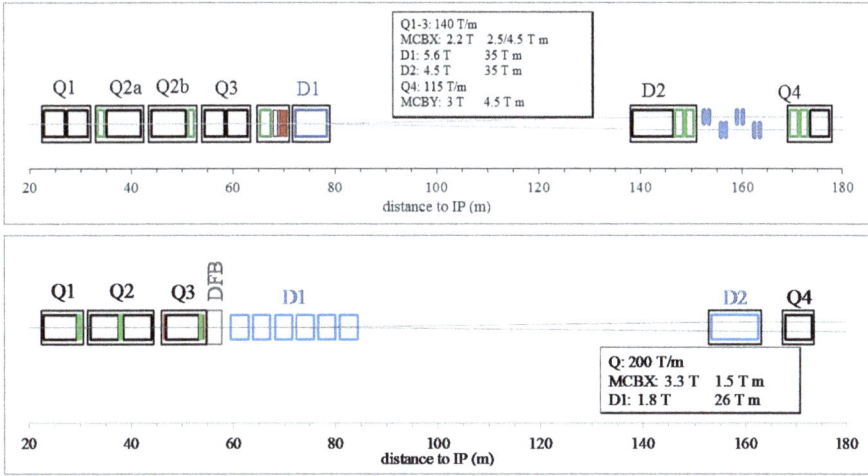

Fig. 1. Comparison of the LHC nominal (bottom) and the HL-LHC baseline layout for the high luminosity insertions (top).

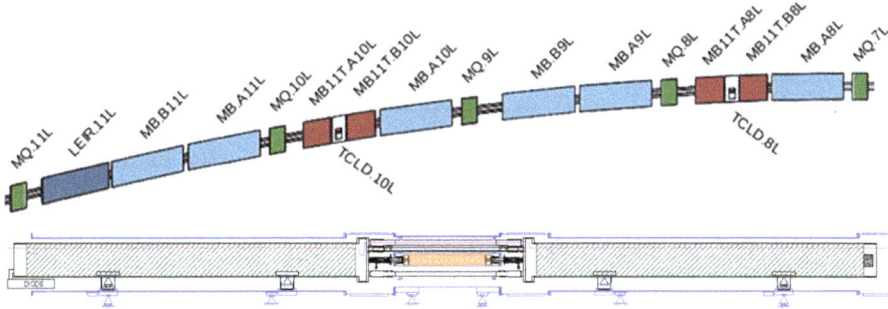

Fig. 2. Design for P7 11 T cryo-assembly per side.

schematic illustration of the required modifications in the Dispersion Suppressor of IR7, which features two collimators per DS instead of the only one collimator per DS between Q9 and Q10 in IR2.

Additional layout modifications are still being examined (e.g. the installation of higher or lower harmonic RF systems and a hollow electron lens for beam halo cleaning in IR4). However, these layout modifications are not yet part of the HL-LHC baseline configuration.

Other layout modifications, like the removal of the power converters from the tunnel area and the change of the feed-boxes for powering the insertion magnets via a superconducting link have no direct impact on the optics and parameter choice for the HL-LHC baseline and will not be deeply discussed here. However, they are vital for improving the LHC efficiency and for achieving the required increase in the average physics fill length for the HL-LHC exploitation.

Table 1. High Luminosity LHC parameters (LHC nominal ones for comparison).

Parameter	Nominal LHC (design report)	HL-LHC 25ns (standard)	HL-LHC 25ns (BCMS)	HL-LHC 50ns
Beam energy in collision [TeV]	7	7	7	7
N_b	1.15E+11	2.2E+11	2.2E+11	3.5E+11
n_b	2808	2748	2604	1404
Number of collisions in IP1 and IP5	2808	2736 [1]	2592	1404
N_{tot}	3.2E+14	6.0E+14	5.7E+14	4.9E+14
beam current [A]	0.58	1.09	1.03	0.89
x-ing angle [μrad]	285	590	590	590
beam separation [σ]	9.4	12.5	12.5	11.4
β^* [m]	0.55	0.15	0.15	0.15
ε_n [μm]	3.75	2.50	2.50	3
ε_L [eVs]	2.50	2.50	2.50	2.50
r.m.s. energy spread	1.13E-04	1.13E-04	1.13E-04	1.13E-04
r.m.s. bunch length [m]	7.55E-02	7.55E-02	7.55E-02	7.55E-02
IBS horizontal [h]	80 -> 106	18.5	18.5	17.2
IBS longitudinal [h]	61 -> 60	20.4	20.4	16.1
Piwinski parameter	0.65	3.14	3.14	2.87
Geometric loss factor R0 without crab-cavity	0.836	0.305	0.305	0.331
Geometric loss factor R1 with crab-cavity	(0.981)	0.829	0.829	0.838
beam-beam / IP without Crab Cavity	3.1E-03	3.3E-03	3.3E-03	4.7E-03
beam-beam / IP with Crab cavity	3.8E-03	1.1E-02	1.1E-02	1.4E-02
Peak Luminosity without crab-cavity [cm^{-2} s^{-1}]	1.00E+34	7.18E+34	6.80E+34	8.44E+34
Virtual Luminosity with crab-cavity: Lpeak*R1/R0 [cm^{-2} s^{-1}]	(1.18E+34)	19.54E+34	18.52E+34	21.38E+34
Events / crossing without levelling and without crab-cavity	27	198	198	454
Leveled Luminosity [cm^{-2} s^{-1}]	-	5.00E+34 [5]	5.00E+34	2.50E+34
Events / crossing (with leveling and crab-cavities for HL-LHC)	27	138	146	135
Peak line density of pile up event [event/mm] (max over stable beams)	0.21	1.25	1.31	1.20
Leveling time [h] (assuming no emittance growth)	-	8.3	7.6	18.0
Number of collisions in IP2/IP8	2808	2452/2524 [7]	2288/2396	0[4]/1404
N_b at SPS extraction [2]	1.20E+11	2.30E+11	2.30E+11	3.68E+11
n_b / injection	288	288	288	144
N_{tot} / injection	3.46E+13	6.62E+13	6.62E+13	5.30E+13
ε_n at SPS extraction [μm] [3]	3.40	2.00	< 2.00 [6]	2.30

[1] Assuming one less batch from the PS for machine protection (pilot injection, TL steering with 12 nominal bunches) and non-colliding bunches for experiments (background studies…). Note that due to RF beam loading the abort gap length must not exceed the 3 μs design value.

[2] An intensity loss of 5% distributed along the cycle is assumed from SPS extraction to collisions in the LHC.

[3] A transverse emittance blow-up of 10 to 15% on the average H/V emittance in addition to the 15% to 20% expected from intra-beam scattering (IBS) is assumed (to reach the 2.5 μm/3.0 μm of emittance in collision for 25 ns/50 ns operation).

[4] As of 2012 ALICE collided main bunches against low intensity. satellite bunches (few per-mill of main bunch) produced during the generation of the 50 ns beam in the injectors rather than two main bunches, hence the number of collisions is given as zero.

[5] For the design of the HL-LHC systems (collimators, triplet magnets,…), a design margin of 50% on the stated peak luminosity was agreed upon.

[6] For the BCMS scheme emittances well below 2.0 μm have already been achieved at LHC injection.

[7] The lower number of collisions in IR2/8 wrt to the general purpose detectors is a result of the agreed filling scheme, aiming as much as possible at a democratic sharing of collisions between the experiments.

2. Alternative Options

As mentioned in the above, HL-LHC project aims at achieving unprecedented peak luminosity and event pile-up per crossing by reducing the IP beta functions, increasing the bunch population and providing crab collisions with crab cavities. In the following three failure scenarios are described together with the possible alternatives to the baseline HL-LHC configuration that, implemented, would allow reaching the desired performance:

2.1. Performance limitations by longitudinal multi-bunch instabilities: These might be mitigated either with a higher harmonic 800 MHz RF system in addition to the nominal 400 MHz LHC RF system or a 200 MHz RF system as a new main RF system [4] and using the existing LHC 400 MHz system as a higher harmonic system. In both cases the RF systems should be operated in bunch shortening mode as this has been experimentally demonstrated in the SPS to be the robust approach for mitigating multi-bunch instabilities. This is in conflict with using the 800 MHz system for bunch lengthening as a means for reducing the peak pile-up density.

2.2. Performance limitations due to the electron cloud effect producing too large heat-load: This might be mitigated by using the 8b+4e filling scheme [5] or longer bunches with a 200 MHz main RF system. The 8b+4e scheme provides larger bunch charge with about 30% fewer bunches. The 200 MHz system might allow to provide bunches as long as 20 cm. Both options show in simulations a suppression of the electron-cloud in the dipoles throughout the full LHC cycle.

2.3. Crab cavities demonstrating not to be operational for hadron beams: SPS tests, machine protection issues, crab cavity impedance, or emittance growth due to RF phase noise might eventually suggest that crab cavities cannot be operated in the HL-LHC. In this scenario it is mandatory to resort to flat optics at the IP. Magnetic or electromagnetic wires [6] might be placed near the separation dipoles in order to compensate for the long-range interactions allowing for a reduction of the crossing angle and therefore increasing the luminous region. A 200 MHz RF system might also help if it allows increasing the bunch intensity. This is expected for single bunch limitations, however multi-bunch instabilities might dominate the performance limitations.

Another set of alternatives to the HL-LHC baseline configuration offer a better luminosity quality by reducing the pile-up density. It has been proposed that lowering the pile-up density might allow for a larger total pile-up and therefore larger luminosity [7]. Three alternatives in this direction follow:

- Peak pile-up leveling with β^*. This alternative does not require any extra hardware and only slows down the baseline β^* leveling to ensure a peak pile-up density below a target value. Since the largest peak pile-up is reached in the baseline only for a short time at the end of the β^* leveling process, it is possible to reduce this largest peak pile-up with little or negligible impact in the integrated luminosity [8].
- Longitudinal bunch profile flattening. Either the use of a lower harmonic 200 MHz or a higher harmonic RF system might be used in conjunction with the existing LHC 400 MHz RF system to lengthen and flatten the longitudinal bunch profile. However it has been remarked that this operational mode implies a substantial hardware upgrade and might be operationally challenging. Alternatively, RF phase modulation has already been success-fully used to slightly flatten the longitudinal bunch profile [9] in the LHC. Further studies of longitudinal bunch profile flattening are required to assess its potential for the HL-LHC. Combining this last option with peak pile-up leveling with β^* offers the lowest possible peak pile-up without significant impact on performance and without any hardware modification to the current baseline.
- Crab kissing [7]. This alternative can be realized in various ways. The initial proposal uses flat bunches, a magnetic or electromagnetic wire to reduce the crossing angle (to lower the crab cavity voltage) and crab cavities in the separation plane to maximize the luminous region. The compensating wire might not be needed if the each crab cavity achieves 5 MV (while the nominal voltage is 3.3 MV). The possibility of doing crab kissing in the crossing plane has also been explored.

Simulations of the fill evolution have been performed taking into account luminosity burn-off, intra-beam scattering and synchrotron radiation. The assumed event pile-up and the expected integrated luminosity per year are shown in Fig. 3 for all the mentioned alternatives. A 50% efficiency over one year of operation (160 days) is assumed. This means that the time in physics plus the time to come back to physics (turn-around time) is 80 days. A turn-around time of three hours has been used in simulations. Beam parameters and further details can be found at [10, 11].

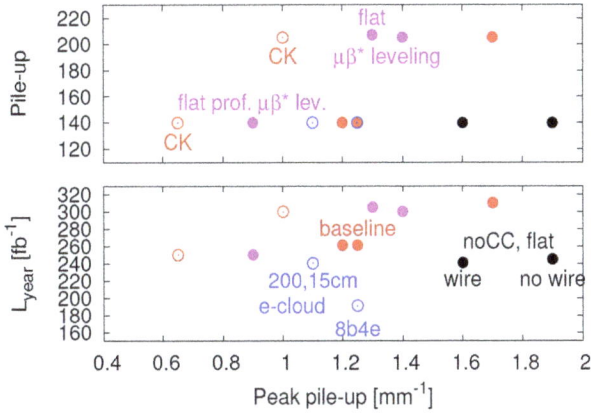

Fig. 3. Performance expectation of different alternatives. The red markers represent the baseline scenarios for different target of total pile-up (140 and 200 event per crossing). Flat optics (no crab cavities) with and without the wire compensator are shown in black. Crab kissing scheme is represented with orange markers. Peak pile-up density leveling with β^* is shown on magenta. In case e-cloud effects need to be mitigated, 200 MHz RF system or the 8b+4e filling scheme (green and blue markers) could be deployed.

3. HL-LHC the Geographical Distribution of the Upgrade Interventions

HL-LHC will require modifying the machine and infrastructure installations of the LHC in several points along the ring. In particular:

- Point 4
- Point 7
- Point 2
- Point 6
- Point 1
- Point 5

Points are listed according to the chronological order foreseen presently for the HL-LHC system installation.

3.1. *Point 4*

Point 4 will be equipped with a new cryogenic plant dedicated to the RF systems (and other cryogenic equipment that might be installed in IR4). The installation will require a warm compressor system on surface and a junction from the surface to the underground installation where a new cold box will be placed. The cold box will then feed a RF dedicated cryogenic distribution line.

3.2. *Point 7*

3.2.1. *The horizontal superconducting links*

In Point 7 two horizontal SC links will be installed in order to electrically feed the 600 A circuits connected to the 2 DFBAs (DFBAM and DFBAN). The related power converters will be installed in the TZ76 and will be connected to the superconducting link via short warm cables. The two superconducting links will then run for about 220 meters in the TZ76 and then enter into the LHC machine tunnel via the UJ76. They will then be routed for about 250 m in the LHC tunnel in order to be connected to the DFBAM and DFBAN (Fig. 4).

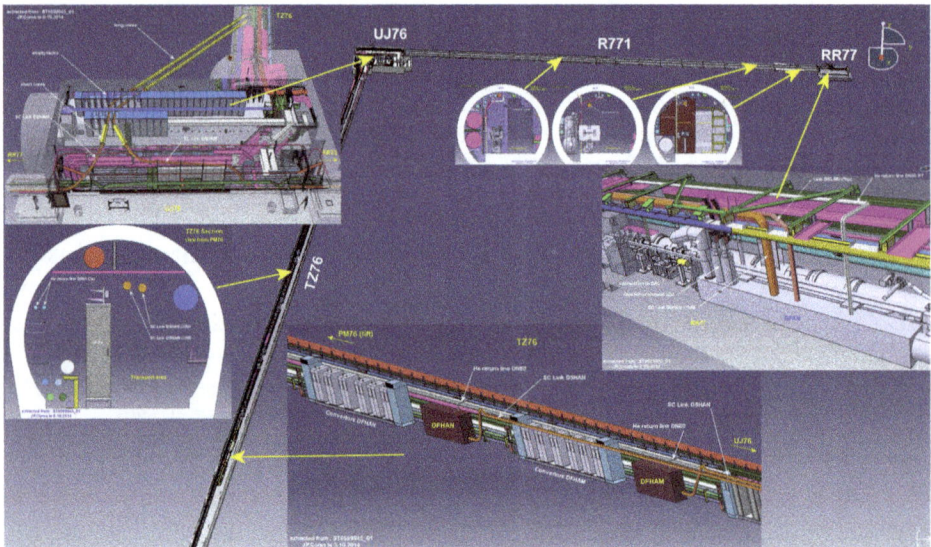

Fig. 4. View of the foreseen installation of the superconducting link system at Point 7. Power converters in the TZ76, routing of the link from the TZ76, via the UJ76 till the RR 77.

3.2.2. *New collimators in the dispersion suppressor*

In order to protect the superconducting magnets (excess heat deposition) from off-momentum proton leakage from the main collimator system itself, some special collimators must be installed in the Dispersion Suppression region, i.e. in the continuous cryostat. The evaluation of the real need of this modification will be completed on the base of the first results of the LHC Run II.

In order to cope with the proton losses in the Dispersion Suppressor area it has been decided to install two collimators on each side of the IP in the slots presently occupied by the Main Bending Magnets MB.B8L7 plus the MB.B10L7 and the symmetric MB.B8R7 plus the MB.B10R7. Each removed dipole will be replaced by a unit composed of two 11 T dipoles separated by a cryogenic by-pass. The collimator will be positioned on the top of the cryogenic by-pass.

3.3. *Point 2*

In order to limit the heat deposition from collision debris in the superconducting magnets during the ion run, collimators in the dispersion suppressor will also be installed in Point 2. In this case the installation will take place only in one slot on each side of the IP replacing the MB.A10L2 and MB.A10R2 main bends.

3.4. *Point 6*

In Point 6 the two quadrupole magnets Q5 will be modified in order to fulfil the needs of the new HL-LHC ATS optics. Two options presently under evaluation lead both to the exchange of the present Q5 with a new and higher gradient Q5.

3.5. *Point 1 and Point 5*

The largest part of the new equipment, required by the HL-LHC performance objectives, will be installed in Point 1 and Point 5. The items to be installed and actions to be carried out are listed below and are applicable to both points if not otherwise specified. The list is organized by geographical areas.

3.5.1. *LHC machine tunnel*

De-installation:

All the machine equipment from the interface with the experimental cavern, starting with the TAS, up to the DFBA (included) need to be removed. The present QRL will be also removed in the same area and a new return module will be installed to allow separating the flows of the coolant coming from the LHC QRL and the one from the new HL-LHC QRL Installation.

- o Installation of the new equipment probably in the following sequence:
  - TAXS
  - Services
  - QRL with related valve and service modules
  - Horizontal superconducting links from the DFM to the magnets
  - Magnets and crab cavity support system
  - Magnets and crab cavity

- TAXN
- Distribution feed boxes for the Q1 to D1 magnet system (DFX) and for the D2 to Q6 magnet system (DFM)

The sequence of installation of the vertical superconducting links to be connected to the DFX and DFM still need to be assessed according to the options retained for its routing.

3.5.2. *Existing LHC tunnel service areas*

The RRs on both sides of Point 1 and Point 5 will need to be re-organized and in particular it will be necessary to: de-install the power converter and other related systems linked to the powering of the removed LHC matching section and then to re-organize the remaining equipment in order to, increase if necessary the radiation shielding.

3.5.3. *New HL-LHC tunnel service areas*

The installation of the new cryogenic plant in Point 1 and Point 5 will have two objectives:

- Provide independent and redundant cooling capacity to feed the final focus and matching sections left and right of each of the two High Luminosity insertions of the LHC.
- Provide redundancy to the cryogenic plant installed to cool the experimental systems.

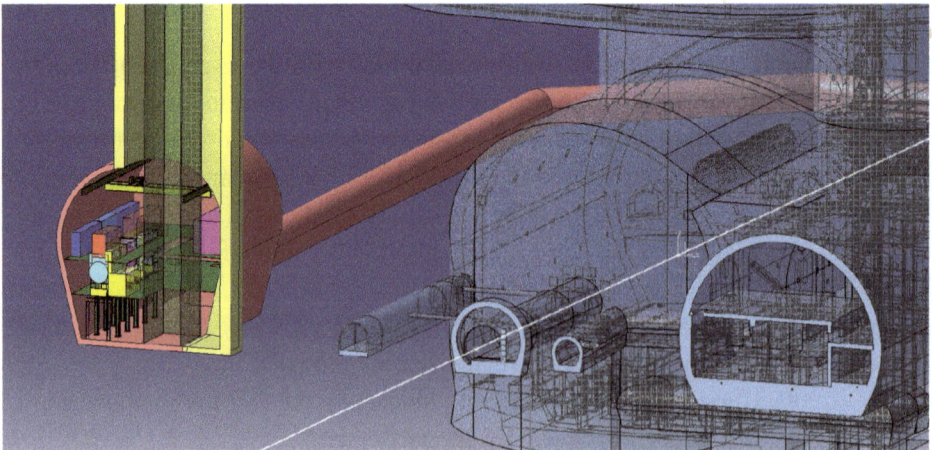

Fig. 5.   Possible option for underground installation of the cryogenic cold box in Point 5.

The cold box shall be installed in underground areas (Fig. 5). Presently the required volume does not exist. Therefore conceptual studies have started in order to identify the best options for building new underground caverns to install this equipment and the related service and control system. Two possible approaches are under study: the baseline corresponds to solutions with magnet power converters on surface, and a second one with power converters in the underground areas.

3.5.4. *New connection from the LHC tunnel and HL-LHC service areas to the surface*

The following connections between the surface and the underground installation shall be made available:

- LHC tunnel, crab cavity area, to the surface. The crab cavities need to be connected to the dedicated RF power system and their control system. The present preferred choice is to install these services in dedicated surface buildings.
- New HL-LHC service area to the surface. These connections are necessary to link the surface part of the cryogenic plant with the cold box installed in the new underground HL-LHC service areas.
- Vertical routing of the superconducting links. In each point at least four superconducting links will need to be routed from the surface to the underground areas.

3.5.5. *New surface installation*

The following installations shall find space on surface in Point 1 and Point 5 and in their proximities:

- Crab cavity RF power and services hosted in two ad hoc surface buildings. They shall be positioned on the surface, vertically directly above the tunnel position where the crab cavities will be installed. There will be two surface buildings for each point, one on the left part of the machine and one on the right part. The surface extremities of the ducts/shaft for the crab cavity coax or shaft shall be housed inside this building.
- Cryogenic installation. On surface the warm compressors and the other part of the cryogenic plant shall be installed.
- Power converters, upper extremities of the superconducting links, protection systems and energy extraction system related to the circuits fed via the

superconducting link. This area shall be possibly located near the surface part of the cryogenic plant and in any case on the top of the surface extremity of the routing of the vertical superconducting link.

4. Schedule

The HL-LHC schedule aims at the installation of the main HL-LHC hardware during LS3, together with the final upgrade of the experimental detectors (so-called upgrade Phase-II). However, a few items like the new cryogenic plant for P4, the 11 T dipole for DS collimation in P2 (for ions), the SC links in P7 and several prototypes for the collimation, beam instrumentation and injection and beam dump systems are already foreseen for LS2.

The HL-LHC schedule is based on the following milestones:

2014: Preliminary Design Report (PDR)
2015: End of Design Phase, release of the Technical Design Report (TDR)
2016: Proof of main hardware components on test benches
2017: Testing of prototypes (including crab cavity test in SPS) and release of TDR_v2
2017–2021: Construction and test of long lead hardware components (e.g. magnets, crab cavities, SC links, collimators)
2018–2019: LS2 — Installation of Cryo-plant P4, DS collimators (11 T) in P2, SC link in P7
2021–2022: String test of inner triplet
2023–2025: LS3 — Main installation (new magnets, crab cavities, cryo-plants, collimators, absorbers, etc.) and commissioning

The present schedule is based on the Project Product Breakdown structure and the HL-LHC lifecycle (Fig. 6). For each one of the components identified there is a simplified schedule that contains the time foreseen for the processes (see Fig. 7):

- Requirements definition
- Functional specification
- Engineering specification
- Acquisition
- Fabrication, assembly and verification
- Installation and commissioning

And the already identified time constrains and dependencies:

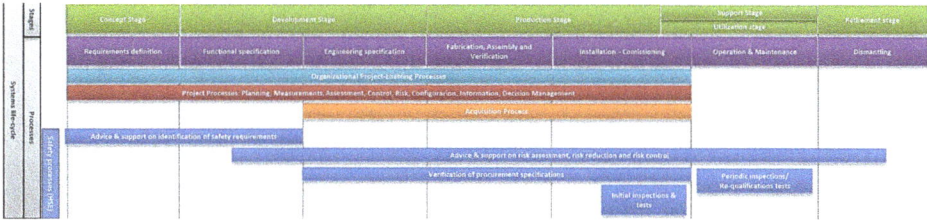

Fig. 6. HL-LHC life cycle processes.

Phase	2014	2015	2016	2017	2018	2019	2020	2021	2022	2023	2024
Requirements definition											
Functional specification											
Engineering specification											
Acquisition Process											
Fabrication, Assembly & Verification											
Installation - Comissioning											

Fig. 7. HL-LHC simplified process schedule for a PBS element.

The schedule also takes in account other general principles such as the reduction of doses taken by the workers during the dismantling of the LHC components maximizing the cold down periods.

The baseline schedule manages also the variants. We call variants the present design alternatives inside the baseline. A variant is for example the installation in surface or underground of the new series of power convertors for the insertion magnets. The variants affect in most cases several components.

The schedule also contains the tasks linked to non-baseline components and their decision parameters.

References

[1] HL-LHC Layout and Parameter Committee, https://espace.cern.ch/HiLumi/PLC/default.aspx.

[2] S. Fartoukh, *Phys. Rev. Spec. Top. Accel. Beams* **16**, 111002 (2013), CERN-ACC-2013-0289.

[3] CERN CDD Drawing number LHCLSXH_0010.

[4] R. Calaga, HL-LHC roadmap, *LHC Performance Workshop*, Chamonix 2014.

[5] H. Damerau *et al.*, LIU exploring alternative ideas, in *Proceeding of RLIUP Workshop*, Archamps, 2013, page 127, https://cds.cern.ch/record/1629486/files/CERN-2014-006.pdf.

[6] A. Valishev and G. Stancari, arXiv:1312.1660.

[7] S. Fartoukh, *Phys. Rev. Spec. Top. Accel. Beams* **17**, 111001 (2014), CERN-ACC-NOTE-2014-0096.

[8] R. Tomás, O. Dominguez and S. White, HL-LHC alternatives, in *Proceedings of RLIUP Workshop*, Archamps, 2013, page 119, https://cds.cern.ch/record/1977364/files/119-126-Tomas.pdf.

[9] E. Shaposhnikova *et al.*, Flat bunches in the LHC, *IPAC 2014*.

[10] G. Arduini *et al.*, Beam Parameters at LHC injection, CERN-ACC-2014-0006.

[11] R. Tomas *et al.*, HL-LHC alternative scenarios, *LHC Performance Workshop*, Chamonix 2014.

Chapter 4

The HL-LHC Accelerator Physics Challenges

S. Fartoukh and F. Zimmermann

CERN, BE Department, Genève 23, CH-1211, Switzerland

The conceptual baseline of the HL-LHC project is reviewed, putting into perspective the main beam physics challenges of this new collider in comparison with the existing LHC, and the series of solutions and possible mitigation measures presently envisaged.

1. Introduction and General Description

The HL-LHC is being designed to deliver an integrated luminosity of at least $250\,\text{fb}^{-1}$/year in each of the two LHC General-Purpose Experiments ATLAS and CMS for proton operation [1], while operating the other two experiments, ALICE and LHCb, at very low and moderate instantaneous luminosity, of about 10^{31} and $1\text{--}2 \times 10^{33}$ cm^{-2}s^{-1}, respectively [2, 3]. The ambitious performance target for ATLAS and CMS cannot be met without pushing both the optics to their extreme, namely β^*, and the nominal parameters of the LHC beam. It relies as well on a number of very challenging new equipment and key innovative technologies, such as:

- new larger aperture super-conducting magnets in order to preserve the transverse acceptance of the two high-luminosity insertions at low β^*, and in particular new inner triplet quadrupoles with a 150 mm coil aperture, more than doubled with respect to the existing Nb-Ti triplet, but still operating at a gradient of 140 T/m (about 30% below the present value) thanks to the Nb$_3$Sn technology,
- crab cavities, which are high-frequency RF transverse deflectors and aim at preserving the luminosity gain with $1/\beta^*$ by ensuring head-on collisions at the interaction point (IP), despite the crossing angle which is needed to separate the two beams after the collision.

Last but not least, the maximum possible peak luminosity is in practice limited by several factors, in particular the number of pile up events per bunch crossing which can rapidly degrade the quality of the data collected for the physics analysis. In this respect, the HL-LHC relies on a constant instantaneous luminosity, not exceeding 5×10^{34} cm^{-2}s^{-1}, and corresponding to approximately 140 events in average per bunch crossing for operation with 25 ns bunch spacing (and more precisely about 2,750 bunches per beam, see Table 1). This is achieved through challenging luminosity leveling techniques, for instance via a gradual reduction of β^* compensating for the proton burn off during the physics coast. In order to sustain such a high luminosity, over typically 10 hours of stable beam, the beam parameters, in particular the total beam current, shall correspond to a so-called virtual luminosity 4 to 5 times higher than the actual (leveled) luminosity. The virtual luminosity would be attained if all the other parameters, for instance β^*, were pushed to their respective limits at the very beginning of a physics fill. The baseline HL-LHC parameters (25 ns version) are listed in Table 1 (see Chapter 3 for more detail), including some key quantities such as the virtual luminosity introduced above. The leveling time of 9 h has been calculated assuming no emittance growth and a total hadron cross-section of 100 mb. The numbers in parentheses in the right column of Table 1 correspond to an ultimate β^* of 10 cm, for which most of the new equipment would be pushed to the limits.

Table 1. Baseline parameters of the HL-LHC (25 ns version) and comparison with the nominal LHC. The numbers in parentheses refer to an ultimate β^* of 10 cm.

Parameters	Nominal LHC	HL-LHC
Energy [TeV]	7	7
Bunch spacing [ns]	25	25
Number of bunches	2,736 (up to 2,808)	2,736 (up to 2,808)
Bunch charge [10^{11}]	1.15	2.2
Total current [A]	0.58	1.11
Bunch length [cm]	7.50	7.50
Energy spread [10^{-4}]	1.20	1.20
Long. emittance [eVs]	2.50	2.50
β^* [cm]	55	15 (10)
Full crossing angle [μrad]	300	590 (720)
Beam separation [σ]	9.9	12.5
Normalized transverse emittance [μm]	3.75	2.5
Peak luminosity (peak w/o crab cavity) [10^{34} cm^{-2}s^{-1}]	1.0	7.4 (7.7)
Virtual luminosity (peak with crab cavity) [10^{34} cm^{-2}s^{-1}]	1.2	21.9 (30.1)
Leveled luminosity [10^{34} cm^{-2}s^{-1}]	NA	5.0
Leveling time [h]	NA	9.0 (10.2)
Pile up event/crossing	≤ 28	135–138

The matchability of such low β^* optics within the specific layout constraints of the LHC, while correcting a series of side effects of chromatic nature (in first instance the linear chromaticity, but also the non-linear chromaticity, the spurious dispersion, and off-momentum β-beating), has been a long-standing problem which was recently solved by the development of a novel optics concept, the Achromatic Telescopic Squeezing (ATS) scheme [4–6] (see Section 1.1). As targeted in Table 1, doubling the beam current and tripling the beam brightness with respect to the nominal parameters represents a challenge in terms of collective effects (instabilities, intra-beam scattering, beam–beam effects, and electron cloud), which will be reviewed in Section 1.2. Finally, the pros and cons of the various possible options for luminosity leveling will be addressed in Section 1.3, while the last section of the chapter will present a summary and conclusions.

1.1. *Optics*

1.1.1. *Optics constraints and challenges*

Reducing the beam sizes at the interaction point, that is acting on β^* at constant transverse emittances, is a key ingredient to boost the performance of any collider. Going in this direction, a series of limitations shall however be overcome, driven by the mechanical aperture available in the inner triplet (IT), and also coming from the rest of the ring. Concerning the demand on the mechanical acceptance of the inner triplet, one can always find a solution based on sufficiently large aperture quadrupoles, by weakening their gradient but making them longer (at more or less constant integrated gradient), regardless of β^* and of the technology chosen for the triplet [7–9]. Indeed decreasing the operational gradient G of the inner triplet at constant β^* and integrated strength, the aperture needed for the beam roughly scales like $1/G^{1/4}$ (or said differently the peak β-function β_{max} reached in the IT is found to increase with $1/\sqrt{G}$), while the coil aperture can in principle be increased with $1/G$ at constant peak field, that is much faster.

The real optics challenge for low β^* is actually elsewhere, i.e. on the "non-triplet" side of the machine [10], where a series of limitations were clearly identified and classified in the framework of the previous upgrade project of the LHC, the so-called Phase I Luminosity Upgrade project [11]. While of very different nature, these limitations can be quantified by the maximum possible peak β-function which is permitted in the inner triplet, namely β_{max}. Indeed, this β_{max} shall then be matched to the regular optics of the arcs within the fixed distance given by the length of the low-β insertion, also known as interaction region (IR), and within the aperture and gradient limits of the IR magnets (quadrupoles of the matching section and of the dispersion suppressor). Finally, a clear strategy shall be established to ensure a proper control of the chromatic aberrations induced, without exceeding the available strength of the lattice sextupoles. The beam observables to be corrected are not only the linear chromaticity Q', which is increasing linearly with β_{max}, but also the non-linear-chromaticities Q'', Q''', ..., the off-momentum β-beating $\partial\beta/\partial\delta$ (i.e. the chromatic variations of the β-functions), and the spurious dispersion induced by the crossing angle in the low-β insertions. Assuming an upgrade which would essentially only rely on the replacement of the existing inner triplet (see e.g. the former upgrade project of the LHC [11]), and no deep conceptual changes in general beam optics for circular colliders, these limitations can rapidly turn into hard limits driven by the existing hardware in large parts of the ring, and given by:

- the mechanical acceptance of the existing matching section,
- the gradient limits of the matching quadrupoles, and
- the strength limits of the arc sextupoles.

As a result, the minimum possible β^* was found to be around 30 cm assuming a Nb-Ti triplet of 120 mm aperture operating at 120 T/m, and only slightly less ($\beta^* \sim 25$ cm), for a triplet of the same aperture but operating with a 50% stronger gradient thanks to the Nb_3Sn technology [10]. These minimum values of β^* were also considered as hard limits at that time, leaving no operational margin, e.g. for fine tuning the optics, the tune, the chromaticity, etc.

1.1.2. *The Achromatic Telescopic Squeezing (ATS) scheme as baseline for the HL-LHC optics*

Concerning the first limitation previously mentioned, the only solution is to equip the LHC matching sections with new two-in-one magnets of larger aperture. Concerning the poor optics flexibility observed at low β^* in the experimental insertions IR1 and IR5, with some quadrupoles being pushed to very low or very high gradients in the matching section and dispersion suppressor, respectively, one possibility is to allow floating matching conditions at the boundaries of these two insertions. More precisely the idea is to maintain the dispersion matching constraints at the entry and exit of the low-β insertions (from Q13.L to Q13.R), but to allow the "auxiliary" insertions on either side (IR8/2 for IR1 and IR4/6 for IR5) to contribute as well to the matching of the β-functions, at least below a certain value of β^*. As a result, β-beating waves are generated in the sectors adjacent to the low-β insertions (sectors 45 and 56 for IR5 and sectors 81 and 12 for IR1). Assuming a phase advance per arc cell strictly matched to $\pi/2$ in these sectors, and if correctly phased with respect to the IP, these waves will reach their maximum at every other sextupole, i.e. at the sextupoles belonging to the same electrical circuit (see Fig. 1).

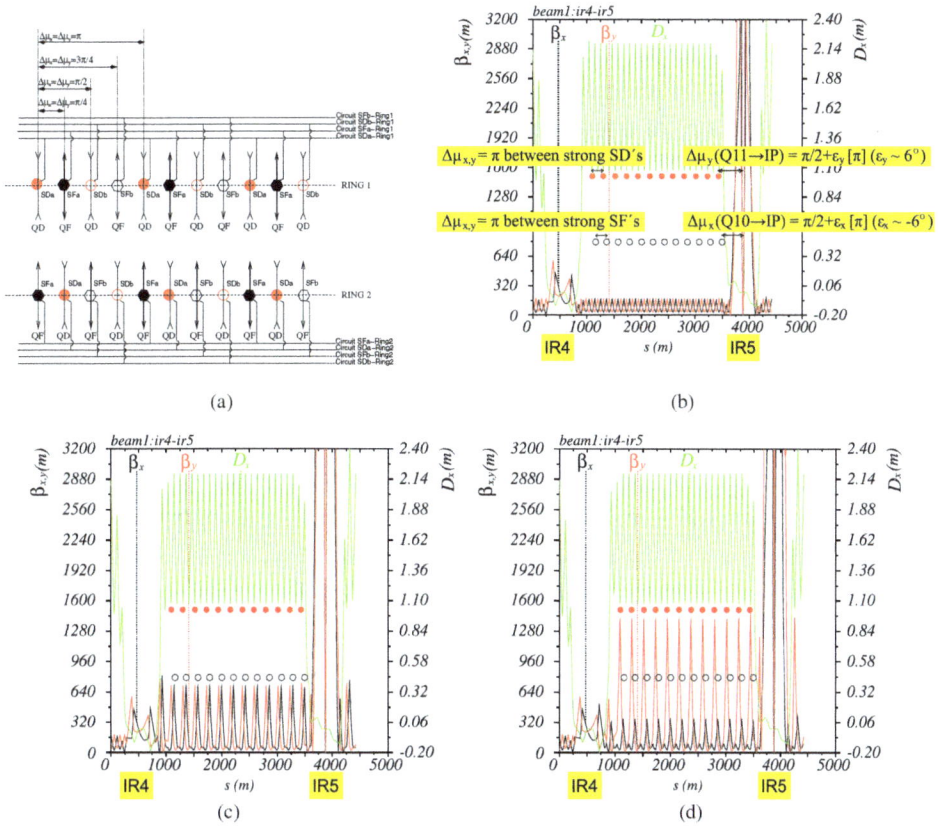

Fig. 1. LHC sextupole powering scheme (a) and zoom of the optics in sector 45: First a pre-squeezed optics (b) with its phasing properties and a typical β^* value of 40 cm, followed by two possible telescopic optics further reducing β^* in a symmetric ("round" optics in (c) with $\beta^*_x = \beta^*_y$) or asymmetric way ("flat" optics in (d) with $\beta^*_x \neq \beta^*_y$). The dispersion function remains matched in the arcs for the telescopic optics, but the β-functions are mismatched reaching their maximum at every other sextupole, i.e. at sextupoles belonging to the same electrical circuit. The relative increase of these maxima with respect to the pre-squeezed optics is inversely proportional to the additional reduction of β^*. As a result, during the telescopic squeeze the chromatic correction of the inner triplet can be achieved at nearly constant sextupole strength.

Consequently, the chromatic correction efficiency of these sextupoles will drastically increase at constant strength which, de facto, will be a definite cure for the third limitation previously mentioned.

This novel approach is particularly well-suited to the LHC for the two following reasons. First, due to the large dynamic range of machine energy, from 450 GeV to 7 TeV, and the reduction in proportion of the transverse emittances during the ramp, the peak β-functions in the arcs could in principle be increased by a factor of about 16 at top energy without exceeding any aperture-related limits (in practice a bit less less since it is advisable to increase the margins at higher energy). Then, at flat top energy, the quadrupole magnets of the so-called "auxiliary" insertions are either moderately pushed, which is the case for the experimental insertions IR8 and IR2 assuming a β^* of a few meters in p-p collision mode, or not pushed at all, in the case of IR4 and IR6 for which the injection optics is kept unchanged during the whole LHC cycle. Therefore all the ingredients are already available in the existing machine to blow up the β-functions in the arcs 81/12/45/56 at 7 TeV, and to implement the principle of the ATS scheme.

A comprehensive description of the scheme can be found in Ref. [6], in particular concerning the constraints imposed on the betatron phases over the left and right side of the low-β insertions, and the optics squeeze which is achieved in a two-stage telescopic mode:

(i) first of all a so-called *pre-squeeze*, which is a more or less standard squeeze, with additional matching constraints, but acting only on the matching quadrupoles of the low-β insertions proper and on the arc sextupoles, till reaching some strength limitations, either for the IR magnets or for the chromaticity sextupole in the arcs,

(ii) then the *telescopic squeeze*, using only the matching quadrupoles of the neighbouring insertions (IR2/8 for squeezing IR1, and IR4/6 for squeezing IR5), and at constant strength for the chromaticity sextupoles of the arcs.

With the exception of the Q5 matching quadrupoles of IR6 (which would need to be made 25% longer for 7 TeV operation), and heavier interventions obviously needed in the matching sections and inner triplets of IR1 and IR5 (see later), the ATS scheme has been found to be fully compatible with the existing LHC hardware and layout, in order to produce and ensure the chromatic correction of the collision optics with extremely small β^*, down to 5–10 cm. ATS optics have been built in practice assuming several possible triplet layouts, e.g. the (120 T/m, 120 mm full aperture) Nb-Ti triplet proposed for the former Phase I LHC upgrade project, two other intermediate triplet layouts with an aperture increased up to 140 mm and compatible with the Nb-Ti or Nb$_3$Sn technology [12] (i.e. with an operating gradient of 100 T/m or 150 T/m, respectively), and more recently with the latest (140 T/m, 150 mm) baseline triplet of the HL-LHC [13]. The main difference between these different cases is the so-called pre-squeezed β^*, ranging from 50 to 40 cm (again with an approximate scaling like $1/\sqrt{G}$), and below which the telescopic techniques of the ATS need to be deployed to reduce β^* further down at constant strength for the chromaticity sextupoles of the lattice. A completely new version of the LHC optics based on the ATS scheme was also developed, strictly compatible with the existing layout of the LHC, in particular with its existing (200 T/m–70 mm) Nb-Ti inner triplets. This allowed testing and successfully

about 31 km. The minimum possible pre-squeezed
case (limited by the strength of the arc sextupoles). This means that a mismatch
of the β-functions by 440% needs to be generated in the arcs 81, 12, 45 and 56
(i.e. with peak β-functions increased by a factor 4.4) in order to build up an ATS
collision optics with $\beta^* = 10$ cm. These
Fig. 2.

A zoomed view of this collision optics is presented in Figs. 3(a)–(c), for the
four experimental insertions of the HL-LHC, namely

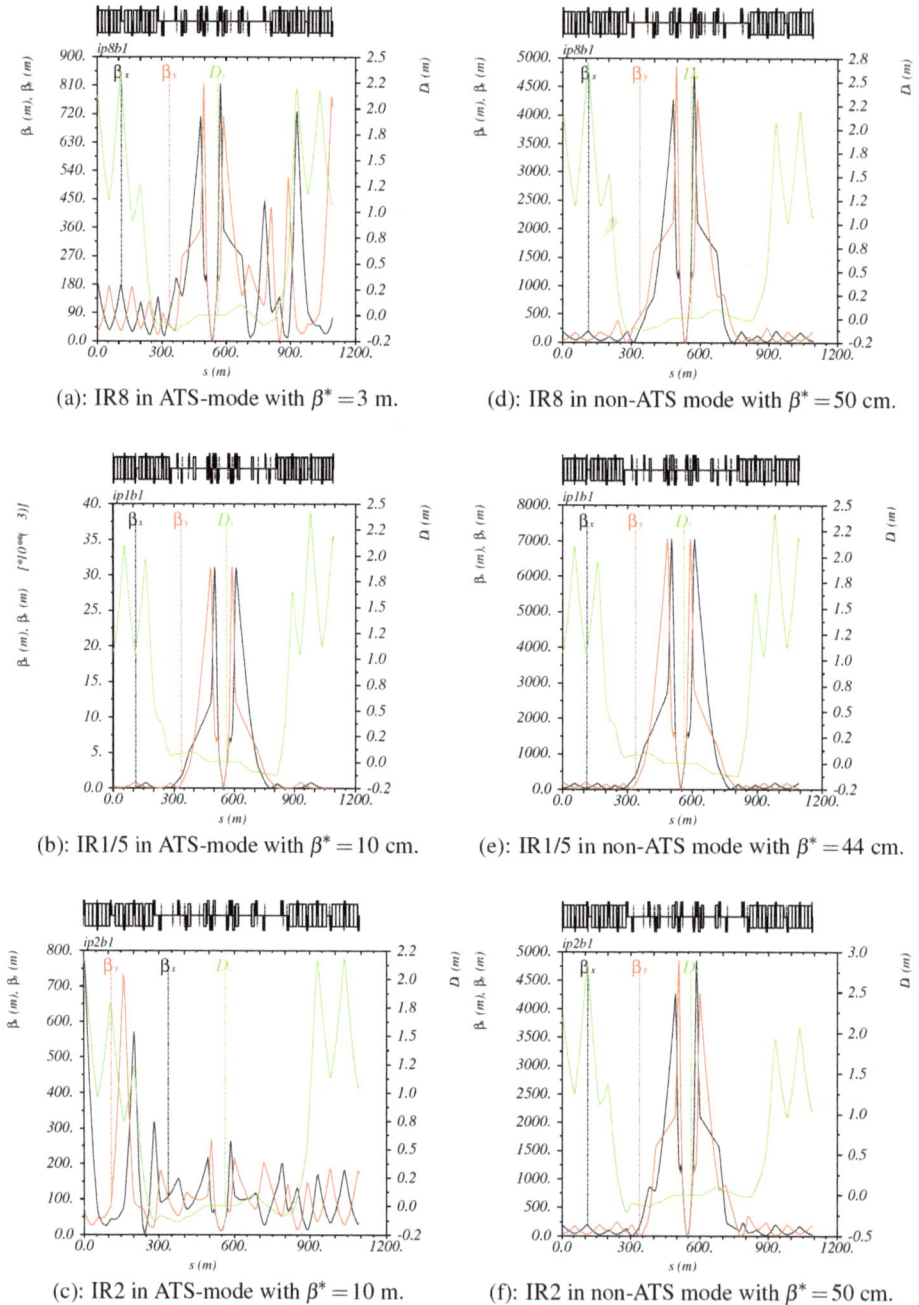

(a): IR8 in ATS-mode with $\beta^* = 3$ m.

(d): IR8 in non-ATS mode with $\beta^* = 50$ cm.

(b): IR1/5 in ATS-mode with $\beta^* = 10$ cm.

(e): IR1/5 in non-ATS mode with $\beta^* = 44$ cm.

(c): IR2 in ATS-mode with $\beta^* = 10$ m.

(f): IR2 in non-ATS mode with $\beta^* = 50$ cm.

Fig. 3.   Zoomed view of collision optics in IR8, IR1/5 and IR2 running the LHC in ATS-mode for proton-proton physics (left pictures), and in non-ATS mode for ion or proton-ion physics (right picture). In ATS-mode β-beating waves are initiated on the right side of IR8 and absorbed on the left side of IR2 (see Figs. (a) and (c)) in order to gain a factor of 4 to 5 in the β^* reduction at IP1.

- LHCb (IR8) squeezed to an intermediate β^* of 3 m which is more than enough for sustaining over about 10 h a luminosity leveled to 1–2×10^{33} cm^{-2}s^{-1} (assuming the HL-LHC beam parameters given in Table 1).
- Alice (IR2) with $\beta^* = 10$ m where halo collisions (with a large beam–beam separation at the IP of more than 5σ) might be requested in p-p mode, without exceeding an instantaneous luminosity of 10^{31} cm^{-2}s^{-1}.
- ATLAS and CMS (IR1 and IR5) with $\beta^* = 10$ cm, where a factor of about 4 in β^* reduction is actually supported by the β-beating waves generated and absorbed by the matching quadrupoles located on the right of IP8 (resp. IP4) and on the left of IP2 (resp. IP6) for ATLAS (resp. CMS).

Starting from the same injection optics as the one used to build up the 10 cm ATS optics described above, the nominal functionality of the LHC experimental insertions is still preserved: in particular IR2 and IR8 can still be squeezed in the usual way down to their nominal β^*, or even slightly below (namely one can achieve $\beta^* = 44$ cm at IP1 and IP5, together with $\beta^* = 50$ cm at IP2 and IP8), assuming ion or proton–ion physics during the HL-LHC runs (see Figs. 3(d)–(f)).

The chromatic properties of ATS optics are particularly interesting to analyze. Pushing β^* down to 10 cm, the chromatic variations of the betatron tunes are only moderately perturbed by a second order chromaticity Q'' showing up over a momentum range of $\delta_p = \pm 10^{-3}$ corresponding to about three times the momentum acceptance of the LHC RF bucket at 7 TeV (see Fig. 4(a)). The chromatic Montague functions (giving the amplitude of the first order chromatic derivative of the β-functions) are nicely vanishing in the collimation insertions IR3 and IR7 and at IP1 and IP5 (see Fig. 4(b)), therefore with no impact on the collimation hierarchy, nor on the machine performance. Another important feature lies in the fact that the off-momentum β-beating waves induced in the two planes by the lattice sextupoles are exactly out of phase by $\pi/2$ with respect to the β-functions themselves, in particular in the triplet and its neighboring magnets. Therefore, no further degradation of the off-momentum mechanical aperture is induced in the arcs, the matching section and the new triplet, except the usual one coming from the contribution of the dispersion, which remains perfectly matched in the ATS scheme. Finally, an extremely important quantity to control is the spurious dispersion induced by the crossing scheme in IR1 and IR5. This dispersion can reach up to 20 m in the new triplets when pushing β^* down to 10 cm with a full crossing angle of 720 μrad (see Table 1). The spurious dispersion is produced by feed-down effects in the inner triplets of one of the two high luminosity insertions, and then exported to the other one. However, thanks to the specific phasing conditions imposed by the ATS scheme, modest H or V orbit bumps of the order of 4–5 mm generated in the sectors adjacent to IR1 and IR5 are found to be sufficient to correct it back to a level of $\sim$0.5–1 m in the inner triplet (see Figs. 4(c) and (d)).

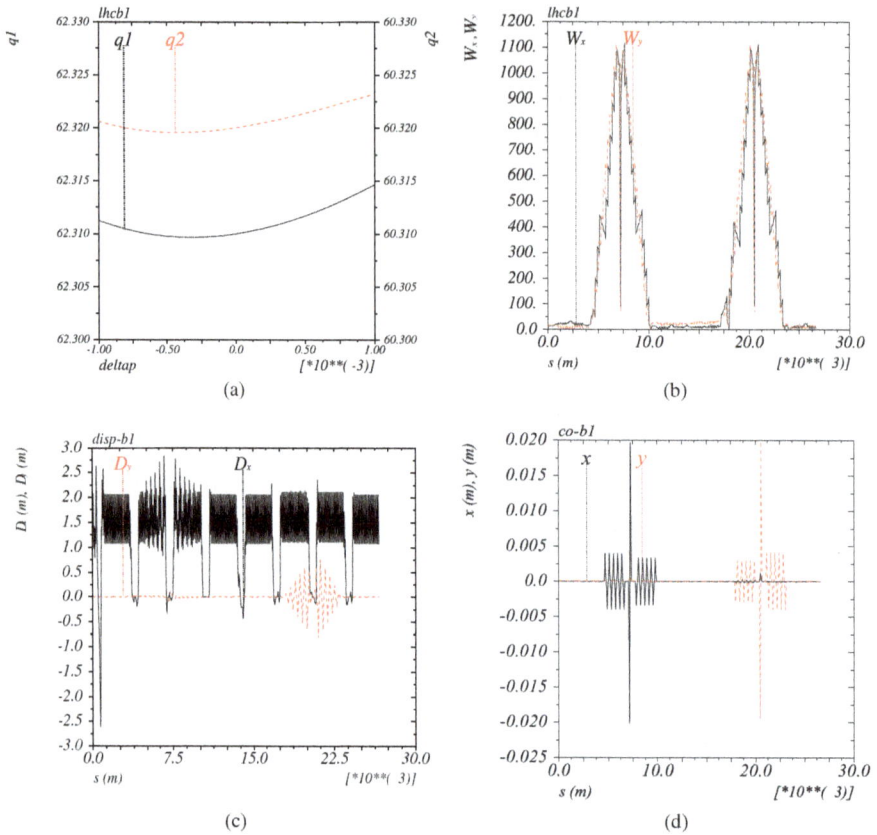

Fig. 4.   Chromatic properties of the 10 cm ATS optics: (a) Chromatic variations of the betatron tunes (assuming the linear chromaticity to be matched to 2 units using standard Q' knobs), (b) Montague functions $W_{x,y}$ around the LHC ring with IP3 chosen as the origin, and (c) residual horizontal and vertical dispersion mismatch induced by the horizontal and vertical crossing angles at IP1 and IP5, but minimized thanks to orbit bumps generated in the arcs 81, 12, 45 and 56 (d). Without this correction, the horizontal and vertical dispersion would be fully mismatched around the ring, in particular in the collimation insertions IR3 and IR7, the RF insertions IR4 and the dump insertion IR6, and may reach up to 20 m in the inner triplet at $\beta^* = 10$ cm, depending on the betatron phase advances between IP1 and IP5.

1.1.3. *Crossing angle and crab cavities*

In order to separate the two beams after the collision, a crossing angle Θ_c is imposed at the interaction point. When reducing β^*, this angle shall be increased in order to guarantee a sufficiently large normalized beam–beam separation d_{bb}, typically of the order of 10σ for the nominal LHC, but which shall be increased up to 12.5σ for the HL-LHC, with longer triplets and therefore a lengthening of the region where the two beams share the same vacuum beam pipe and continue to interact with each other:

$$\Theta_c = d_{bb} \times \sqrt{\frac{\varepsilon}{\beta^*}} = \frac{d_{bb}\,\sigma^*}{\beta^*}, \tag{1}$$

where ε and $\sigma^* = \sqrt{\varepsilon\beta^*}$ denote the 1σ physical beam emittance and the r.m.s. spot size at the IP, respectively. Increasing the crossing angle, however, affects directly the so-called Piwinski angle ϕ_w which characterizes the overlap of the two collid-ing beam distributions, and therefore the luminosity in the presence of a non-zero crossing angle:

$$\phi_w \equiv \frac{\Theta_c\,\sigma_z}{2\,\sigma^*} = \frac{\beta_w^*}{\beta^*}, \tag{2}$$

where σ_z is the r.m.s. bunch length and β_w^* is a characteristic β^* defined by

$$\beta_w^* \equiv \frac{d_{bb}\,\sigma_z}{2}. \tag{3}$$

The potential gain of luminosity with $1/\beta^*$ saturates rapidly below this character-istic β^*, of the order of $\beta_w^* \sim$ 40–50 cm for typical (HL)-LHC parameters:

$$\mathcal{L}(\beta^*) \propto \frac{1}{\beta^*\,\sqrt{1+\phi_w^2}} = \frac{1}{\beta^*\,\sqrt{1+(\beta_w^*/\beta^*)^2}} \xrightarrow{\beta^* \ll \beta_w^*} \frac{1}{\beta_w^*}. \tag{4}$$

A possible mitigation measure consists of using flat optics, with a β^* as small as possible in the parallel separation plane, if possible down to $\beta_\parallel^* \sim \sigma_z$, and of the order of $\beta_X^* \sim \beta_w^*$ in the crossing plane [4]. Indeed, this strategy still leads to an in-crease of the luminosity with $1/\sqrt{\beta_\parallel^*}$, until saturation occurs due to the hourglass effect when $\beta_\parallel^*$ becomes comparable to the r.m.s. bunch length. Another possibility to fully restore the luminosity gain with $1/\beta^*$ is the use of so-called crab cavities (deflecting RF dipoles) which aim at maximizing the overlap of the two beam dis-tributions despite the crossing angle, as illustrated in Fig. 5. The HL-LHC project has chosen the second option, which is more challenging in terms of hardware but also more performing.

The so-called compact crab cavities [16], running at the main RF frequency of $\omega_{CC}/(2\pi) = 400$ MHz, are planned to be installed in between the recombina-tion dipole D2 and the first quadrupole Q4 of the matching section [17], at about ± 150 m on either side of the IP where the beam separation has reached its nominal value of 194 mm. The β functions are still relatively high at this location, of the order of $\beta_{CC} \sim$ 3.5–4 km for $\beta^* = 15$ cm, and the phase advance with respect to the IP is still very close to $90°$. This in turn reduces the required voltage of the RF deflecting kick, but also and mainly ensures the quasi-closure of the RF bumps on

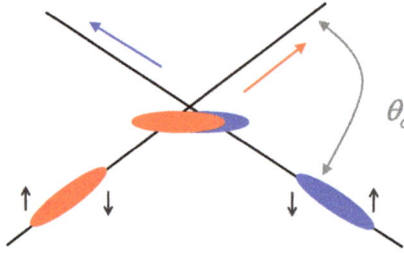

Fig. 5. Bunch rotation induced by crab cavities, in order to maximize the overlap of the two beam distributions at the interaction point, and therefore the luminosity, in the presence of a non-zero crossing angle.

either side of the IP. The total RF deflecting voltage required per beam and per IP side, V_{CC}, is given by

$$\frac{V_{CC}}{E} = \frac{c}{\omega_{CC}} \times \frac{\Theta_c}{2\sqrt{\beta_{CC}\beta^*}},$$ (5)

where E denotes the beam energy [eV] and c is the speed of light. As an example, $V_{CC} \sim 12.5$ MV is needed in order to restore head-on collisions in the presence of a crossing angle set to $\Theta_c = 590\,\mu$rad at $\beta^* = 15$ cm (see Table 1).

In order to minimize the beam loading in the crab cavities, a strict control of the closed orbit is also demanded at the position of the crab cavities. At this location, however, the crossing and parallel separation bumps are not yet closed in the LHC, inducing in particular a non-zero closed orbit which can substantially vary depending on the operation mode of the machine (injection, ramp, squeeze, Vernier scans,...). It was however found that the closure of these bumps upstream of the crab cavities could actually be achieved by means of new strong orbit correctors which would be installed on the non-IP side of D2, and by reinforcing the triplet orbit corrector MCBX located on the non-IP side of Q3 (see Fig. 6 and Ref. [17] for more details).

1.1.4. Layout and mechanical aperture

After having introduced the two main ingredients, namely the ATS optics for producing and properly controlling the chromatic aberrations of very low β^* collision optics, and the crab cavities to maximize the efficiency of a reduced β^* in terms of luminosity performance, the need and the basic parameters (in particular the mechanical aperture) of new HL-LHC magnets and various equipments can be defined. First of all, it is clear that for the target β^* values ranging from 15 to 10 cm (see Table 1), that is a factor of 4 to 5 below the nominal collision β^* of the LHC, the peak β-functions reached in the inner triplet increase in proportion. Since the beam sizes go with the square root of the β-functions, the aperture of the

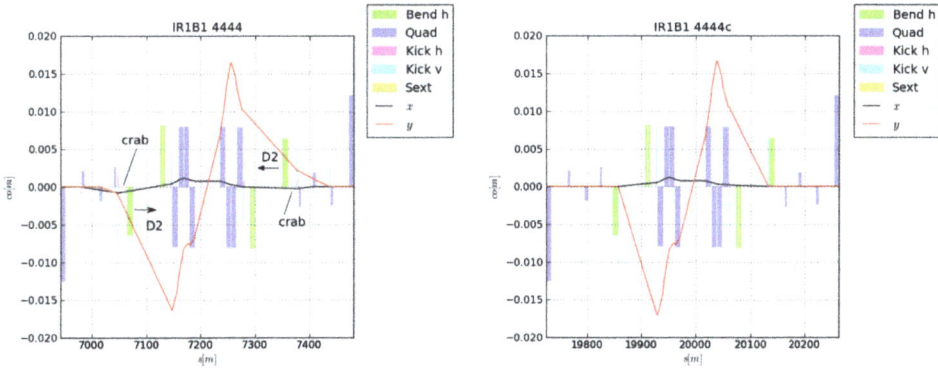

Fig. 6. Nominal (left) and new (right) crossing (red) and parallel separation (black) bumps. In both cases, the recombination dipole D2 has been displaced towards the IP in order to accommodate crab cavities on the IP-side of Q4. In the first case, these bumps are closed at Q6, as in the existing LHC. In the second case, the bumps can be closed before the crab cavities using the triplet orbit correctors (in particular the one on the non-IP side of Q3, see Section 1.1.4), but also additional orbit correctors installed on the non-IP side of D2 [17] (Courtesy of Riccardo de Maria).

new triplet shall be more or less doubled (with respect to the 70 mm coil aperture of the existing low-β quadrupoles), similarly for the TAS (which is an absorber in front of the first low-β-quadrupole Q1), and the separation dipole D1. Detailed aperture calculations demonstrated as well the need to replace the recombination dipole D2 and the first two quadrupoles of the matching section, namely Q4 and Q5, with magnets of larger aperture [4, 5, 12, 13]. In this respect, the coil aperture of the new triplet was fixed to 150 mm, which is compatible with an operating gradient of 140 T/m for the Nb_3Sn technology (see Chapter 6). While in the existing machine the separation dipole D1 is a normal conducting dipole made of six consecutive modules in IR1 and IR5, the new D1 will be superconducting in the high-luminosity insertions, with an aperture of 150 mm as for the new triplets. The aperture of the recombination dipole D2 will also be increased, from 80 mm to 105 mm, that of Q4 from 70 mm to 90 mm, and that of Q5 from 56 mm to 70 mm (see Table 2).

A sketch of the layout of the new triplet up to D1 is shown in Fig. 7, with a zoomed view of the triplet corrector package on the non-IP side of Q3. The main features of this layout, and in particular its modifications and similarities with respect to the existing one, are emphasized below:

- The triplet remains symmetric, that is with the same magnetic length for Q1 and Q3 ($L_{Q1} \equiv L_{Q3} \approx 8$ m, to be compared with 6.4 m for the existing magnets). Q1 and Q3 are however split into two 4-m-long magnets, for technical reasons. Also, contrary to the existing triplet layout, Q2a and Q2b are hosted in two sep-

Table 2. Aperture and performance specifications for the main HL-LHC equipments foreseen to be installed in IR1 and IR5. The numbers in parentheses refer to the nominal LHC.

Equipment	Coil aperture [mm] (gap for the existing D1)	Aperture separation [mm] for 2-in-1 equipment	Performance
TAS	60 (34)	-	-
Triplet (Q1-2-3)	150 (70)	-	140 T/m (205 T/m)
D1	150 (63)	-	35 T·m (27 T·m)
D2	105 (80)	186 (188)	35 T·m (35 T·m)
Crab cavity	84	194	12.5 MV/beam/IP side
Q4	90 (70)	194 (194)	550 T/m×m (550 T/m×m)
Q5	70 (56)	194 (194)	770 T/m×m (770 T/m×m)

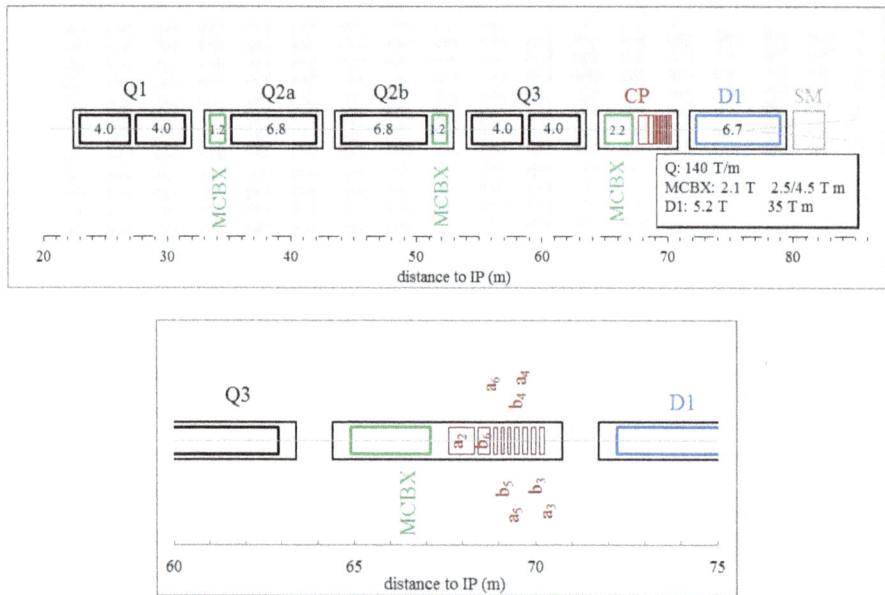

Fig. 7. Sketch of the layout of the new high-luminosity insertions from the IP up to D1 (top picture, courtesy of Ezio Todesco) and zoom into the triplet corrector package installed on the non-IP side of Q3 (bottom picture, courtesy of R. de Maria).

arate cryostats and have a magnetic length $L_{Q2a} \equiv L_{Q2b} \approx 6.8$ m, to be compared with 5.5 m for the existing Q2's.

- Nested horizontal and vertical orbit correctors, so-called MCBX, are installed on the non-IP side of Q2b and IP-side of Q2a, with an integrated strength which is rather similar to the one available in the existing MCBX magnets, but of course with a coil aperture of 150 mm. The integrated strength (or more precisely the length) of the MCBX installed in a corrector package on the non-IP side of Q3 (see below) is however almost doubled in order to cope with the requirement of the new crossing scheme. The main specifications of these orbit correctors are given in Table 3, together with the new orbit correctors which are needed on the non-IP side of D2 in order to close the crossing and separation bumps before the crab cavities.

- A series of multipole correction coils, for all orders from $n = 1$ to $n = 6$, normal and skew (except b_2), are hosted in a separated cryostat, the so-called triplet corrector package (CP) installed on the non-IP side of Q3. Compared to the nominal LHC, only the normal and skew decapole (namely a_5 and b_5), and the skew dodecapole (namely a_6) magnets are new types of triplet correctors. They were indeed found to be fundamental in order to preserve the dynamic aperture of the machine in the presence of field imperfections in the new triplet and D1 (see Section 1.1.5 and specification in Table 4).
- Finally, as already mentioned, the separation dipole D1 will be superconducting in the HL-LHC, with a coil aperture of 150 mm, a length of 6.7 m, and an integrated strength of about 35 T·m.

Table 3. Performance specification for the new triplet and D2 orbit correctors [13].

Corrector	Location	Aperture type	Integrated strength [T·m]
MCBX2a	IP-side of Q2a	Triplet	2.5 (in both planes)
MCBX2b	Non-IP side of Q2b	Triplet	2.5 (in both planes)
MCBX3	Non-IP side of Q3	Triplet	4.5 (in the crossing plane) 2.5 (in the parallel sep. plane)
MCBRD	Non-IP side of D2	D2	7.0 (in the crossing plane) 2.0 (in the parallel sep. plane)

Table 4. Performance specification for the triplet multipole correctors [18].

Corrector	Multipole	Aperture type	Integrated strength [mT·m] at 50 mm
MQSX	a_2	Triplet	1000
MCSSX	a_3	Triplet	63
MCSX	b_3	Triplet	63
MCOSX	a_4	Triplet	46
MCOX	b_4	Triplet	46
MCDSX	a_5	Triplet	25
MCDX	b_5	Triplet	25
MCTSX	a_6	Triplet	17
MCTX	b_6	Triplet	86

Concerning the layout of the matching section, D2, Q4 and Q5 are planned to be replaced by magnets of larger aperture. D2 will in addition be shifted by about 15 m towards the IP in order to liberate sufficient room for the integration of the crab cavities, and Q5 displaced by about 10 m towards the arcs, which was found to be optimal in terms of optics flexibility with the new longer HL-LHC triplet. On the other hand, from a strictly qualitative point of view, the conceptual layout of the matching section does not need to be further modified for the HL-LHC, thanks to the telescopic squeeze possibilities offered by the ATS.

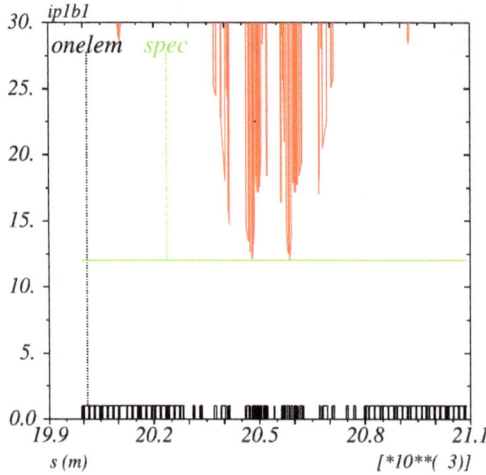

Fig. 8. Normalized aperture [σ] of the new high luminosity insertion (from Q7.L to Q7.R) calculated at 7 TeV, for a normalized emittance of 3.5 μm, a typical HL-LHC collision optics ($\beta^* = 15$ cm and a full crossing-angle of 590 μrad). Regardless of the beam emittance, the normalised settings of the collimators are defined for the above-mentioned reference emittance of 3.5 μm. The minimum aperture is located in the inner triplet, still compatible with the target of 12σ defined by the collimation and machine protection requirements (see Chapters 12 and 13), but assuming challenging mechanical tolerances (1.6 mm radially, including ground motion) and a budget for the linear optics distortions based on the LHC experience (2 mm for the closed orbit variations, and a β-beating not exceeding 10 %). The aperture model used for the new triplet and D1 is described in more detail in Chapter 10, including the specific shielding which is foreseen to maximize the lifetime of these new equipments. Concerning the other magnets, the beam-screen inner diameter has been taken equal to a certain fraction (typically 83%) of the coil diameters defined in Table 2.

The normalized aperture of the new high luminosity insertions (from Q7.L to Q7.R) is shown in Fig. 8, including some mechanical tolerances and a tolerance budget for the beam based on the LHC operational experience so far (i.e. 2 mm closed orbit and 10% β-beating). A minimum of 12σ is reached in the inner triplet at $\beta^* = 15$ cm, which is considered to be sufficient for machine protection and collimation (see Chapters 12 and 13).

1.1.5. *Dynamic aperture*

In the presence of non-linear field imperfections, both in the existing LHC magnets and in the new ones foreseen for the HL-LHC, the preservation of the dynamic aperture (DA), that is the region of the transverse phase space where the particle motion remains stable after a very large number of turns, is a fundamental input driving both the design of the optics, of the layout (in terms of need for correctors) and, of course, of the magnets themselves. The target dynamic aperture has been fixed to 10.5σ in collision for the baseline β^* of 15 cm, with a scaling like β^* for smaller β^* [19], i.e. reaching the 7σ extension of the secondary halo (see Chapter 13) for the ultimate β^* of 10 cm. More precisely, although the new HL-LHC magnets do not seem to degrade the DA for the injection optics at 450 GeV [20], with still quite relaxed assumptions concerning the field quality which is expected for these new magnets, the situation is much more critical in collision when the optics is squeezed. This fact is both related to the features of the collision optics, with a net increase of the β-functions in the arcs imposed by the ATS scheme, and of course the very large beam sizes reached in the inner triplet and downstream magnets at very small β^*.

Concerning the first item, a potential strong limitation of the ATS scheme is indeed its detrimental impact on the DA, due to the net increase of the peak β-functions in the arcs during the telescopic squeeze, combined with the non-linear field imperfections of the main dipoles and quadrupoles, and with the strength imposed on some sextupole families for the chromatic correction. On the other hand, thanks to the tremendous efforts which were deployed for the LHC main magnets during the construction and installation phases of the machine, both in terms of field quality specifications [21] and monitoring [22, 23], and in terms of sorting strategy [24], the existing machine was found compatible with the ATS scheme for this aspect, although not designed for it. More precisely,

- within an increase of the peak β-functions in the arcs by a factor of 4 in both planes (corresponding to a β^* of about 10 cm starting from an optics pre-squeezed to $\beta^* \approx 40$ cm), or up to 8 in one plane, but only 2 in the other plane (corresponding to a flat optics with at typical β^* aspect ratio of 4, e.g. with $\beta^* \approx 20/5$ cm in the crossing/separation planes, alternated between IR1 and IR5, when starting from a 40 cm pre-squeezed optics [4]),
- considering only in a first step the field imperfections of the arc magnets as measured and installed in the existing LHC ring (i.e. assuming new IR magnets for the HL-LHC which are ideal from the field quality point of view), and
- ensuring that the sextupole families participating in the chromatic correction do contain an even number of magnets (for a two-by-two compensation at π of their contribution to the third order resonance driving terms), the (HL)-LHC

dynamic aperture was found to be about 40 beam σ for the pre-squeezed optics, but dropping to about 15σ and 11σ for the ATS round and flat optics, described above [17, 25]. This net degradation determines the limit of the ATS scheme in terms of β^* reach. Fortunately, the β^* value achievable lies within (and even beyond) the HL-LHC target range.

Concerning the impact of the new matching section quadrupoles Q4 and Q5, and separation-recombination dipoles D1 and D2, and based on first estimates of their expected field quality, the situation is still relatively comfortable for the baseline β^* of 15 cm of the HL-LHC [26]. Preliminary designs of the new D2 magnet, however, seem to indicate that its field quality could be much worse than initially thought. Therefore, dedicated studies are ongoing to improve the D2 design, in order to decide whether or not this magnet needs to be equipped with dedicated spool-piece correctors. The latter would definitely solve the problem.

Implementing the triplet corrector package described in Section 1.1.4, and assuming the field quality presently expected for the inner triplet [27] (see Table 5), the target DA of 10.5σ seems to be within reach for $\beta^* = 15$ cm in collision (see Fig. 9(a)). However, dedicated studies and iterations with the magnet builders are still ongoing in order to identify, and if necessary to act on, the most dangerous multipoles [28], so as to clearly meet the target DA at $\beta^* = 15$ cm, and to further improve the situation in order to gain margin in view of the ultimate β^* of 10 cm (see Fig. 9(b)).

Table 5. Field imperfections expected at high current in the new 150 mm aperture inner triplet quadrupoles [27], expressed in terms of relative normal and skew multipole errors in units of 10^{-4} at a reference radius of 50 mm. A reduction (by at most 50%) is recommended for some multipoles [28] (see numbers in parenthesis). The superscripts S, U and R stands for systematic, uncertainty (maximum possible average deviation with respect to the expected systematic over the production), and random (giving the r.m.s fluctuation from magnets to magnets over the production).

n	a_n^S	a_n^U	a_n^R	b_n^S	b_n^U	b_n^R
3	0	0.800	0.800	0	0.820	0.820
4	0	0.650	0.650	0	0.570	0.570
5	0	0.430	0.430	0	0.420	0.420
6	0	0.310	0.310	0.800	1.100 (0.550)	1.100 (0.550)
7	0	0.190 (0.152)	0.190 (0.095)	0	0.190 (0.095)	0.190 (0.095)
8	0	0.110 (0.088)	0.110 (0.055)	0	0.130 (0.065)	0.130 (0.065)
9	0	0.080 (0.064)	0.080 (0.040)	0	0.070 (0.035)	0.070 (0.035)
10	0	0.040	0.040 (0.032)	0.150 (0.075)	0.200 (0.100)	0.200 (0.100)
11	0	0.026	0.026 (0.021)	0	0.026 (0.021)	0.026 (0.021)
12	0	0.014	0.014	0	0.018 (0.014)	0.018 (0.014)
13	0	0.010	0.010	0	0.009 (0.007)	0.009 (0.007)
14	0	0.005	0.005	-0.040 (-0.020)	0.023 (0.012)	0.023 (0.012)

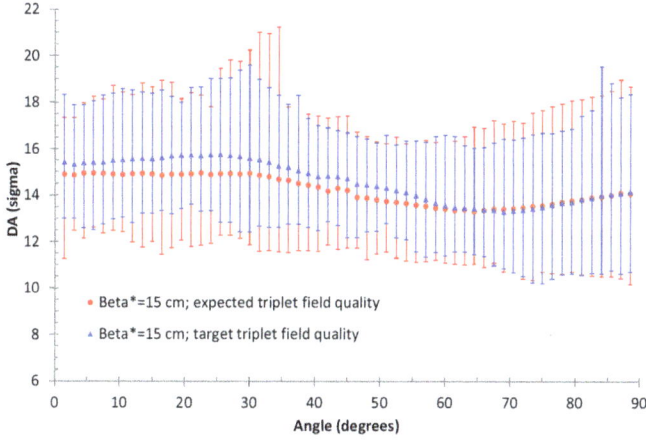

(a): $\beta^* = 15$ cm.

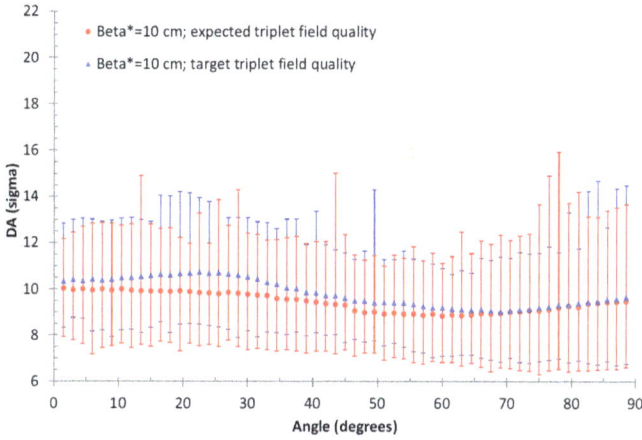

(b): $\beta^* = 10$ cm.

Fig. 9.   100,000 turns dynamic aperture [σ] of the HL-LHC in collision as a function of the phase space angle: minimum and maximum (error bars) and average (markers) obtained over 60 possible different configurations of the machine (seeds) in terms of field quality of the new inner triplets. The top and bottom pictures refer to an ATS collision optics where β^* has been matched to 15 cm and 10 cm, respectively, with the corresponding crossing angle given in Table 1. For both optics, the tracking has been performed with the expected and targeted triplet error tables of Table 5 (courtesy of M. Giovannozzi).

1.2. *Collective effects*

Collective beam effects arise from the electromagnetic interaction of the beam particles among themselves, with their environment (including electron cloud) and with the other beam. The two times higher than nominal beam current of the

HL-LHC (see Table 1) will enhance such interactions, so that collective effects may ultimately limit the performance of the HL-LHC. Depending on the charge per bunch, on the bunch length, and on the bunch filling pattern around the machine, electromagnetic interactions with the beam surroundings give rise to parasitic losses (overheating of sensitive components due to the excitation of trapped modes), can cause beam instabilities, or may degrade the beam quality by inducing emittance growth and poor lifetime of all or of some specific bunches. For the nominal LHC a systematic review of single beam collective effects was given in Ref. [29], and a comprehensive updated summary in Ref. [30].

The next section summarizes the current knowledge of the longitudinal and transverse impedance in the actual LHC, while Section 1.2.2 presents an impedance forecast for the HL-LHC, and Section 1.2.3 discusses some consequences of these impedances for Landau damping. Section 1.2.4 then reviews synchrotron radiation and electron cloud, Section 1.2.5 intrabeam scattering, and Section 1.2.6 Touschek scattering. Lastly, Section 1.2.7 addresses beam–beam effects at the HL-LHC.

1.2.1. *Present LHC Impedance*

The longitudinal impedance of the present LHC is known to be compatible with the LHC design specifications, at least the inductive part [31]. Measurements of the frequency shift of bunch-length oscillations with bunch intensity have yielded a limit consistent with the design impedance budget of $|Z/n| \sim 0.1\,\Omega$. Observations of the loss of Landau damping during acceleration (without controlled emittance blow up) have confirmed this estimate [32].

Concerning the transverse impedance of the LHC as built, the total transverse effective impedances can be estimated from the measured tune slopes and bunch parameters [33] to be between 4 and 12 MΩ/m at injection (larger in the vertical plane than in the horizontal), and between 30 and 50 MΩ/m at 4 TeV (here larger in the horizontal plane than in the vertical), assuming average β functions of $\langle \beta_x \rangle \approx 66$ m and $\langle \beta_y \rangle \approx 72$ m around the LHC ring [34].

The measurements at 4 TeV were performed with the so-called 'tight collimator settings' of 2012 (collimators in IR3 and 7 closed in towards the end of the ramp) prior to the squeeze [35]. Therefore, the collimator settings at the time of the 4-TeV measurement were almost exactly identical to those used during physics operation, the only significant difference being the settings of the special absorbers (TCLs) located downstream of the high luminosity IPs. The latter were retracted during these beam measurements, but this should not make any significant difference for impedance since, firstly, the TCLs are made from copper and, secondly, the β^* in IP1 and IP5 was 11 m.

Within the measurement accuracy ($\pm 2 \times 10^{-4}$ per 10^{11} protons for the tune variation with intensity), the horizontal and vertical betatron tunes were found insensitive to the rms bunch length at injection, when the latter was varied between 8.6 and 11.6 cm. The rms bunch lengths at 4 TeV were then adjusted in between 7.1 and 8.1 cm, and, also here, no noticeable change of tune was observed within the range of the bunch-length variations.

Both at injection and at 4 TeV beam energy the measured tune shifts — and hence also the associated effective impedances — are up to three times larger than expected from the LHC impedance model [36]. The LHC transverse broadband impedance is thought to be dominated by the collimation system.

1.2.2. *HL-LHC impedance*

Longitudinal impedance

The HL-LHC longitudinal impedance is composed of several significant contributions, the largest of which, over a wide frequency range, up to about 100 MHz, is the resistive wall of the arc beam screen, as is illustrated in Fig. 10.

Another important part of the machine impedance, both longitudinally and transversely, is high-Q resonances, normally related to higher-order modes (HOMs) in cavity-like objects. As for the LHC, such modes, either damped or un-damped, will be present in the HL-LHC, from the 400 MHz main RF cavities, the transverse damper system, and/or the vacuum chambers of the four main experiments. New for the HL-LHC will be an extensive crab cavity system, with up to 16 novel compact RF cavities per beam, with a total transverse RF voltage of about 20 MV (similar to the main RF system), and possibly as well additional 200 MHz (6 MV) or 800 MHz (8 MV) lower or higher harmonic RF cavities. Higher- or lower-order modes in these new cavities will significantly add to the set of higher-order modes in the LHC. Also the inductive impedance is expected to at least double due to these new equipments.

The large contribution of the crab cavities to the overall longitudinal impedance is evident from Fig. 10 (the orange portions). The longitudinal effective inductive impedance for 16 crab cavities is $(Z/n)_{\text{eff}} \approx 0.04\,\Omega$, which corresponds to an increase by about 50%, in the total longitudinal inductive impedance, compared with the present LHC, with possible implications for bunch lengthening and loss of longitudinal Landau damping.

Transverse impedance

Figure 11 compares the transverse impedance models for the LHC and the HL-LHC. As it is already the case for the LHC, the impedance of the HL-LHC will be dominated by the collimation system. The effective impedance is sensitive to

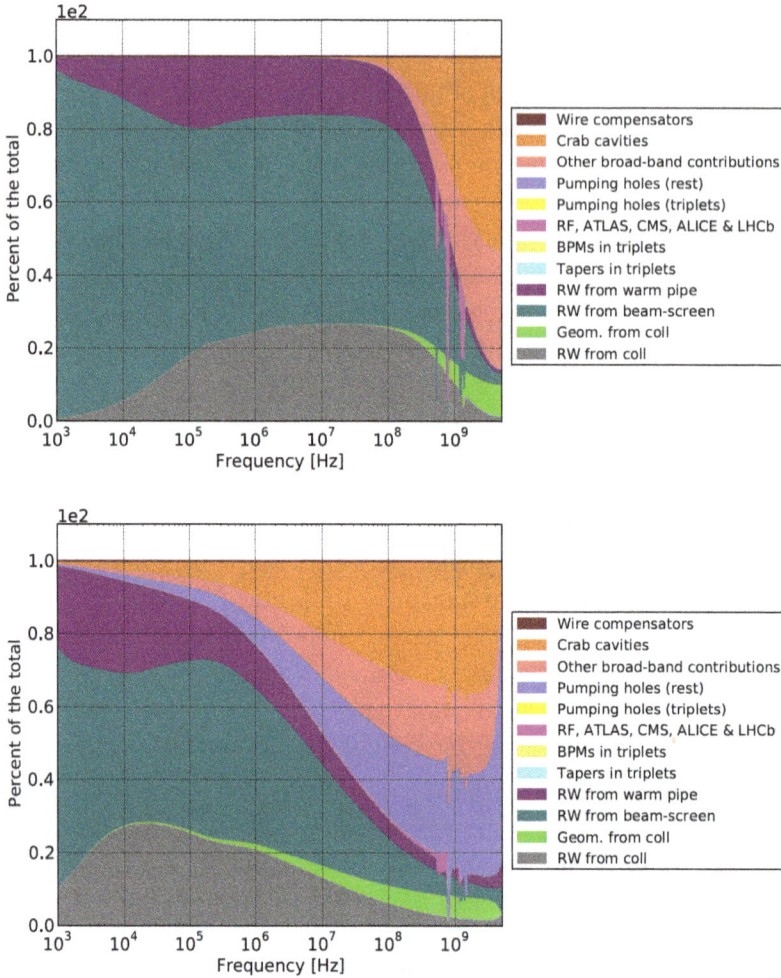

Fig. 10. Composition of the real (top) and imaginary part of the longitudinal impedance (bottom) for the HL-LHC at 7 TeV as a function of frequency [37] (Courtesy N. Mounet).

the normalized aperture of the secondary collimators, i.e. expressed in units of the transverse rms beam size, as well as to the collimator material and its possible coatings. The transverse impedance scales inversely with (slightly less than) the third power of the normalized aperture of the collimator jaws, and, therefore, the effective impedance, i.e. the impedance weighted by the local beta function, depends only weakly on the beta value at the collimator.

Figure 12 illustrates that over a wide frequency range the collimators contribute about 90% of the total transverse impedance in the HL-LHC.

Figure 13 (top picture) reveals that more than half of this transverse impedance is due to the secondary collimators ("TCSG"), which have fairly long jaws placed

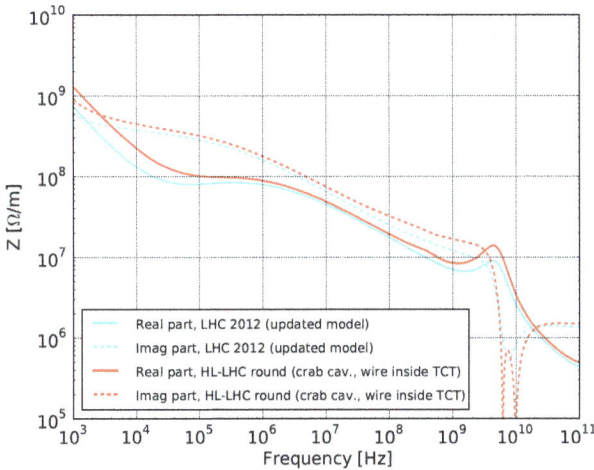

Fig. 11. Comparison of the transverse impedance models for the LHC and HL-LHC as a function of frequency. The LHC and HL-LHC beam power spectrum at top energy extends up to about 1 GHz (more precisely 6×10^8 Hz for a Gaussian beam, where the beam spectrum would drop by $1/e$) [38] (Courtesy N. Mounet).

close to the beam. As part of the LHC upgrade it is considered to replace 22 secondary collimators in the betatron cleaning insertion of IR7, and 8 secondary collimators in the momentum cleaning section of IR3, by new collimators using jaws made from molybdenum-graphite (Mo-Gr) instead of the presently employed carbon-fibre reinforced carbon (CFC). The CFC had been chosen for its robustness in case of a beam impact, but it exhibits a rather low conductivity. The bottom picture of Fig. 13 shows the expected overall impedance reduction thanks to the replacement of the secondary collimators in IR7. For frequencies between a few 100 kHz (a few tens of revolution harmonics) and a few GHz (beyond the cutoff of the bunch spectrum) the impedance will be reduced by about a factor of two, which should ensure beam stability at the (twice the nominal) bunch current of the HL-LHC, provided the normalized aperture of the collimators remains unchanged. The latter, in turn, depends on the apertures of the final triplet and of the matching-section magnets, as well as on the $\beta^*_{x,y}$ values.

As in the longitudinal plane, high-Q resonances can also be a significant part of the transverse machine impedance, with, again, many of these resonances related to higher-order modes (HOMs) in cavity-like objects. Such modes will be present in the 400 MHz main RF cavities, the transverse damper system, and/or the vacuum chambers of the four main experiments. Similar to the longitudinal plane, the HL-LHC crab-cavity system will significantly increase the set of higher-order modes. Namely, the 16 crab cavities per beam will add about 25% to the total effective

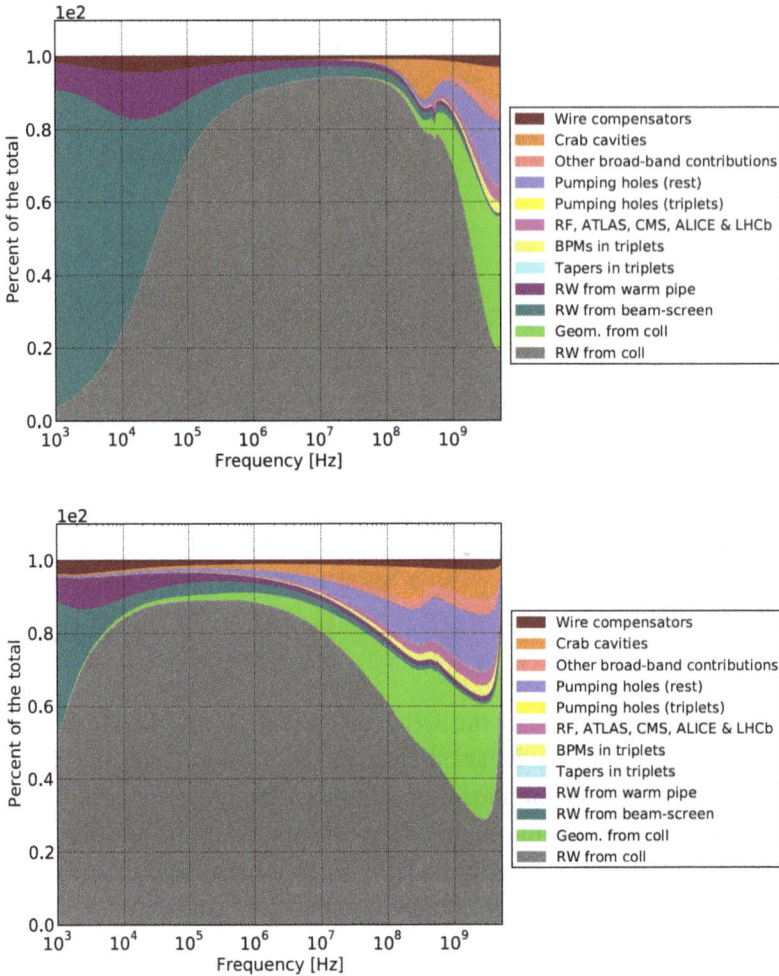

Fig. 12. Composition of the real (top) and imaginary transverse impedance (bottom) for the HL-LHC at 7 TeV as a function of frequency. The resistive-wall impedance of the collimators is shown in gray, the geometric collimator impedance in green [38] (Courtesy N. Mounet).

transverse broadband impedance. Their large contribution to the overall transverse impedance is illustrated in Fig. 12 (the orange portions).

The impedance of other additional devices, such a long-range wire compensators and, especially, (hollow) electron lenses could potentially be large and harmful, and requires a careful investigation and strict impedance control during design and fabrication. Other possibly harmful devices are the kickers and monitors currently under study as part of a stochastic cooling implementation for the HL-LHC ion operation.

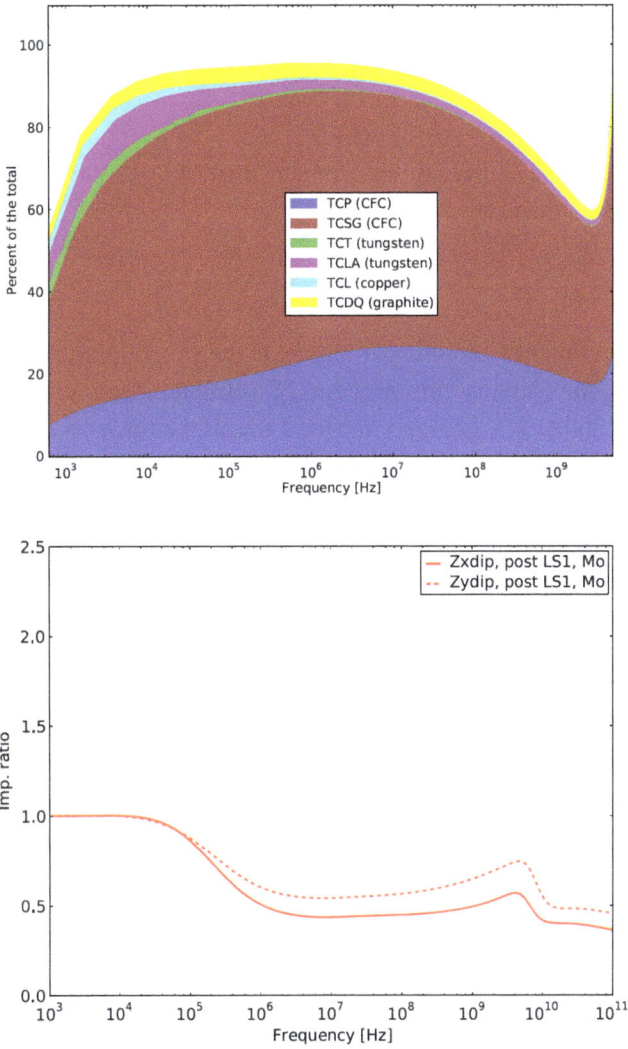

Fig. 13. The contributions of the resistive-wall impedance of collimator families to the total imaginary dipolar vertical impedance (top), and the reduction of the modulus of the total transverse dipolar impedance (bottom) thanks to a replacement of all secondary collimators in IR7 by coated collimators, with 50 μm of Mo on top of the CFC (carbon reinforced carbon) as a function of frequency [39] (Courtesy N. Mounet).

The transverse resistive wall impedance is another potential concern. At twice the nominal LHC bunch charge the resistive wall instability rise time for the most unstable rigid coupled-bunch mode will shrink to about 150 turns or 13 ms. About half of this growth rate comes from the Cu-coated arc beam screen, and the other half from the collimator impedance at 8 kHz (which is not much affected by the planned collimator modifications). A growth rate of 150 turns should however be easily damped by the transverse damper system.

Space-charge

Considering a Gaussian longitudinal bunch distribution, the incoherent space-charge tune shift is given by

$$\Delta Q^{\text{inc}}_{\perp,sc} = -\frac{N_b r_p C}{4\pi \beta \gamma^2 \varepsilon_n \sqrt{2\pi}\sigma_z}. \tag{6}$$

Here N_b is the number of protons per bunch, C the ring circumference, σ_z the rms bunch length, r_p the classical proton radius, $\beta = v/c$ the beam velocity normalized to the speed of light, $\gamma = (1 - \beta^2)^{-1/2}$ the Lorentz factor, and ε_n the normalized transverse beam emittance. At injection energy, assuming the same bunch length as for the nominal LHC, the incoherent space charge tune shift for HL-LHC bunch intensity and emittance is -3.6×10^{-3} and the direct space charge impedance [40] $i 8.9$ MΩ/m. These values are significantly larger than for the nominal LHC, with possible consequences including (enhanced) Landau damping [41] and (reduced) dynamic aperture [42]. Reassuringly in beam studies at the SPS much larger space-charge tune shifts (less than -0.1) did not significantly affect beam lifetime or stability [43]. However, any effect of direct space charge arises from its interplay with the nonlinear magnetic fields of a storage ring, which are different in the superconducting LHC and the warm SPS. At 7 TeV the incoherent HL-LHC space-charge tune shift and direct space-charge impedance become -2×10^{-5} and $i 0.62$ MΩ/m, respectively.

1.2.3. Landau damping

Loss of longitudinal Landau damping has been observed in the present LHC on different parts of the cycle (flat bottom, ramp and flat top) for bunches with small longitudinal emittances [32]. For bunch intensities of around 1.5×10^{11} protons, injection phase oscillations are not damped on the flat bottom with longitudinal emittances (defined as $4\pi\sigma_t\sigma_E$, with σ_E the rms energy spread), of less than 0.5 eVs. Quadrupole (or non-rigid dipole) instability has been observed during acceleration for emittances below 0.4 eVs (ramp in 2010) and on the flat top below about 0.7 eVs. As a mitigation measure, in normal operation the beam is stabilized by controlled emittance blow-up during the ramp [44]. A possible additional approach considered for the HL-LHC is to use a second RF system, operating at the second or third harmonic frequency of the fundamental system (i.e. at 800 MHz or 1.2 GHz), which could more than double the threshold for the loss of longitudinal Landau damping [45].

Landau damping in the transverse plane is provided by existing dedicated octupole magnets located in the arcs (MO), as well as, during physics, primarily by the large tune spread resulting from the head-on collisions. On the other hand, during the optics squeeze the tune spread induced by the long-range (LR) beam–beam interactions increases with $1/\beta^{*^2}$ (assuming the crossing angle to be kept constant during this process, as in 2012). Therefore, depending on the MO polarity, the LR beam–beam tune spread can add up or, on the contrary, compensate the amplitude detuning induced by the Landau octupoles themselves. In addition, during the collapsing of the parallel separation bumps for bringing the two beams into collision, the aforementioned effect is further amplified because the degree of parallel separation at the main IP also affects the beam–beam distance at the long-range beam–beam encounters. Finally at small parallel separation, the tune spread due to the beam–beam forces at the main IP changes sign, and Landau damping can be lost completely if the tune spread provided by the octupoles is insufficient, or if (for the wrong sign of the octupole polarity) at a certain transverse separation the tune shift due to the beam–beam force cancels the effect of the octupoles, resulting in essentially zero tune spread and onset of instability [46]. All together, for the negative MO polarity, and in the presence of long-range beam–beam interactions, the overall tune spread may first vanish on the anti-diagonal of the tune diagram towards the end of the squeeze, and on the diagonal during the collapsing process of the separation bumps [46] (see Fig. 14). For this reason the LHC octupole polarity was changed from negative to positive in the middle of the 2012 run [46], although the positive polarity is known to be non-optimal with regard to single beam stability [47] as, in the LHC, the impedance induced tune shifts have negative real parts.

This effect can also be analyzed from the point of view of Landau stability diagrams, as is illustrated in Fig. 15, which shows stability limits in complex tune plane computed separately for each of the following components for the nominal LHC parameters [48]: (1) the Landau octupoles, (2) the long-range beam–beam interactions, and (3) head-on collisions at the interaction points 1 and 5.

For the 2012 LHC parameters, combining the above three contributions, and analyzing the situation for both octupole polarities, Fig. 16 shows the negative imaginary part of the tune shift, $-\mathrm{Im}(\Delta Q)$, related to the Landau damping rate through $1/\tau_{\mathrm{damp}} = -2\pi f_{\mathrm{rev}}\mathrm{Im}(\Delta Q)$, as a function of the real part of the tune shift and of the beam–beam separation at the two main collisions points IP 1 and 5, but ignoring any collisions in IPs 2 and 8 [48]. This figure highlights the lack of stability for a transverse beam–beam separation around $2\sigma^*$, especially for the case where the Landau octupole magnets are powered at negative polarity (left picture). Comparing the left and the right pictures, this figure illustrates as well the aforementioned effect during the last part of the collision making

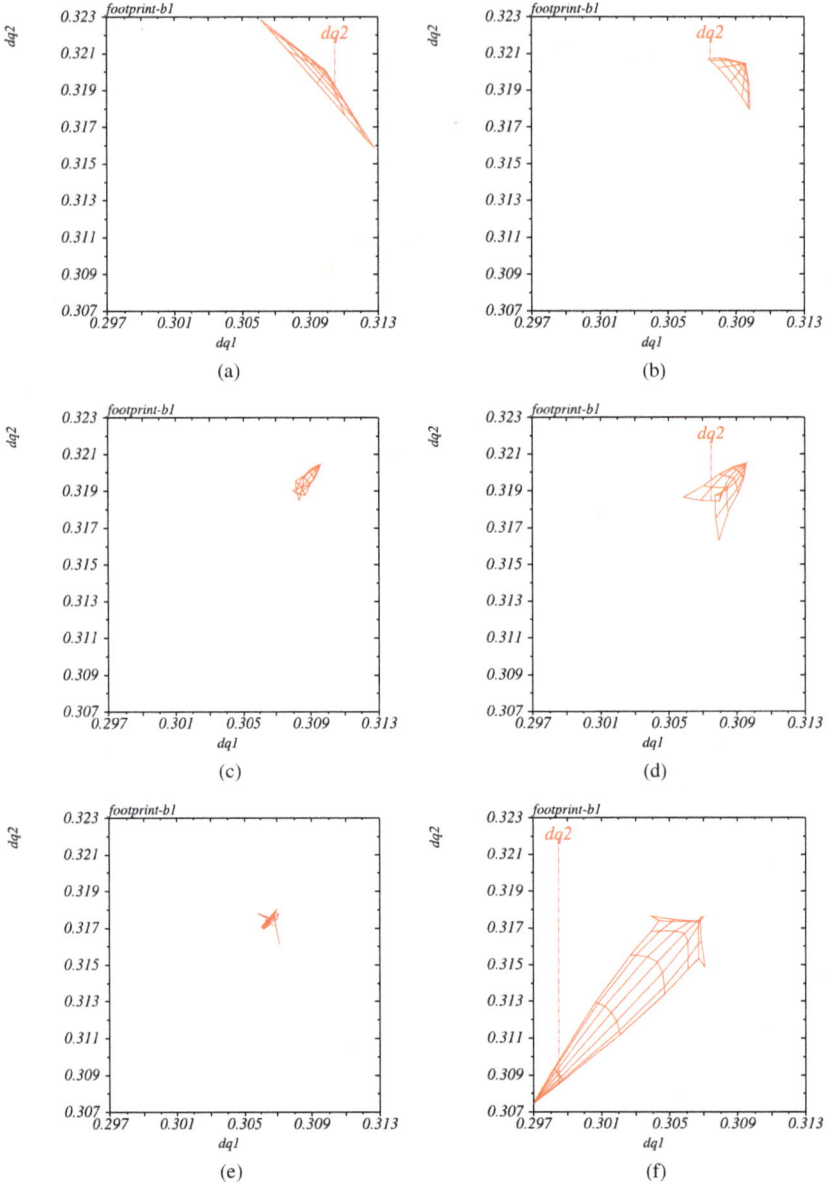

Fig. 14.   Tune footprint up to transverse amplitude of 6σ for the 2012 LHC parameters (4 TeV/beam, $N_b = 1.5 \times 10^{11}$, $\gamma\varepsilon = 2.5\,\mu$m) for bunches experiencing the maximum number of long-range interactions in IR1 and IR5. The Landau octupoles are assumed to be powered with a negative polarity (i.e. $\partial Q_x/\partial J_x \equiv \partial Q_y/\partial J_y < 0$ and $\partial Q_x/\partial J_y \equiv \partial Q_y/\partial J_x > 0$). The different pictures correspond to: (a) the end of the ramp; the end of the squeeze with the full (b) and an intermediate parallel separation (c), corresponding to $\delta x^* \sim 70\sigma^*$ and $\delta x^* \sim 15\sigma^*$, respectively; the very end of the collision making process with a small (d) and or very small (e) parallel separation of $\delta x^* \sim 5\sigma^*$ and $\delta x^* \sim 1.8\sigma^*$, respectively; (f) the beginning of the physics coast with head-on collisions at IP1 and IP5.

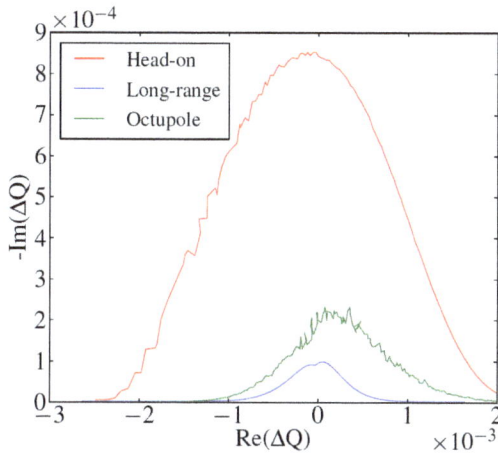

Fig. 15. Comparison of stability diagrams from either octupoles powered with at maximum nega-tive excitation (500 A magnet current), long-range collisions in IPs 1 and 5, or head-on collisions in IPs 1 and 5, for the nominal LHC parameters [48] (Courtesy X. Buffat).

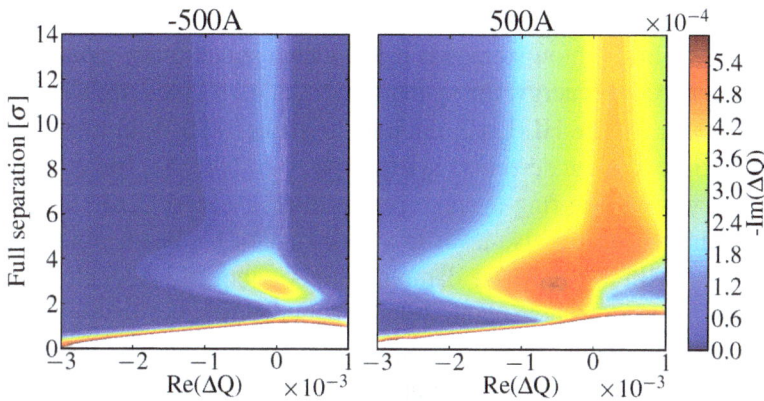

Fig. 16. Stability diagram as a function of beam separation in IPs 1 and 5 for a bunch experiencing the maximum number of long-range interactions assuming LHC parameters from 2012, and consid-ering both polarities of the octupoles [48] (the left and the right picture; Courtesy X. Buffat).

process, i.e. for parallel separations below $10 - 15\sigma^*$ where, for the 2012 LHC beam parameters and a negative octupole polarity, the direct tune shifts with am-plitude induced by the octupoles were fully compensated by the LR beam–beam effects.

The Landau damping scales with the transverse geometric emittance, and with the normalized octupole strength, both of which decrease by a factor $4/7 \approx 0.57$ at the design beam energy of 7 TeV, compared with the 4 TeV from 2012. In

addition the bunch charge will be about 40% higher for the HL-LHC beam (2.2 ×
10^{11} p/bunch) than what it was for the 50 ns LHC beam in 2012 (1.5–1.6 × 10^{11}
p/bunch). In total without collision and prior to the final β^* squeeze, the HL-LHC
beam may be about four times more unstable transversely than it has been the
case in 2012 (where, for the negative octupole polarity, already many fills were
aborted due to transverse instabilities). A possible countermeasure could be to
ramp and squeeze with colliding beams, thereby stabilizing the beam thanks to the
larger beam–beam tune spread. Also, after the squeeze, the ATS optics greatly
increases the effective strength of the Landau octupoles, by almost a factor of
4 for $\beta^*_{x,y} = 15$ cm, thanks to the beta wave introduced in the arcs around IP1
and IP5. Finally long-range beam–beam compensators based on electro-magnetic
wires [49] could locally correct the tune spread induced by the long-range beam–
beam interactions, and therefore restore the full freedom concerning the choice of
the octupole polarity.

1.2.4. Synchrotron radiation and electron cloud

The LHC is the first proton storage ring for which synchrotron radiation is a noticeable effect. At top energy, the synchrotron radiation gives rise to a significant
heat load, which for the HL-LHC amounts to about 3886 W in total, and which
is intercepted by a beam screen at a temperature of 5–20 K. The critical photon
energy is 43 eV and the ring-average photon flux 3.4×10^{21} cm^{-2}s^{-1} per beam.
The synchrotron radiation also leads to a shrinkage of the beam emittance during
physics stores, with an emittance damping time of about 13 h longitudinally, and
twice longer in the transverse plane. When operating at 3.5–4 TeV in 2011–12 the
damping has been about ten times weaker. At injection energy, the synchrotron
radiation stays negligible in terms of heat load (though still usable for beam diagnostics purposes), with a total radiated power of order 0.1 W.

Table 6 summarizes the various heat loads induced by the circulating beam on
the arc beam screen for the HL-LHC at 7 TeV. The corresponding values for the
nominal LHC are also shown for comparison. The total value of these conventional
heat loads determines the cooling capacity remaining for any additional source of
heat, such as due to incident electrons discussed in the following.

Namely, seed electrons created by ionization of the residual gas at injection or
photo-electrons liberated by the large number of hard U.V. synchrotron radiation
photons at 7 TeV are pulled towards the positively charged LHC proton bunches.
When they hit the opposite wall, they generate secondary electrons which can in
turn be accelerated by the next bunch if they are slow enough to remain in the
vacuum chamber until the next bunch arrives.

Table 6.   Summary of beam-induced heat loads on the arc beam screen at 7 TeV beam energy for the nominal LHC and for the HL-LHC (assuming the same bunch length). The three columns show the various sources together with the corresponding heat loads for the two machines in units of mW/m.

Heat Source	Power [mW/m] LHC at 7 TeV	Power [mW/m] HL-LHC at 7 TeV
Synchrotron Radiation [50]	170	330
Ohmic Losses [51]	110	400
Pumping Slots [52]	~10	~35
Welds [29]	~10	~35
Shielded Bellows [29]	≤ 30	≤ 100
Total	~330	~900

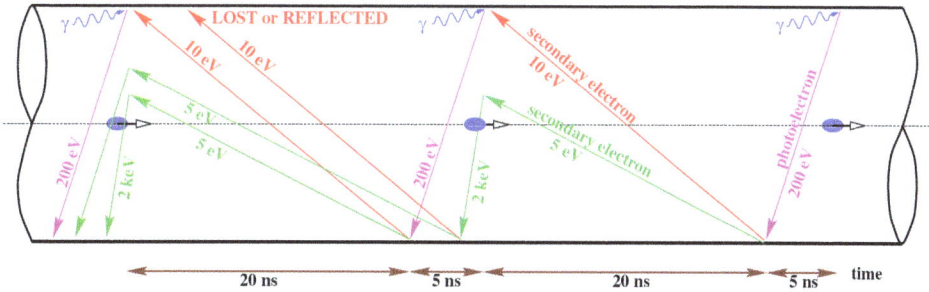

Fig. 17.   Illustration of the beam induced multipacting process leading to the build up of an electron cloud in the LHC beam pipe. The horizontal axis is time, illustrating the bunch spacing. Blue curvy lines represent synchrotron radiation photons hitting the chamber wall at the moment of a bunch arrival. Emitted photoelectrons (red) gain energy in the electric field of the passing bunch. When hitting the opposite side of the beam pipe, these are either reflected (still red) or generate lower-energy secondary electrons (green), which are then again accelerated transversely during the subsequent bunch passage (Courtesy F. Ruggiero).

This mechanism is illustrated in Fig. 17. It can lead to the fast build up of an electron cloud, with potential implications for beam stability, emittance growth, and heat load on the cold beam screen in the arcs. Electron cloud effects have been actively investigated at CERN since 1997 by analytic estimates, simulations, and experimental tests [53–55]. The electron cloud can ultimately cure itself since the secondary emission yield of the chamber surface decreases as a result of the continuing bombardment with electrons of sufficiently high energy (e.g. above about 30 eV). The variation of the maximum secondary emission yield, δ_{max}, with electron dose measured in the laboratory for the colaminated copper surface of the LHC beam screen (at perpendicular incidence of the primary electrons) is illustrated in Fig. 18. The LHC strategy to overcome electron-cloud related limitations

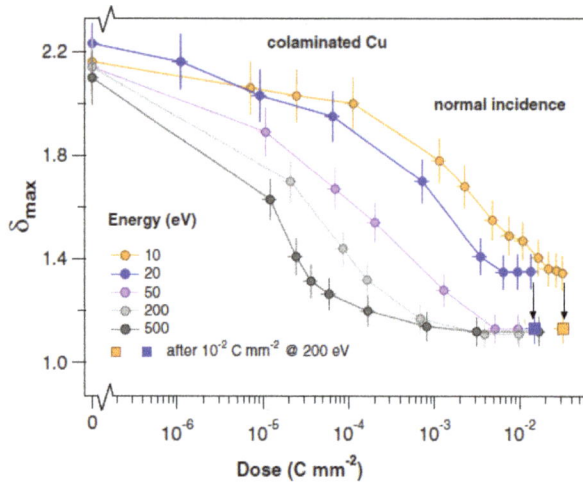

Fig. 18.   Variation of the secondary emission yield of copper with the incident electron dose for different energies of the impinging electrons at normal incidence on colaminated Cu of the LHC beam screen. The squares represent the δ_{max} after an additional dose of 10^{-2} C mm^{-2} at 200 eV [56] (Courtesy R. Cimino).

relies on this process of surface conditioning (intentionally enhanced in dedicated "scrubbing runs").

Electron-cloud effects and a reduction of the secondary emission yield with time have indeed been observed in the first years of LHC operation, where operation with 50-ns spacing has become possible thanks to rapid surface improvement (lowering of the secondary emission yield) in dedicated "scrubbing" runs with 25-ns beams [57–59]. Figure 19 presents the LHC surface conditioning observed during the 2011–12 LHC run for an uncoated field-free region, Fig. 20 the one for the cold arcs in 2011.

Arc heat-load observations in 2012, together with associated 'benchmarking' simulations, have, however, given rise to a concern that further scrubbing at 25-ns spacing may not decrease the secondary emission to a level sufficiently low to completely avoid electron-cloud build up during physics operation with 25-ns beams. Particularly critical are the arc quadrupole magnets, for which the electron-cloud threshold is lower than for dipoles or field-free regions (see Fig. 21) [59]. In addition at beam energies of 6.5 or 7 TeV the amount of photoelectrons will be much higher than at the 3.5–4 TeV energies of 2011–13. Also the generation of "UFOs" (micro-size dust particles falling into the beam, sometimes leading to very fast losses and consequently triggering a beam abort [60]) could be related to the presence of an electron cloud, as both occur more strongly, or more frequently, at the 25-ns bunch spacing. One conjecture is that the electron-cloud build-up could lead to a charging of macro-particles residing inside the vacuum system, thereby making them more likely to move under the influence of the electric field of the proton beam. For all these reasons, HL-LHC physics operation with 50-ns spacing (implying lower luminosity or higher pile up) is still retained as a back-up

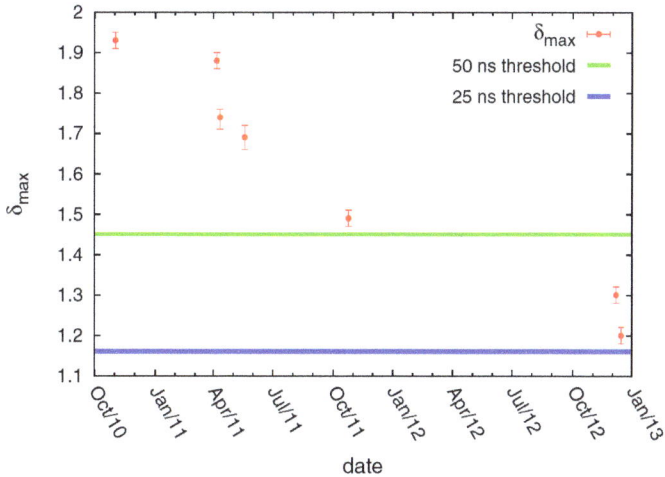

Fig. 19. LHC surface conditioning in an uncoated field-free region around the LHC vacuum gauge VGI.141.61.4.B, from 2010 to 2013, inferred by benchmarking of simulations with pressure-rise observations. The symbol δ_{max} denotes the maximum secondary emission yield, measurements of which were shown in Fig. 18. The calculated multipacting thresholds for bunch spacings of 50 and 25 ns are marked for reference (Courtesy O. Dominguez).

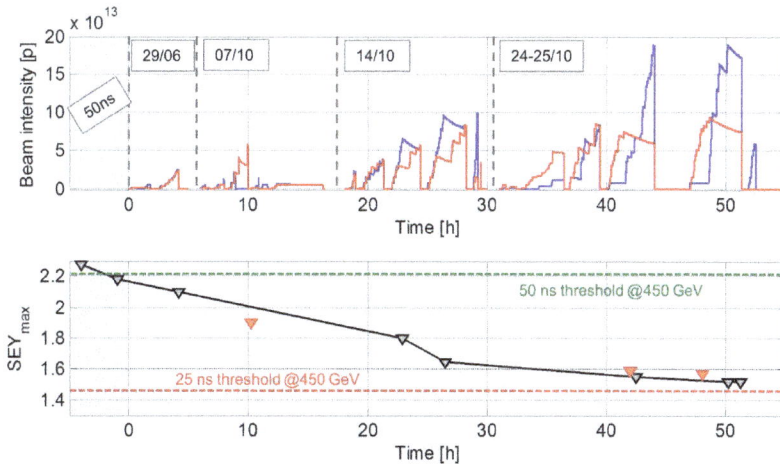

Fig. 20. LHC surface conditioning in the cold arcs during 2011: beam intensity (top) and estimated evolution of the maximum secondary emission yield (bottom) as inferred by benchmarking of simulations with cryogenic heat-load measurements (Courtesy G. Iadarola and G. Rumolo).

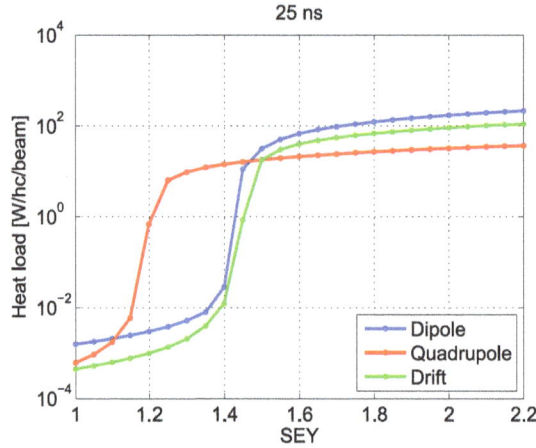

Fig. 21. Simulated electron-cloud heat load in dipoles, field-free regions and quadrupole magnets of the LHC cold arcs, for a single beam of 2,808 bunches (25 ns bunch spacing) as a function of maximum secondary emission yield [59] (Courtesy G. Iadarola, G. Rumolo).

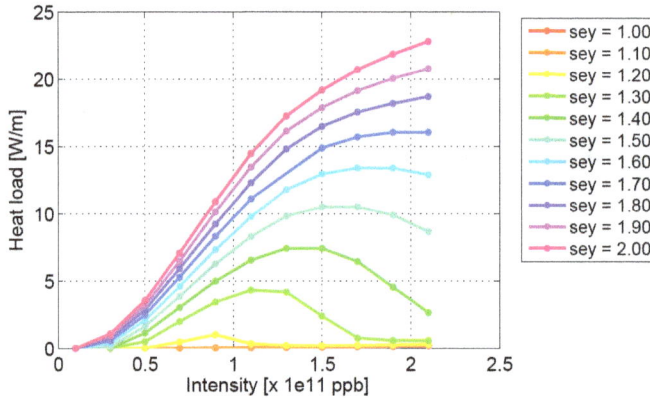

Fig. 22. Heat load per unit length inside an arc quadrupole for a single beam of 2,808 bunches as a function of bunch intensity simulated by the PyECLOUD code; the different curves refer to various values for the maximum secondary emission yield as indicated [62] (Courtesy G. Rumolo, G. Iadarola).

option. With 50-ns electron cloud is predicted not to be a problem at the HL-LHC either [61]. Fortunately, at the higher bunch intensity of the HL-LHC the electron cloud build up and related heat load inside the quadrupole magnets are expected to improve [62], as is illustrated in Fig. 22, so that for the HL-LHC the 25-ns operation should be more easily feasible than for the present LHC.

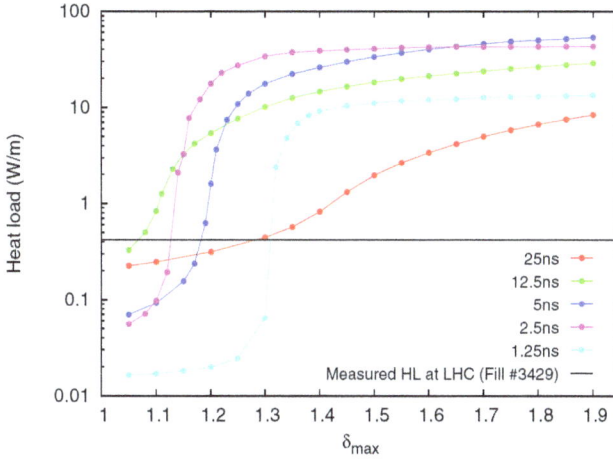

Fig. 23. Simulated heat load per unit length inside an arc dipole at 7 TeV beam energy as a function of the maximum secondary emission yield δ_{max}; the different curves refer to various values of the bunch spacing as indicated; the total beam current is held constant at 1.12 A; a low-energy electron reflectivity of $R \approx 0.7$ and a characteristic incident electron energy $\varepsilon_{max} = 332$ eV (specifically, ε_{max} refers to the energy at which the maximum yield δ_{max} is attained) have been assumed [58] (Courtesy O. Dominguez).

The HL-LHC could conceivably also operate with shorter bunch spacing. The variation of the electron-cloud heat load for shorter spacing is illustrated in Fig. 23, where the total beam current is held constant. This figure indicates that a bunch spacing equal to half the nominal, ~ 12.5 ns is about the worst case in terms of the onset of multipacting, while the heat load decreases again for even shorter distances between bunches, approaching the limit of a coasting continuous beam with static electric potential (in which the electrons cannot gain any energy).

1.2.5. *Intra-beam scattering*

In the dispersive regions of the ring, small-angle particle–particle collisions within a bunch couple the horizontal and longitudinal particle oscillations and, above transition energy, give rise to an irreversible emittance growth in both planes. The horizontal emittance growth rate caused by this intra-beam scattering (IBS) is roughly proportional to $N_b/(\varepsilon_\perp^{2.2}\varepsilon_z^{0.75})$, which increases by a factor 4.7 when transiting from the LHC to the HL-LHC beam parameters. The longitudinal IBS growth rate roughly scales as $N_b/(\varepsilon_\perp^{1.2}\varepsilon_z^{1.75})$, which rises only by a factor 3.1 (assuming the same longitudinal emittance, bunch profile, and RF voltage). In addition, the IBS growth rates also depend on the optics. Especially for the ATS collision optics [63] deviations from the simple scaling are expected due to the large β-beating waves induced in four of the arcs, which are not part of the LHC design optics. At 7 TeV beam energy, for the nominal LHC the longitudinal IBS emittance growth time is 57 h and the horizontal growth time is 102 h. For the HL-LHC, with the ATS optics, these growth times become 19 and 18 h, i.e., they decrease by a factor of about 6 horizontally (the difference from the aforementioned factor of 4.7 is attributed mainly to the change in optics) and by a factor of about 3.0 longitudinally. For constant normalized emittances the IBS growth times are approximately independent of the beam energy. However, the longitudinal emittance is a factor 2.5 smaller at injection than in collision (as a result of a controlled blow up on the LHC ramp) which leads to larger IBS growth rates at lower energy. As a result, for the HL-LHC the IBS rise times at injection are 5 h and 7.5 h, respectively, in the longitudinal and in the horizontal plane. Therefore, an emittance growth of about 10% and 7% is expected to occur longitudinally and transversely, respectively, during 30 minutes on the injection plateau of the HL-LHC. This emittance growth could be reduced by means of a lower frequency capture RF system (e.g. 200 MHz), allowing for a larger longitudinal emittance at injection, which enters roughly quadratically or linearly into the longitudinal and transverse IBS rise times, respectively. Another possible mitigation could be longitudinal bunch shaping using two RF systems (e.g., 200 and 400 MHz, or 400 and 800 MHz): A flat bunch of total length $2\sqrt{\pi}\sigma_z$ has a factor $\sqrt{2}$ lower IBS rate than a Gaussian bunch with rms length σ_z of the same charge [64].

The IBS growth rates cited above were computed with the latest version of MAD-X [65] considering either the design optics, for the nominal LHC, or the ATS optics (SLHCV3.1b [12]), with $\beta^*_{x,y} = 15$ cm in IPs 1 and 5 for the HL-LHC collision scenarios and with $\beta^*_{x,y} = 11$ m at injection energy.

1.2.6. *Touschek scattering*

Touschek scattering refers to particle–particle collisions within a bunch, by which enough energy is transferred from the transverse into the longitudinal oscillations such that the scattered particles leave the stable RF bucket. The Touschek loss rate for the nominal LHC has been estimated in [66]. According to this study, for the nominal LHC coasting beam is produced at a rate per proton of $2 \times 10^{-4}\,\mathrm{h}^{-1}$ during injection and of $8 \times 10^{-5}\,\mathrm{h}^{-1}$ at 7 TeV. Since the loss rate per proton due to Touschek scattering is linear in the bunch population and roughly inversely proportional to the square of the transverse emittance, it will be about four times higher at the HL-LHC, with a coasting beam production rate per proton of $8 \times 10^{-4}\,\mathrm{h}^{-1}$ during injection and of $2 \times 10^{-4}\,\mathrm{h}^{-1}$ at 7 TeV. In addition, the total number of protons circulating in the ring is doubled. Therefore, the absolute Touschek scattering rate is expected to be about one order of magnitude higher at the HL-LHC compared to the nominal LHC.

Once the protons are outside the RF bucket, they lose energy due to synchrotron radiation. If the collimators provide an energy aperture of 3.9×10^{-3}, a scattered proton is lost after about 390 h at injection or after 6.5 min at top energy, respectively. While the energy drift due to synchrotron radiation is not noticeable at injection (which means that most protons coasting outside of the RF bucket are lost during a short time interval at the start of the energy ramp), at 7 TeV it gives rise to a steady-state coasting beam component of about 10^{-5} for the present LHC and of 10^{-4} for the HL-LHC. This unavoidable coasting beam component, which will also populate the beam abort gap, could be reduced (linearly) by collimating more tightly in energy.

1.2.7. *Beam–beam effects*

Beam–beam effects include incoherent effects, such as betatron tune spreads associated with the nonlinear head-on and long-range collisions as well as a reduction of dynamic aperture in case of insufficient beam separation at the parasitic encounters, and coherent effects affecting orbit, tunes, and chromaticities of the different bunches (depending on their different collision schedules) or coherent oscillation modes.

For the HL-LHC both long-range and head-on beam–beam effects can be more severe than for the nominal LHC.

The effect of the long-range collisions is enhanced compared with the nominal LHC, since both the charge per bunch is almost doubled and the number of relevant long-range encounters (till the entrance of the separation dipole D1) is increased, due to the longer final triplet, from about 16 to 19 encounters for the latest layout of new insertions ATLAS and CMS [13]. Figure 24 illustrates that a crossing angle of 590 μrad (12.5σ) is required to obtain a short-term dynamic aperture around 7σ for a β^* of 15 cm (round). This choice of crossing angle is confirmed by Fig. 25, which compares the tune footprints up to 6σ computed for crossing angles of 475 μrad (10σ normalized separation) and for 590 μrad (12.5σ), the former case appearing "pathological". These results are also consistent with a scaling law for the dynamic aperture induced by the long-range beam–beam interaction, as first observed by Irwin [67].

The residual effect of long range beam–beam encounters can be mitigated by electromagnetic wires [49], which are being investigated for a possible deployment at the HL-LHC.

The head-on beam–beam tune shift at one IP with crab crossing and round colliding beams is given by

$$\Delta Q_{bb} = \frac{r_p N_b}{4\pi \gamma \varepsilon_{x,y}} \, , \tag{7}$$

where r_p denotes the "classical proton radius," N_b the bunch population and $(\gamma \varepsilon_{x,y})$ the normalized transverse emittance. For the HL-LHC beam parameters (see Table 1) and full crabbing at IP1 and IP5 (or collision at small Piwinski angle as in IR8), $\Delta Q_{bb} \approx 0.011$ per IP. With collisions at three IPs, the total tune shift becomes $\Delta Q_{tot} \approx 0.033$, which is more than three times higher than for the nominal LHC. Some simulation results indicate that this value may be close to the beam–beam limit of the LHC [69]. At the beam–beam limit, once it is reached, the luminosity can however be further optimized by increasing bunch length and crossing angle [71], essentially at the beginning of the store when the bunch charge is maximal. Another means for mitigating the total head-on beam–beam tune shift is to assume a luminosity leveling through a variation of the parallel separation at IP8 rather than with β^*, as successfully deployed in 2012. Finally an alternative which emerged very recently but has not yet been approved for the HL-LHC baseline is the so-called "crab-kissing" scheme [72], which consists in leveling the luminosity (and the peak line density of pile-up events, see Subsection 1.3) by acting on

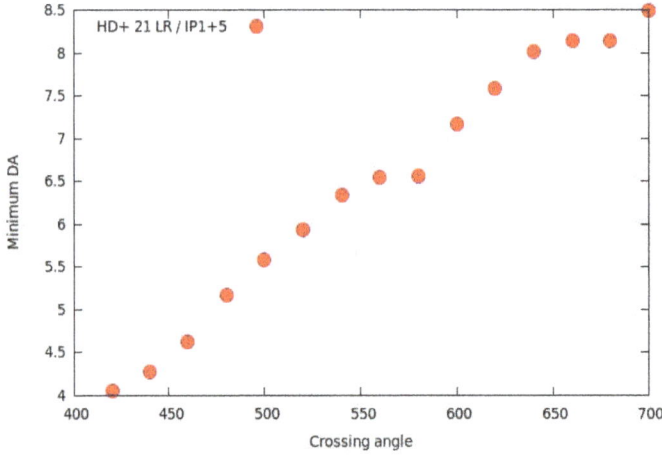

Fig. 24. Simulated dynamic aperture from tracking over 100,000 turns as a function of full crossing angle in IP1 and IP5 (and collisions in these IPs only), using the 25 ns HL-LHC beam parameters of Table 1 ($\beta^*_{x,y} = 15$ cm), with lattice sextupoles, long-range and head-on beam–beam interactions, but without field imperfections (Courtesy S. White). This simulation was performed at maximum beam current and minimum β^* (15 cm) with head-on collisions at IP1 and IP5 (with only one slice at the IP to maximize the beam–beam tune shift), but without any collision in IP8. It therefore corresponds to the full virtual luminosity of about 2×10^{35} cm^{-2}s^{-1} delivered to ATLAS and CMS, to a worst case in terms of long-range beam–beam effects, while the total head-on beam–beam tune shift of 0.022 in this simulation is lower than the maximum possible value, which would be obtained with head-on collisions at low Piwinski angle at 3 IPs. However, the above conditions lead to a worst case in terms of dynamic aperture. Indeed a series of other simulations, considering various more realistic situations, in particular ones with β^* leveling, have confirmed the choice of the crossing angle inferred from the study shown here [68].

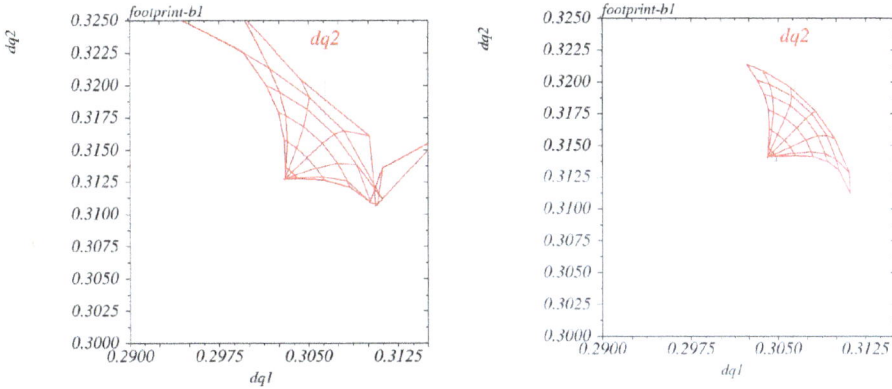

Fig. 25. Tune footprint up to a transverse amplitude of 6σ assuming 21 long-range encounters per IP side (for the longest possible inner triplet ever envisaged for the HL-LHC), as well as head-on collisions at IP1 and 5, for the 25 ns HL-LHC beam parameters of Table 1, with $\beta^*_{x,y} = 15$ cm and a full crossing angle of $475\,\mu$rad (10σ) or $590\,\mu$rad (12.5σ) for the left and right picture, respectively.

additional (rotated) crab cavities providing an RF dipole field in the separation plane of IR1 and IR5, so as to generate a time-dependent parallel separation at the IP and, therefore, to simultaneously reduce the head-on beam–beam tune shift at IP1 and IP5 [73].

Coming back to the present baseline (β^* leveling at IP1, IP5 and IP8) and for the HL-LHC beam parameters given in Table 1, Fig. 26 presents a beam–beam tune footprint and frequency map for a worst case scenario [74], assuming collisions with a very small Piwinski angle at IP8 (relatively small luminosity and large β^*), and head-on fully crabbed collisions at IP1 and IP5 with $\beta^*_{x,y}$ equal to 40 cm and a crossing angle of 360 μrad (12σ separation at this value of β^*, compared to 590 μrad for the nominal HL-LHC β^* of 15 cm). The tune footprint straddles resonances of 7th, 10th and 13th order, all of which are known to degrade the beam lifetime in collision, according to past experience at the $S p\bar{p}S$ collider [75]. Figure 27 shows the corresponding plots for a later time in the store, including changes due to luminosity leveling, namely with a smaller β^* of 15 cm, a lower bunch population of 1.06×10^{11} and a larger (nominal HL-LHC) crossing angle of 590 μrad (12.5σ separation) [74]. Resonance effects have all but disappeared from the frequency maps.

Other important considerations related to beam–beam effects are the bunch to bunch variation of closed orbit, betatron tune and chromaticity, arising from the fact that different bunches do not experience the same number of long-range (LR) beam–beam interactions depending on their relative position in the trains. Neglecting the LR interactions in IR2 and IR8, this effect is vanishing in terms of tune shift assuming an alternated horizontal-vertical crossing angle in IR1 and IR5 [76]. The

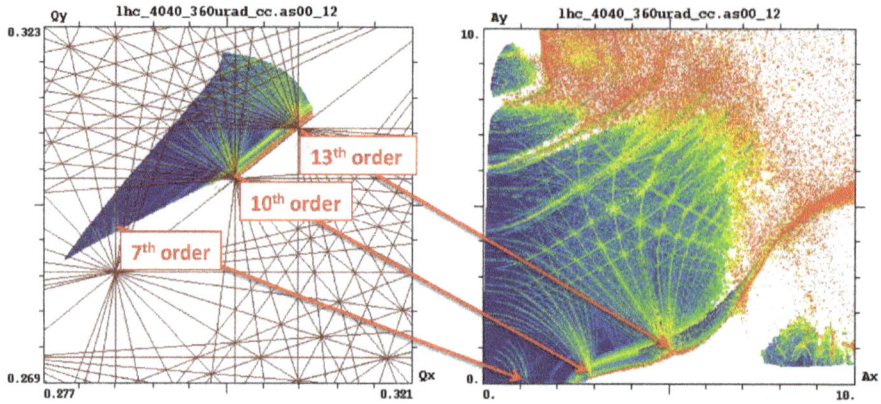

Fig. 26.   Tune footprint (left) and frequency map (right) for the HL-LHC with fully crabbed col-
lisions in 3 IPs, considering $\beta^*_{x,y} = 40$ cm, $N_p = 2.2 \times 10^{11}$ protons per bunch, $\theta_c = 360$ μrad full
crossing angle, normalized long-range separation of 12σ, crab voltage of 7.7 MV per side of IP, and
head-on beam–beam tune shifts of $\xi_x = 0.0306$ and $\xi_y = 0.031$ in total for the 3 IPs, as obtained in
weak-strong simulations with the code LIFETRAC [74] (Courtesy A. Valishev).

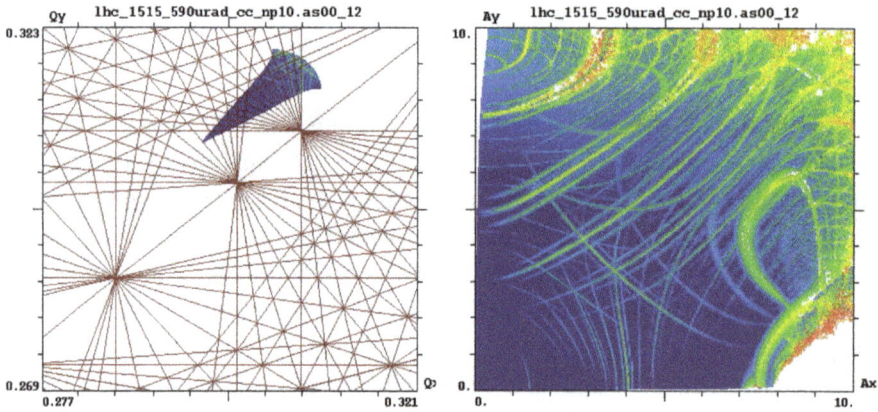

Fig. 27.   Tune footprint (left) and frequency map (right) for the HL-LHC with fully crabbed col-
lisions in 3 IPs, considering $\beta^*_{x,y} = 15$ cm, $N_p = 1.1 \times 10^{11}$ protons per bunch, $\theta_c = 590$ μrad full
crossing angle, normalized long-range separation of 12.5σ, crab voltage of 12.5 MV per side of IP,
and head-on beam–beam tune shifts of $\xi_x = 0.015$ and $\xi_y = 0.016$ in total for the 3 IPs, as obtained
in weak-strong simulations with the code LIFETRAC [74] (Courtesy A. Valishev).

bunch to bunch variations of chromaticity are also marginal for the ATS optics, and
essentially driven by the sextupole-like component of the LR beam–beam interaction combined with an almost vanishing spurious dispersion in the low-β insertion
(see Subsection 1.1.2). The most visible effect concerns the bunch to bunch variation in terms of closed orbit, scaling as follows with the bunch charge N_b, the

normalized emittance $\gamma\varepsilon$, the normalized crossing angle d_{bb}, and the total number of LR beam–beam interactions: N_{LR}:

$$\delta x[\sigma] \propto \frac{N_{LR} N_b}{\gamma\varepsilon d_{bb}}. \qquad (8)$$

In a worst case scenario without β^* leveling, in particular when starting the store with a minimal beam–beam separation of 12.5σ (see Table 1), this effect would be almost tripled for the HL-LHC, reaching about $\pm 0.3\sigma$, to be compared with $\pm 0.1\sigma$ for the nominal LHC [77]. The first and last bunches of each train may then collide with the other beam with a relative offset of about $\delta x^*_{b1} - \delta x^*_{b2} \sim 0.5\sigma^*$ at the IP depending on the optics of both beams. This offset will be visible in terms of luminosity (5% luminosity loss for these specific collisions), but should remain marginal in terms of integrated performance over the overall collision schedule. Both simulations for the HL-LHC [70] and LHC operational experience so far [78] indicate that the emittance growth due to static collision offsets at the level of a few tenths of the rms beam size is insignificant compared with the emittance growth caused by the nonzero crossing angle. Even this small effect could be further mitigated by an appropriate adjustment of the phase advance between IP1 and IP5 for the clockwise rotating beam, and between IP5 and IP1 for the other beam.

1.3. Dealing with pile up limits

1.3.1. Detector limitations

The detector technology sets limits on the total number of events per crossing (e.g. for calorimetry), as well as on the longitudinal event line density (for tracking of the primary vertices) and, possibly, also on the number of events per unit time during the collision. The nominal LHC parameters correspond to a peak pile up of about 20 events per crossing. During the 2012 LHC run the average pile up already was about 21 for ATLAS and CMS, and the maximum pile up in physics runs about 40 (i.e. twice the design). Higher peak values, close to 80 events per crossing, were reached in dedicated machine studies with a few bunches. The new ATLAS and CMS detectors are being designed for an average pile up of $\mu_{tot} = 140$ events per bunch crossing, with tails up to 200. At 25 ns bunch spacing, with about 2,800 bunches per beam, and considering a total inelastic cross-section of 85 mbarn an average pile up of 140 events per crossing corresponds to a luminosity of 5×10^{34} cm^{-2}s^{-1} (more precisely 5.2×10^{34} cm^{-2}s^{-1} for 2,808 bunches per beam). By contrast, the new LHCb detector is designed for an instantaneous luminosity not exceeding 2×10^{33} cm^{-2}s^{-1}, corresponding to about 5.2 visible pile up events on average per bunch crossing [3] (for an inelastic cross-section of 70 mb, which is visible by this detector, and assuming 2,400 collisions at IP8, compared with about 2,800 at IP1 and IP5).

In addition to the limit of $\mu_{tot} = 140$ in the new ATLAS and CMS detectors, the average peak pile-up line density should not exceed 1.3 events per mm, with tails up to 1.8 events per mm [80]. The latter condition determines the minimum extent of the longitudinal "luminous region", which is related to the bunch length, the value of β^*, the normalised crossing angle and the voltage assumed in the crab cavities (maximum voltage for full crabbing as in the baseline, or partial crabbing). Assuming transverse and longitudinal bunch distributions to all be of Gaussian shape, and denoting the effective full crossing angle (including the possible effect of crab cavities) by θ_c the rms length of the luminous region can be written as

$$\sigma_{lum} = \frac{\sigma_z}{\sqrt{2}} \frac{1}{\sqrt{1 + \left(\frac{\sigma_z \theta_c}{2\sigma_\perp}\right)^2}} = \frac{\sigma_z}{\sqrt{2}} \frac{1}{\sqrt{1 + \phi_w^2}}, \tag{9}$$

where ϕ_w signifies the Piwinski angle. For the baseline parameters of Table 1 at $\beta^* = 15$ cm, the extension of the luminous region is expected to be 4.4 cm r.m.s., and the peak line density of pile-up is calculated as 1.27 event/mm at $\mu_{tot} = 140$, including the degradation due to the hourglass effect and due to the RF curvature of the crab-cavity field [72].

A limit on the number of events per unit time during a bunch collision may also exist, essentially for forward physics, but has not yet been specified.

Finally, it is expected that bunch spacings shorter than 25 ns cannot be handled by either the present or the upgraded LHC detectors.

1.3.2. *Luminosity leveling*

The HL-LHC upgrade project aims at achieving a 'virtual' peak luminosity that is considerably higher than the maximum value imposed by the acceptable event pile up and to deploy a controlled reduction of the peak luminosity during operation ('luminosity leveling') so that the operational luminosity can be sustained over a significant length of time.

This luminosity leveling during a physics store can be accomplished in a number of ways: (1) dynamic β^* squeeze, (2) crossing angle variation, (3) changes in the crab RF voltage, (4) dynamic bunch-length reduction, or (5) controlled variation of the transverse distance between the two colliding beams[a].

---

[a]Historically, leveling with β^* variation was mentioned for the LHC ion programme around 2000 [81]. For ions the history of this proposal goes back further than this, probably to about 1995 [82]. Leveling for *pp* collisions in the context of the LHC luminosity upgrade was first proposed in February 2007

Due to proton consumption in the collisions, the total beam intensity, N_{tot}, decays as $dN_{tot}/dt = -n_{IP}\sigma_{tot}L$, where n_{IP} denotes the number of high-luminosity IPs ($n_{IP} = 2$ for HL-LHC), σ_{tot} the total hadron cross-section ($\sigma_{tot} \approx 100$ mbarn), and L the luminosity at IP1 and IP5. Setting L equal to the leveled luminosity, L_{lev}, the effective beam lifetime is

$$\tau_{eff} = \frac{N_{tot}(t \equiv 0)}{n_{IP}\sigma_{tot}L_{lev}} . \tag{10}$$

Next, introducing the ratio of virtual peak luminosity and leveled luminosity, $k = \hat{L}/L_{lev}$, we can express the maximum leveling time as

$$t_{lev} = \tau_{eff}\left(1 - \frac{1}{\sqrt{k}}\right) \equiv \tau_{eff}K , \tag{11}$$

where $K \equiv (1 - 1/\sqrt{k})$ designates the ratio of leveling time and effective beam lifetime. For the general case, where the physics run is extended beyond the end of the leveling period by a certain decay time t_{dec} (see Fig. 28), the time-averaged luminosity becomes [86]

$$L_{ave} = L_{lev}\frac{t_{lev} + t_{dec}\tau_{eff}/(t_{dec} + \tau_{eff})}{t_{dec} + t_{lev} + t_{ta}} , \tag{12}$$

with t_{ta} denoting the average turnaround time, i.e. the time between the end of one physics run and the start of the next one (time needed for magnet ramp down, injection, acceleration, β^* squeeze, collimation set up, etc.).

The average luminosity assumes a maximum value, $L_{ave,opt}$, for t_{dec} equal to the 'optimum decay time' [86]:

$$t_{dec,opt} = \frac{\tau_{eff}}{1 + K}\left(-K + \sqrt{(K^2 + (1+K)t_{ta}/\tau_{eff}}\right) . \tag{13}$$

The larger the turnaround time t_{ta} is, compared with the effective lifetime τ_{eff}, the longer is the optimum decay time. The resulting optimum total length of a single run is $t_{run,opt} = (t_{lev} + t_{dec,opt})$.

---

[83]. Both dynamic β^* squeeze and bunch-length variation were analysed as examples. A subsequent discussion concluded that "Luminosity leveling in IP1 and 5 is a very attractive solution for constant luminosity throughout the runs, achievable by i.e. a continuous β^* squeeze. Emittance blow-up effects remain to be assessed. Difficulties of changing the bunch length with RF systems is stressed with very high RF voltage required. Changing the β^* is the most promising option, although it is probably difficult to use in practice because of the behaviour of superconducting magnets." A few months later leveling by crossing angle variation in the so-called "early separation scheme" was proposed as another attractive option [84]. Leveling using the crab RF voltage was suggested as well [85]. The former two options have however the drawback of inducing very high line pile-up density at the beginning of the physics fill (about 5 event/mm). Very recently, the so-called crab-kissing scheme was introduced [72], assuming crab cavities in the parallel separation plane in anti-phase between Beam 1 and Beam 2, and offering a powerful luminosity leveling tool while reducing simultaneously the peak pile up line density and the head-on beam–beam tune shift. This scheme, however, is still under discussion, essentially due to its additional need in terms of hardware, and therefore will not be discussed any further here.

For comparison, at the nominal LHC without leveling the optimum run time is $t_{run,nol} = \sqrt{t_{ta}\tau_{eff}}$, yielding the average luminosity of $L_{ave,nol} = \hat{L}/\left(1 + \sqrt{t_{ta}/\tau_{eff}}\right)^2$, where $\hat{L}$ denotes the initial peak luminosity.

The variation of the beam–beam tune shift during a physics store depends on the leveling scheme [87]. In case of β^* variation, the beam–beam tune shift is maximized at the beginning, but then decreases during the store. When leveling via the bunch length, crossing angle or crab voltage the tune shift is minimized at the beginning but then increases, by a factor of 2 to 3. When leveling with the transverse offset the beam–beam tune shift changes sign and its modulus can increase even more strongly during the leveling process, by up to an order of magnitude [87].

The integrated annual luminosity for LHC and HL-LHC can simplistically be estimated by multiplying the total time scheduled for physics production T, the machine availability A (time without hardware failures divided by total time scheduled), and the average luminosity over the time periods without any hardware failure, as

$$L_{int} \equiv \int_{year} L(t)dt = T_{tot}AL_{ave} . \tag{14}$$

For the HL-LHC target parameters we require L_{int} to be 250 fb^{-1}, consider $T_{tot} = 160$ days (per year). We can then use the above equation to deduce the minimum availability required to be about 45%, which may be compared with an actual LHC availability of 73% in the year 2012.

Defining the machine efficiency, E, as the time spent in physics divided by the total allocated calendar time, we can also estimate the minimum needed efficiency, as

$$E \approx A\, t_{run}/(t_{run} + t_{ta}) . \tag{15}$$

For example, assuming the estimated minimum machine availability A of 45% deduced previously, an optimum HL-LHC run (leveling) time t_{run} of 9 h, and an average turnaround time of 5 h, the minimum needed efficiency E becomes 29%, which is lower than the efficiency of 36.5% obtained during LHC 4-TeV operation in 2012, and, therefore, appears an achievable target.

More refined estimates of integrated luminosities or necessary efficiencies might be obtained by considering a realistic random run-time distribution of (prematurely aborted) physics stores.

Figure 28 illustrates the HL-LHC luminosity time evolution for a single fill with and without leveling. The curve for the nominal LHC is also included for comparison. Figure 29 displays the luminosity evolution with and without leveling over several successive fills, where the fill length for either case has been optimized for maximum luminosity, assuming a turnaround time of 5 h. Without leveling the

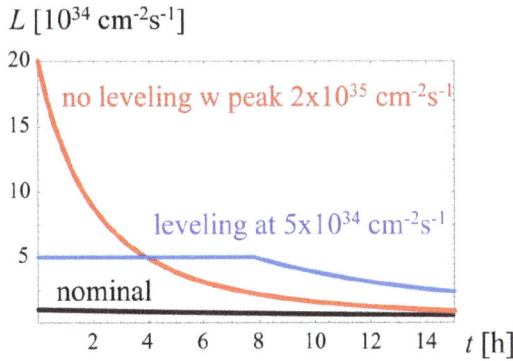

Fig. 28. HL-LHC luminosity evolution as a function of time, for a single fill, without (red), and with leveling at a pile up of 140 events per crossing (blue), compared with the LHC design (black). An inelastic pile-up cross-section of 85 mbarn is assumed for the mapping between number of pile-up events and luminosity, while a total cross-section of 100 mbarn is adopted for evaluating the proton burn off rate during the store.

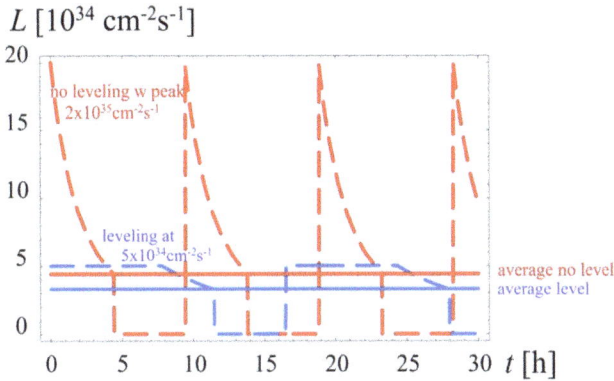

Fig. 29. HL-LHC luminosity evolution as a function of time, during several length-optimized fills over 30 h, without (dashed red), and with leveling at a pile up of 140 events per crossing (dashed blue). A turnaround time of 5 h has been assumed. The corresponding time-averaged luminosity values are indicated by the solid red and blue lines.

optimized fill length would be less than half the fill length of the leveled case, implying a significant increase in the fraction of time spent for turnaround without any luminosity. The solid lines indicate the time-averaged luminosities with and without leveling. The difference is only some 25% while the peak luminosity differs by 400%. Finally, Fig. 30 shows the integrated luminosity as a function of time. It is about 4 fb^{-1} per day. With an efficiency of 40% and considering 160 days of days scheduled for physics per calendar year, the luminosity delivered per year will exceed 250 fb^{-1}.

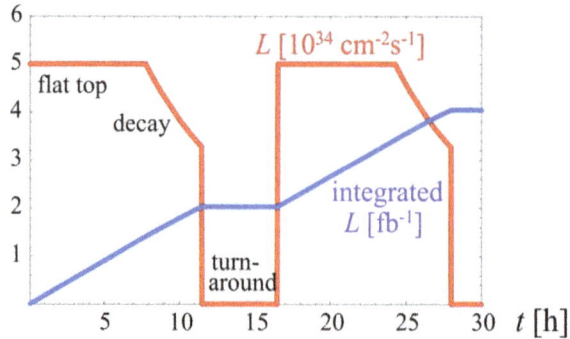

Fig. 30. Bare (red) and integrated HL-LHC luminosity (blue) as a function of time, during two length-optimized fills over 30 h, with leveling at a pile up of 140 events per crossing (dashed blue).

1.4. Summary and conclusions

The yearly performance targeted by the HL-LHC ($\sim$250 fb^{-1}) is a factor of 5 to 6 larger than the one expected for the nominal LHC. In order to reach this goal, both the optics (β^*) and the beam parameters shall be pushed by substantial factors in order to achieve a so-called virtual luminosity of about 2×10^{35} cm^{-2}s^{-1}, which will enable long fills at a leveled luminosity of 5×10^{34} cm^{-2}s^{-1} (limited by the number of pile up events per bunch crossing), to be compared with the LHC design peak luminosity of 10^{34} cm^{-2}s^{-1} (see Table 1). The generation and chromatic correction of the collision optics with extremely small β^* requires the deployment of a novel scheme, the so-called Achromatic Telescopic Squeezing (ATS) scheme, but also a series of new magnets of larger aperture including a challenging 150 mm aperture inner triplet (but not only). Novel RF devices, the so-called compact crab cavities, are also mandatory in order to maximize the luminosity at small β^* despite the crossing angle. Due to the increased bunch charge and smaller emittance, collective effects may ultimately limit the performance of the HL-LHC. Some of them are however still expected to be well under control (e.g. IBS), and/or with clear mitigation measures already identified (e.g. head-on and long-range beam–beam effects). On the other hand, as for the nominal LHC and pending more understanding and operational experience at 7 TeV per beam, some uncertainties remain concerning the beam stability in the transverse plane, electron-cloud effects for 25-ns bunch spacing, UFOs, the attainable machine availability and efficiency, as well as machine protection (e.g. crab-cavity failure modes).

Acknowledgments

The authors would like to warmly thank many collaborators of the HiLumi Design Study for various inputs and materials, in particular X. Buffat, O. Dominguez, J. Esteban Muller, M. Giovannozzi, G. Iadarola, R. de Maria, N. Mounet, Y. Nosochkov, G. Papotti, G. Rumolo, B. Salvant, E. Shaposhnikova, E. Todesco, A. Valishev, and S. White.

We also would like to express our gratitude to L. Rossi and O. Brüning for inviting us to address this very interesting aspect of the HL-LHC project, and for their diligent proofreading of an earlier draft version.

References

[1] L. Rossi, LHC Upgrade Plans: Options and Strategy, in *Proceedings of the 2nd International Particle Accelerator Conference 2011*, San Sebastiàn, Spain, 4–9 September 2011, pp. 908.

[2] The ALICE Collaboration, Upgrade of the ALICE Experiment, Letter Of Intent, CERN-LHCC-2012-012 (2013).

[3] The LHCb Collaboration, LHCb VELO Upgrade Technical Design Report, CERN-LHCC-2013-021; LHCB-TDR-013 (2013).

[4] S. Fartoukh, Towards the LHC Upgrade using the LHC well-characterized technology, *SLHC-Project-Report 49*, CERN, Geneva, Switzerland, 2010.

[5] S. Fartoukh, Breaching the Phase I optics limitations for the HL-LHC, in *Proc. Chamonix 2011 Workshop on LHC Performance*, Chamonix, France, 24–28 Jan 2011, CERN-ATS-2011-005, pp. 302–316.

[6] S. Fartoukh, Achromatic telescopic squeezing scheme and its application to the LHC and its luminosity upgrade, *Phys. Rev. ST Accel. Beams* **16**, 111002 (2013).

[7] O. Brüning, R. De Maria, Low gradient triplet layout options, in *Proc. 3rd CARE-HHH-APD Workshop: Towards a Roadmap for the Upgrade of the LHC and GSI Accelerator Complex*, Valencia, Spain, 16–20 Oct 2006, pp. 12–19.

[8] J. P. Koutchouk, L. Rossi, E. Todesco, A solution for phase-one upgrade of the LHC low-beta quadrupoles based on Nb-Ti, LHC Project Report 1000, 2007.

[9] O. Brüning, R. De Maria, R. Ostojic, Low gradient, large aperture IR upgrade options for the LHC compatible with Nb-Ti magnet technology, LHC Project Report 1008, CERN, Geneva, Switzerland, 2007.

[10] S. Fartoukh, Optics Challenges and Solutions for the LHC Insertion Upgrade Phase I, in *Proc. Chamonix 2010 Workshop on LHC Performance*, Chamonix, France, 25–29 Jan 2010, CERN-ATS-2010-026, pp. 262–290.

[11] R. Ostojic, Overview of IR upgrade scope and challenges, in *Proc. Chamonix 2010 Workshop on LHC Performance*, Chamonix, France, 25–29 Jan 2010, CERN-ATS-2010-026 (2010), pp. 253–257.

[12] S. Fartoukh, R. De Maria, Optics and Layout Solutions for HL-LHC with Large Aperture NB3SN and NB-TI Inner Triplets, in *Proc. 3rd International Particle Accelerator Conference 2012*, New Orleans, LA, USA, 20–25 May 2012, pp. 145.

[13] R. De Maria, S. Fartoukh, A. Bogomyakov, M. Korostolev, HLLHCV1.0: HL-LHC layout and optics models for 150 mm Nb3Sn triplets and local crab cavities, to be published in *Proc. 4th International Particle Accelerator Conference 2013*, Shanghai, China, 12–17 May 2013.

[14] S. Fartoukh, First demonstration with beam of the Achromatic Telescopic Squeezing (ATS) scheme, in *Proc. Chamonix 2012 Workshop on LHC Performance*, Chamonix, France, 6–10 Feb 2012, CERN-ATS-2012-069, pp. 128–134

[15] S. Fartoukh *et al.*, The 10 cm beta* ATS MD, CERN-ATS-Note-2013-004 MD, CERN, Geneva, Switzerland, 2013.

[16] R. Calaga, R. Tomas, F. Zimmermann, LHC crab-cavity aspects and strategy, in *Proceedings of the 1st International Particle Accelerator Conference 2010*, Kyoto, Japan, 23–28 May 2010, pp. 1240.

[17] R. De Maria, S. Fartoukh, Optics and layout for the HL-LHC upgrade project with a local crab cavity scheme, SLHC-Project-Report 55, CERN, Geneva, Switzerland, 2011.

[18] M. Giovannozzi, S. Fartoukh, R. De Maria, Specification of a System of Correctors for the Triplets and Separation Dipoles of the LHC Upgrade, to be published in *Proc. 4th International Particle Accelerator Conference 2013*, Shanghai, China, 12–17 May 2013.

[19] M. Giovannozzi *et al.*, Dynamic Aperture Performance for Different Collision Optics Scenarios for the LHC Luminosity Upgrade, to be published in *Proc. 4th International Particle Accelerator Conference 2013*, Shanghai, China, 12–17 May 2013.

[20] R. De Maria, S. Fartoukh, M. Giovannozzi, Specifications of the Field Quality at Injection Energy of the New Magnets for the HL-LHC Upgrade Project, to be published in *Proc. 4th International Particle Accelerator Conference 2013*, Shanghai, China, 12–17 May 2013.

[21] S. Fartoukh, O. Brüning, Field quality specification for the LHC main dipole magnets, LHC-Project-Report 501, CERN, Geneva, Switzerland, 2001.

[22] E. Todesco, Estimates of the LHC magnetic optics versus requirements, in *Proc. LHC Performance Workshop, Chamonix XIV*, CERN-AB 2005-014 ADM (2005), pp. 228–234.

[23] E. Todesco, B. Bellesia, L. Bottura, A. Devred, V. Remondino, S. Pauletta, S. Sanfilippo, W. Scandale, C. Vollinger, E. Wildner, Steering field quality in the main dipole magnets of the Large Hadron Collider, in *Proceedings of the MT-18 Conference, IEEE Trans. Appl. Supercond.* **14** 177–80 (2004).

[24] S. Fartoukh and M. Giovannozzi, Dynamic aperture computation for the as-built CERN Large Hadron Collider and impact of main dipoles sorting, *Nucl. Instrum. & Methods A* **671**, 10 (2012).

[25] R. De Maria, S. Fartoukh, SLHCV3.0: layout, optics and long term stability, SLHC-Project-Report-0050, CERN, Geneva, Switzerland, 2010.

[26] Y. Nosochkov, Y. Cai, M.-H. Wang, S. Fartoukh, M. Giovannozzi, R. de Maria, E. McIntosh, Evaluation of Field Quality for Separation Dipoles and Matching Section Quadrupoles for the LHC High Luminosity Lattice at Collision Energy, to be published in *Proc. 4th International Particle Accelerator Conference 2013*, Shanghai, China, 12–17 May 2013.

[27] G. Sabbi, E. Todesco, Requirements for Nb3Sn Inner Triplet and Comparison with Present State of the Art, HILUMILHC-MIL-MS-33, CERN, Geneva, Switzerland, 2012.

[28] Y. Nosochkov, Y. Cai, M.-H. Wang, S. Fartoukh, M. Giovannozzi, R. de Maria, E. McIntosh, Optimization of Triplet Quadrupoles Field Quality for the LHC High Luminosity Lattice at Collision Energy, to be published in *Proc. 4th International Particle Accelerator Conference 2013*, Shanghai, China, 12–17 May 2013.

[29] F. Ruggiero, Single-beam collective effects in the LHC, in *Proc. Workshop on Collective Effects in Large Hadron Colliders*, Montreux, 1994, eds. E. Keil and F. Ruggiero *Part. Accel.* **50**, 83–104 (1995).

[30] O. S. Brüning *et al.* (eds), LHC Design Report, v.1, Chapter 5: Collective Effects, CERN-2004-003-V-1 (2004).

[31] E. Shaposhnikova, private communication (2014).

[32] E. Shaposhnikova, Loss of Landau Damping in the LHC, in *Proc. IPAC'11 San Sebastian* (2011), pp. 211.

[33] T. Argyropoulos *et al.*, Probing the LHC impedance with single bunches, CERN-ATS-Note-2013-001 MD (2013).

[34] B. Salvant, private communication (2014).

[35] R. Alemany-Fernandez *et al.*, Operation and Configuration of the LHC in Run 1, CERN-ACC-Note-2013-0041 (2013).

[36] N. Mounet *et al.*, Beam Stability with Separated Beams at 6.5 TeV, in *Proc. 4th Evian Workshop on LHC Beam Operation*, 17–20 December 2012, Evian-les-Bains, France, CERN-ATS-2013-045 (2013), pp. 95.

[37] N. Mounet, B. Salvant, E. Metral, HL-LHC impedance model, http://impedance.web.cern.ch/impedance/HLLHC.htm (2013).

[38] N. Mounet, First Estimates of Intensity Limitations from HL-LHC Transverse Impedance, HiLumi WP2 Task 2.4 meeting, CERN, 22 January 2014.

[39] N. Mounet, E. Métral, B. Salvant, R. Bruce, S. Redaelli, B. Salvachua, G. Valentino, Collimator Impedance, Collimation Review, CERN, 30 May 2013.

[40] A. Chao, *Physics of Collective Beam Instabilities in High Energy Accelerators* (John Wiley, New York, 1993).

[41] E. Metral, F. Ruggiero, Stability Diagrams for Landau Damping with Two-Dimensional Betatron Tune Spread from both Octupoles and Nonlinear Space Charge Applied to the LHC at Injection, in *Proc. EPAC 2004*, Lucerne (2004), pp. 1897.

[42] F. Ruggiero, F. Zimmermann, Consequences of the Direct Space Charge Effect for Dynamic Aperture and Beam Tail Formation in the LHC, in *Proc. PAC'99*, New York (1999), pp. 2628.

[43] H. Burkhardt, G. Rumolo, F. Zimmermann, Investigations of Space Charge Effects in the SPS, in *Proc. PAC 2003*, Portland (2003), pp. 3041.

[44] P. Baudrenghien *et al.*, Longitudinal Emittance Blow-up in the LHC, in *Proc. IPAC'11*, San Sebastian (2011), pp. 1819.

[45] T. Linnecar and E. Shaposhnikova, An RF System for Landau Damping in the LHC, CERN LHC Project Note 394 (2007).

[46] S. Fartoukh, The sign of the LHC octupoles, *141th LHC Machine Committee*, 11 June 2012, https://espace.cern.ch/lhc-machine-committee/Minutes/1/lmc_141.pdf.

[47] J. Gareyte *et al.*, Landau Damping, Dynamic Aperture and Octupoles in LHC, CERN LHC Project Report 91 (1997).

[48] X. Buffat, W. Herr, N. Mounet, T. Pieloni, and S. White, Stability Diagrams of Colliding Beams in the Large Hadron Collider, to be submitted for publication in *Phys. Rev. ST Accel. Beams*.

[49] J.-P. Koutchouk, Correction of the long-range beam-beam effect in LHC using electromagnetic lenses, CERN-SL-2001-048-BI (2001) and *Proc. IEEE Particle Accelerator Conference (PAC2001)*, Chicago, IL, USA, 18–22 June 2001, eds. P. Lucas and S. Webber (IEEE, Piscataway, NJ, 2001), pp. 1681–1683.

[50] F. Zimmermann, Synchrotron Radiation in the LHC Arcs — Monte-Carlo Approach, LHC Project Note 237 (2000).

[51] F. Caspers, M. Morvillo, F. Ruggiero, J. Tan, and H. Tsutsui, Surface Resistance Measurements of LHC Dipole Beam Screen Samples, in *Proc. EPAC 2000*, Stockholm, and CERN LHC Project Report 410 (2000).

[52] A. Mostacci and F. Ruggiero, Pumping slots and thickness of the LHC beam screen, CERN LHC Project NOTE 195 (1999).

[53] F. Zimmermann, A Simulation Study of Electron-Cloud Instability and Beam-Induced Multipacting in the LHC, CERN-LHC-Project-Report-95 (1997).

[54] O. Gröbner, Beam Induced Multipacting, *PAC 97*, Vancouver, and CERN-LHC-Project-Report-127 (1997). J. S. Berg, Energy Gain in an Electron Cloud During the Passage of a Bunch, CERN-LHC-Project-Note-97 (1997). G. Stupakov, Photoelectrons and Multipacting in the LHC: Electron Cloud Build-up, CERN-LHC-Project-Report-141 (1997). O. Brüning, Simulations for the Beam-Induced Electron Cloud in the LHC beam screen with Magnetic Field and Image Charges, CERN-LHC-Project-Report-158 (1997). F. Caspers, J.-M. Laurent, M. Morvillo, and F. Ruggiero, multipacting tests with a resonant coaxial setup, CERN-LHC-Project-Note-110 (1997). F. Ruggiero, Electron Cloud in the LHC, CERN-LHC-Project-Report-166 (1998). M. Furman, The Electron Cloud Effect in the Arcs of the LHC, CERN-LHC-Project-Report-180 (1998). L. Vos, Electron Cloud: an Analytic View, CERN-LHC-Project-Note-150 (1998). V. Baglin *et al.*, Beam-Induced Electron Cloud in the LHC and Possible Remedies, *EPAC 1998*, Stockholm, and CERN-LHC-Project-Report-188 (1998). F. Zimmermann, Electron Cloud Simulations: An Update, in *Proc. Chamonix 2001*, CERN-SL-2001-003 DI (2001). See also World-wide web page on Electron Cloud in the LHC at the address http://wwwslap.cern.ch/collective/electron-cloud/.

[55] F. Zimmermann, Electron Clouds – Operational Limitations and Simulations, in *Proc. Chamonix 2003*, CERN-AB-2003-008-ADM (2003).

[56] R. Cimino, M. Commisso, D.R. Grosso, T. Demma, V. Baglin, R. Flammini, and R. Larciprete, Nature of the Decrease of the Secondary-Emission Yield by Electron Bombardment and Its Energy Dependence, *Phys. Rev. Lett.* **109**, 0604801 (2012).

[57] O. Dominguez *et al.*, First Electron-Cloud Studies at the Large Hadron Collider, *Phys. Rev. ST Accel. Beams* **16**, 011003 (2013).

[58] O. Dominguez, Electron Cloud Studies for the LHC and Future Proton Colliders, PhD thesis, EPFL Lausanne (2013).

[59] G. Rumolo, G. Iadarola, Update on Electron Cloud Studies for the LHC, 169th Meeting of the LHC Machine Committee, 18 September 2013.

[60] T. Baer *et al.*, UFOs in the LHC: Observations, studies and extrapolations, in *Proc. IPAC12*, New Orleans, 20–25 May 2012 (2012), pp. 3936.

[61] H. Maury Cuna, J.G. Contreras, F. Zimmermann, Simulations of Electron-Cloud Heat Load for the Cold Arcs of the CERN Large Hadron Collider and its High-Luminosity Upgrade Scenarios, *Phys. Rev. ST Accel. Beams* **15**, 051001 (2012).

[62] G. Rumolo, G. Iadarola, presented by R. De Maria, How to Maximize the HL-LHC Performance, in *Proc. 2013 RLIUP Workshop*, CERN, Geneva, Switzerland (2013).

[63] M. Schaumann *et al.*, Influence of the ATS optics on intra-beam scattering for HL-LHC, in *Proc. IPAC'13*, Shanghai (2013), TUPFI024.

[64] F. Ruggiero, G. Rumolo, F. Zimmermann, Y. Papaphilippou, Beam Dynamics Studies for Uniform (Hollow) Bunches or Super-bunches in the LHC: Beam-Beam Effects, Electron Cloud, Longitudinal Dynamics, and Intrabeam Scattering, Presented at the International Workshop on Recent Progress in Induction Accelerators (RPIA2002), KEK, Tsukuba, Japan, October 29–31, 2002, and LHC-Project-Report-627 (2002).

[65] F. Antoniou, F. Zimmermann, Revision of Intrabeam Scattering with Non-Ultrarelativistic Corrections and Vertical Dispersion for MAD-X, CERN-ATS-2012-066 (2012).

[66] F. Zimmermann and M.-P. Zorzano, Touschek Scattering in HERA and LHC, CERN-LHC-Project-Note-244 (2000).

[67] J. Irwin, Diffusive Losses From Ssc Particle Bunches Due To Long Range Beam-Beam Interactions, *SSC-233* (1989).

[68] A. Valishev, Preliminary Estimates of Beam-Beam Effects, HiLUMI LHC Milestone Report, CERN-ACC-2014-0066 (2014).

[69] K. Ohmi, Study of Beam-Beam Effect at Various Collision Schemes in LHC, in *Proc. EPAC'08* (2008), pp. 2593.

[70] K. Ohmi and F. Zimmermann, Beam-Beam Limit in Hadron Colliders, to be submitted to *Phys. Rev. ST Accel. Beams*.

[71] F. Ruggiero and F. Zimmermann, Luminosity optimization near the beam-beam limit by increasing bunch length or crossing angle, CERN-SL-2002-005-AP (2002) and *Phys. Rev. ST Accel. Beams* **5**, 061001 (2002).

[72] S. Fartoukh, Pile up density at HL-LHC with new shaping and leveling techniques, ECFA High Luminosity LHC Experiment Workshop, 1–3 October 2013, Aix-Les-Bains, http://indico.cern.ch/conferenceOtherViews.py?view=standard&confId=252045.

[73] S. Fartoukh, Pile up management at the High Luminosity LHC and introduction to the crab-kissing concept, CERN-ACC-2014-0076 (2014), CERN, Geneva, Switzerland.

[74] A. Valishev and S. Fartoukh, Weak-Strong Beam-Beam Simulations with Crab Kissing Scheme, Presentation at LHC-CC13 workshop, CERN, 10 December 2013.

[75] K. Cornelis, Beam-Beam Effects in the SPS Proton-Antiproton Collider, in *Proc. Workshop on Beam-Beam Effects in Large Hadron Colliders*, Geneva, Switzerland, 12–17 April 1999, CERN-SL-99-039-AP (1999), pp. 2–5.

[76] W. Herr, Is there an alternative to alternating crossing schemes for the LHC?, CERN-SL/93-45 (AP), also in CERN-LHC-Note-258 (1993), CERN, Geneva, Switzerland.

[77] H. Grote, Self-consistent orbits with beam-beam effects in the LHC, CERN-LHC-Project-Report-404 (2000), CERN, Geneva, Switzerland.

[78] R. Alemany *et al.*, Test of Luminosity Levelling with Separated Collisions, CERN-ATS-Note-2011-028 MD (2011).

[79] G. Papotti *et al.*, Experience with Offset Collisions in the LHC, in *Proc. 2nd International Particle Accelerator Conference 2011*, San Sebastian, 4–9 September 2011 (2001), pp. 1858.

[80] A. Ball, B. Di Girolamo, S. Fartoukh, L. Rossi, Pile-up and its density at HL-LHC: a common view from the machine and experiments, 3rd Joint HiLumi LHC-LARP Annual Meeting, Daresbury, 11–14 November 2013, Daresbury, UK, https://indico.cern.ch/conferenceDisplay.py?confId=257368.

[81] D. Brandt, Review of the LHC Ion Programme, CERN-LHC-Project-Report-450, Section 3.3 (2000).

[82] Proposal of A. Morsch, private communication by J. Jowett (2013).

[83] F. Zimmermann, presentation at CERN PAF/POFPA meeting on 13.02.2007, available at http://care-hhh.web.cern.ch/care-hhh/Literature, pages 31–39 (2007); also http://edms.cern.ch/file/820244/1/PAFPOFPA-Min13Feb07rev1.pdf.

[84] G. Sterbini, J.-P. Koutchouk, CERN-LHC-Project-Note-403 (2007).

[85] W. Scandale, F. Zimmermann, Scenarios for sLHC and vLHC, presented at Hadron Collider Physics Symposium, La Biodola, Italy, 20–26 May 2007, CARE-Conf-07-011-HHH; published in *Nucl. Phys. B, Proc. Suppl.* **177-178** (2008), pp. 207-211.

[86] O. Brüning, F. Zimmermann, Parameter Space for the LHC High-Luminosity Upgrade, in *Proc. 3rd International Particle Accelerator Conference 2012*, New Orleans, LA, USA, 20–25 May 2012 (2013), pp. 127.

[87] F. Zimmermann, HL-LHC: Parameter Space, Constraints & Possible Options, in *Proc. Chamonix 2011 Workshop on LHC Performance*, Chamonix, France, 24–28 January 2011, CERN-2011-005 (2011), pp. 317–330.

Chapter 5

Interface with Experimental Detector in the High Luminosity Run

H. Burkhardt

CERN, BE Department, Genève 23, CH-1211, Switzerland

This chapter describes the upgrade of the interaction regions as relevant for the experiments.

1. Introduction

The machine upgrade for high luminosity requires major changes on the machine side. Key ingredients for the luminosity increase are larger apertures in the focusing sections around the experiments and higher beam intensities.

The experiments are upgraded for reduced inner beam pipes with more powerful vertex detectors. This is important for physics and essential for the increased pile-up. Other key design considerations for the upgraded LHC detectors include longevity at increased radiation levels and minimization of activation.

As is often the case in machine-detector interfacing, these are to some extend conflicting requirements, which require a coherent planning for experiments and machine together.

2. Overview of the Main Changes Relevant for the Experiments

In this chapter, we discuss more generally hardware changes of relevance to the experimental regions, with respect to the original design of the LHC as described in the LHC design report [1]. The changes required for the high-luminosity upgrade and the requests and planning of the experiments for the future running of the LHC have been discussed in two joint machine-experiments workshops [2, 3].

Figure 1 shows the schematic layout of the LHC with its four interaction regions.

The HL-LHC design is for four experiments, the two high-luminosity experiments ATLAS and CMS at IR1 and IR5, and the ALICE and LHCb experiments installed at IR2 and IR8.

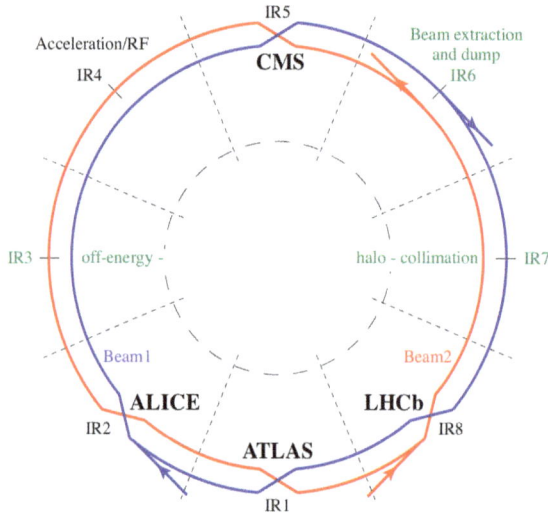

Fig. 1. Schematic layout of the LHC with its four interaction regions which provide collisions to the ALICE, ATLAS, CMS and LHCb experiments.

Table 1 shows the target luminosities for the experiments in proton–proton collisions in the LHC as originally designed, and for the high-luminosity upgrade. The main luminosity upgrade is for the interaction regions IR1 and IR5 and will be implemented in the long shutdown LS3.

The ALICE and LHCb experiments installed in IR2 and IR8 will have their most significant detector upgrades during LS2 scheduled in 2018/2019 and will continue to run after LS3.

LHCb has asked for a luminosity increase up to $2 \times 10^{33}\,\mathrm{cm^{-2}s^{-1}}$. This is possible without changes to the magnet layout in IR8 and the required detector and vacuum beam pipe upgrades can be implemented in the long shutdown LS2.

Table 1. Target luminosities $\mathscr{L}$ for p-p operation for the LHC and HL-LHC. For the HL-LHC, the ATLAS and CMS target luminosities include luminosity leveling which will allow for constant luminosities for the first hours during a fill.

IR	LHC $\mathscr{L}$, cm^{-2}s^{-1}	HL-LHC $\mathscr{L}$, cm^{-2}s^{-1}	Experiment
1	1×10^{34}	5×10^{34}	ATLAS
2	1×10^{31}	1×10^{31}	ALICE
5	1×10^{34}	5×10^{34}	CMS
8	4×10^{32}	2×10^{33}	LHCb

It should be accompanied by an improved shielding (including a minimal TAN at D2), to minimize the impact of the increase in radiation and heating of cold machine elements [4].

The low target luminosity for ALICE in p-p operation will require collisions with large transverse offsets.

The experimental programs of other smaller experiments (LHCf, TOTEM) do not extend beyond LS3.

In discussion with all experiments in the HL-LHC coordination working group during 2013, it was confirmed, that the LHC machine upgrade design can be considered as dedicated to high-luminosity and should not be constraint by other modes of operation, which can be completed before LS3. High-β^* ($\gg 100$ m) operation is not planned after LS3 and far detectors (ZDCs, roman pots, Hamburg beam-pipes) are not foreseen in IR1 and IR5 after LS3. While new proposals for extra detectors or special running conditions are not *a priori* excluded, they are not included in the HL-LHC design considerations and should not limit the high-luminosity performance of the HL-LHC.

The magnet layout in IR1 and IR5 will change significantly. This is shown schematically in Fig. 2 for the first 80 m from the interaction point and discussed in more detail in the following chapters [5]. Details like the exact length of the new triplet magnets are still under discussion. The distance of the first quadrupole magnet (Q1) from the IP will remain the same (23 m) as before the upgrade.

The most relevant machine modification for the experiments will be the installation of the new large aperture triplet magnets Q1–Q3 in IR1 and IR5. The inner coil diameter of these triplet magnets will increase by roughly a factor of two from 70 mm to 150 mm.

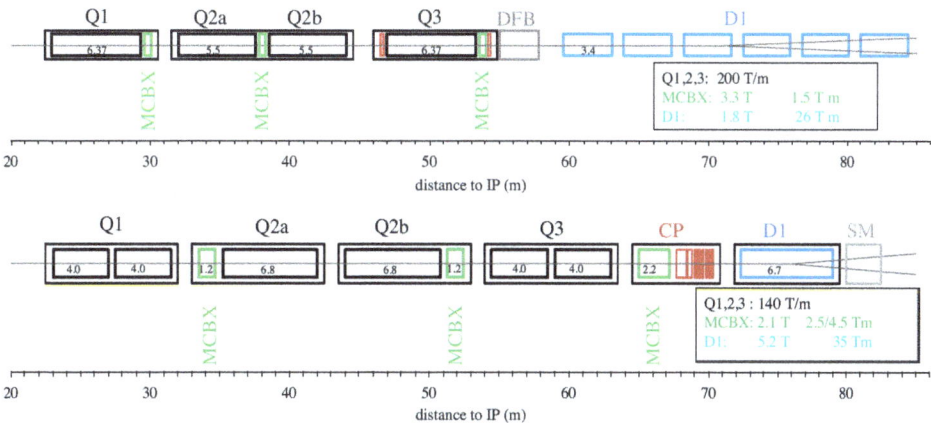

Fig. 2.   Schematic magnet layout for the current LHC (top) and the HL-LHC in IR1 and IR5 (bottom) up to first separation magnet D1.

H. Burkhardt

As presently the case, the magnet layout will be the same for IR1 and IR5, and also remain approximately left/right anti-symmetric with respect to the interaction points.

3. Experimental Beam-pipes

The four large experiments have asked for reductions of the diameter of the central beam pipes. Table 2 summarizes the original and reduced inner beam pipe radii [7]. For ATLAS and CMS the new beam pipes are installed during LS1. The reduction for ALICE has been requested but not yet approved from the machine side. The LHCb VELO is movable. It is only closed in stable physics to the value shown in the table, and retracted to 30 mm otherwise [6].

Table 2.   Original and reduced inner beam pipe radii at the IPs.

IP	Original r_{min} mm	Reduced r_{min} mm	Experiment	When
1	29	23.5	ATLAS	LS1
2	29	18.2	ALICE	LS2
5	29	21.7	CMS	LS1
8	5	3.5	LHCb, VELO	LS2

The reduction of the central beam pipes for IR1 and IR5 was approved on the machine side for operation up to LS3.

4. TAS, TAN

The high-luminosity interaction regions IR1 and IR5 are equipped with 1.8 m long copper absorbers called TAS at 19 m from the interaction points, located in front of the first superconducting quadrupoles Q1, see Fig. 3.

Their primary function is to reduce the energy flow from collision debries into the superconducting quadrupole triplet magnets [8]. In addition, the TAS also acts as a passive protection. It reduces the flux of particles into the inner detectors of ATLAS and CMS in case of abnormal beam losses. The inner radius of the TAS as presently installed is 17 mm both in IR1 and IR5. This is significantly less than the central beam pipe radius of ATLAS and CMS. The radius of the reduced central beam pipes installed in LS1 was chosen such that they still remain in the shadow of the TAS, including alignment tolerances and sagging. This is of direct relevance for high-β^* operation in the LHC, where the beam size is approximately constant throughout the experimental regions.

For the HL-LHC upgrade, the inner coil diameter of the triplet magnets will increase from 70 mm to 150 mm. The inner TAS radius will also have to be increased from 17 to approximately 27 mm, which will be significantly larger than the radius of the central beam pipes. The TAS material, length and outer dimensions may remain as originally designed. Additional shielding will be installed around the beam screens in the triplet region. The energy deposition for the enlarged TAS to the triplet magnets has been determined by simulations and remains within specifications, see Chapter 10.

Increasing the central beam-pipes after LS3 in the same proportion as the inner TAS radius would compromise the vertex detector performance. Optimal vertex resolution for ATLAS and CMS is essential to deal with the increased pile-up after LS3. The working strategy is that we assume that the beam pipe radii will remain after LS3 at the reduced values given in Table 2, and to assure by detailed tracking simulations including all relevant failure cases and by fast active protection (beam loss detection and dump) that LHC operation will remain safe for the experiments at the HL-LHC, in-spite of increased intensities and apertures.

There will also be major changes further outside in IR1 and IR5. The D2 magnet and neutral absorber TAN which is located in front of the D2 magnet will move by 13 m closer to the interaction points, to make space available for the installation of the crab cavities. The β-functions at the TAN will increase and require a larger TAN aperture. The half-crossing angle will roughly double for the HL-LHC (from typically 142.5 to 295 μrad) and move the neutral cone from collision debries closer to the beam aperture of the TAN. The TAN surrounds both beams and also acts as passive absorber for the incoming beam. The increase in TAN aperture results in a reduction of passive protection compared to the present LHC, which can be minimized by closer matching of the holes through the TAN to

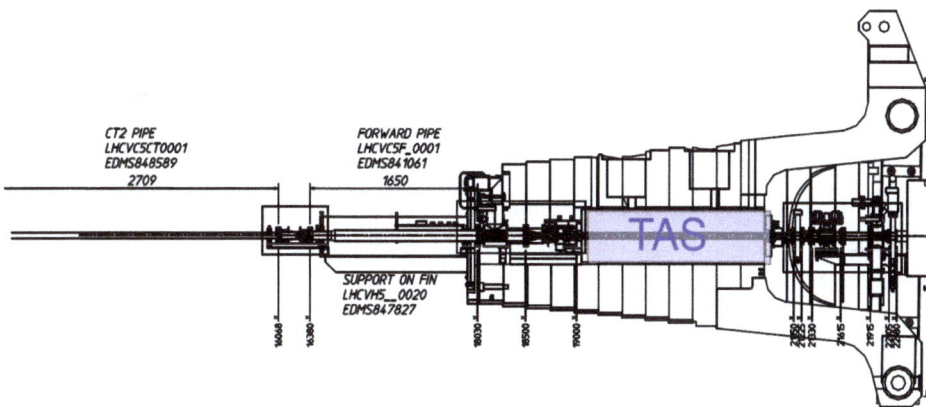

Fig. 3.   Layout right of IR5 (CMS) with the TAS.

the beam geometry and by addition of movable collimators. Details remain to be worked out.

It appears un-avoidable, that the ATLAS and CMS detectors will be more exposed to accidental beams losses and machine induced backgrounds after LS3. Understanding, minimizing and mitigating any un-avoidable negative impact of the machine upgrade to the experiments is a key objective for the machine detector interface for the HL-LHC [2].

5. Failure Scenarios and Experiments Protection

Active machine protection, based on continuous beam loss monitoring (BLM) and fast beam dump (within 3 turns) has already been proven to be essential and reliable for the present LHC [9]. It will be even more important for the HL-LHC. In addition to the protection of the machine elements which will be discussed in a later chapter, we will have to rely on active protection for the experiments. This implies, that we have to identify all relevant failure scenarios which may result in significant beam losses to the experiments, and to make sure that these abnormal beam losses can be detected sufficiently fast and beams be dumped before they cause any significant damage to the experiments.

Detailed studies with particle tracking have started.

Most critical for experiments protection is the operation at top energy with squeezed beams. The potentially most relevant failures scenarios for the HL-LHC are:

- Crab cavity failures [10, 11];
- Asynchronous beam dumps [12];
- Mechanical non-conformities, i.e. objects which accidentally reduce the aperture (example rf-fingers) or UFO's (dust particles falling through the beam) resulting in showers with local production of off-momentum and neutral particles around the experiments [14].

Other more or less dangerous scenarios do exist and will also be followed up, together with any changes which may result in increased beam losses into the experiments:

- D1 magnet failures. The present 6 warm D1 magnets at either side of IP1 and IP5 will be replaced with single superconducting D1 magnets. It is expected that this will result in longer time constants in case of D1 trips, which would leave more time to safely dump the beams in case of D1 failures.

- Injection (kicker) failures and grazing beam impact on injection elements (TDI).
- Check that any new equipment and in particular moving objects are compatible with experiments protection. It was already decided to not use fast vacuum sector valves close to the experiments.
- Experiments protection for ALICE in IP2 and LHCb in IP8 with reduced beam pipes for HL-LHC beam parameters.

Figure 4 shows a schematic view of IR5 with beam envelopes and apertures. Beam-pipe apertures are shown as lines as implemented in present simulations, both for the LHC as originally build for RUN 1 (dark blue lines), as well as for the HL-LHC after LS3 (green). The colored bands show 5σ beam envelopes for $\beta^* = 15$ cm as relevant for the HL-LHC. Two off momentum tracks with $\Delta p/p = -20\%$ and -30% are also shown. A -30% track originating at 150 m from the interaction point (originating by collisions of beam particles with dust particles, for example) will pass through the enlarged HL-LHC apertures and directly hit the central experimental beam pipe.

Figure 5 illustrates the beam envelope growth induced by an immediate 90° phase jump on a single crab cavity.

Early simulations suggest a growth of amplitudes by 14% within 5 turns. This would be a very fast, and potentially dangerous failure scenario. Detailed simulations also predict where the particles will be lost. As can be generally expected, the

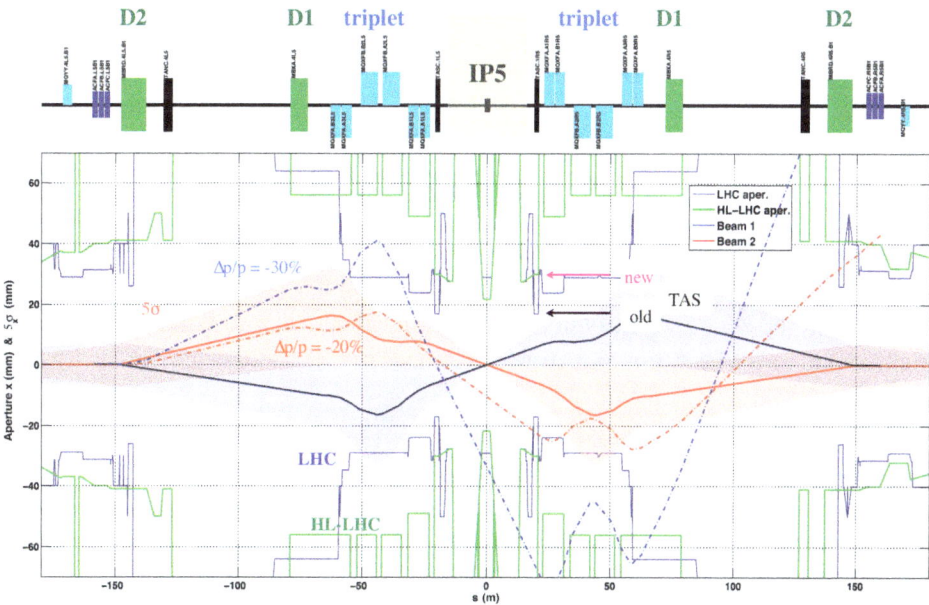

Fig. 4.   Schematic view of IR5 with beam envelopes and apertures.

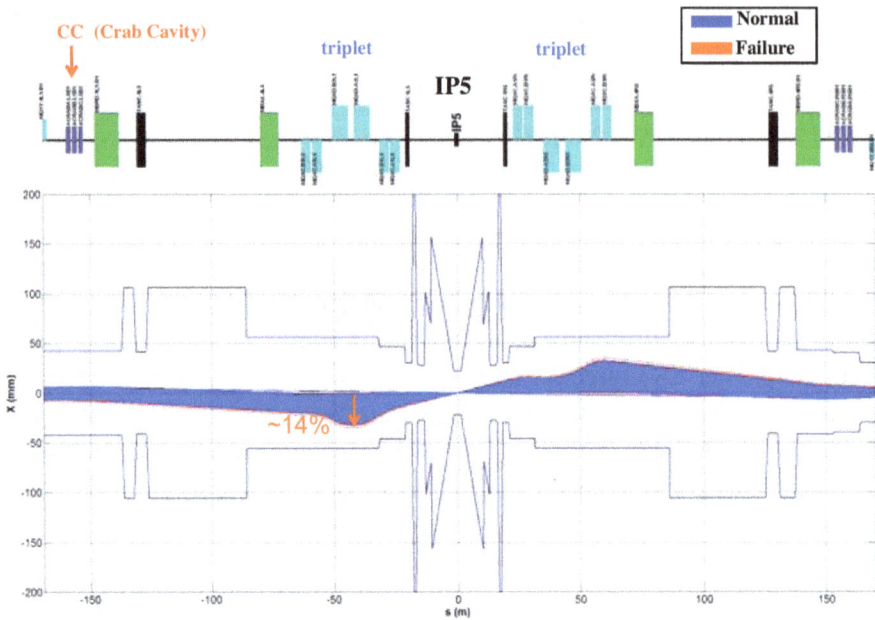

Fig. 5.   Schematic view the beam envelope growth induced by a crab cavity failure, resulting in a growth of 14% within 5 turns.

first simulations with crab cavity failures indicate that the losses will be mostly in the collimation sections and that only a small fraction will reach the experiments.

More detailed and complete simulations are being prepared. It is planned to develop a detailed model of the transient behavior of crab cavities. It is also planned to provide for fast detection of failures directly at the crab cavities and to reduce the effect on the beams, for example by turning cavities off on both sides of the interaction region in case of a single cavity trip. See also the chapters on crab cavity developments and machine protection. A close collaboration between the machine and experiments teams has started to re-design the beam pipe-layout in the experimental section. The CT2 chamber of CMS at approximately 15 m from the IP will have to be enlarged. Easy access and reduced activation (using preferentially lighter elements like Aluminium) are also important, for access and repair.

6. Machine Induced Backgrounds

Machine induced backgrounds in the LHC are dominated by beam gas scattering. Beam gas backgrounds scale with the beam intensity and vacuum pressure and are to a large extend generated locally in the straight section and dispersion suppressors around the experiments. Under normal conditions, they depend only weakly on optics details and collimator settings.

Background conditions have generally been very good in RUN 1 of the LHC [13]. Signal to background ratios of the order of 10^4 were observed in good running conditions in ATLAS and CMS in RUN 1.

For ALICE, which operates at much lower luminosity, machine induced backgrounds were already significant in proton–proton running even under good running conditions. Excellent vacuum conditions (pressures below 5×10^{-9} mbar) are essential for ALICE. During part of the proton–proton operation in 2012, machine induced backgrounds in ALICE were too high to permit data taking. This was related to heating in the injection absorber (TDI) region and is expected to improve in RUN 2 after hardware modification including increased pumping in IR2, implemented in LS1.

There are several reasons why machine induced backgrounds could increase a lot in future LHC operation. Continued efforts to monitor, understand and minimize backgrounds are important for all experiments.

Detailed tracking simulations for the HL-LHC have recently been started and will have to be updated as details of the new layout become available. From what we know so far, the detectors will be more exposed to machine induced backgrounds by the increase in apertures. The possible increases of backgrounds from geometry are expected to be moderate, below an order of magnitude and may partially be compensated by going to lighter structures in the central detector region, which reduce the number of secondary particles produced. Since luminosities will increase as well, the signal to background ratio should still remain comfortable for the high-luminosity experiments.

An increase in residual gas pressure would directly translate into increased backgrounds. Potential reasons for an increase are:

- synchrotron radiation, main step here will be the increase from 4 to 6.5 TeV at beginning of RUN 2;
- electron cloud due to reduced bunch spacing, main step 50 ns $\to$ 25 ns at beginning of RUN 2;
- electron cloud due to increased bunch intensities;
- local heating from increased intensities.

Two major changes, the increase in beam energy and the reduction in bunch spacing will already happen at the beginning of RUN 2. Observing and understanding backgrounds in RUN 2 by comparison with simulations and experience from RUN 1 should provide us with very valuable information for the optimisation of the experimental conditions at the HL-LHC.

References

[1] O. Brüning *et al.* (ed.), LHC design report, Vols. 1–3, CERN-2004-003-V-1–3.

[2] H. Burkhardt, D. Lacarrere and L. Rossi, Executive summary of the 1st Collider-Experiments Interface Workshop on 30 Nov. 2012.

[3] ECFA High Luminosity LHC Experiments Workshop, Aix-Les-Bains, 1–3 Oct. 2013.

[4] L. Esposito *et al.*, in *Proc. IPAC 2013*, tupfi022.

[5] Magnets for Insertion Regions, HL-LHC work package 3.

[6] M. Ferro-Luzzi, presentation in LMC 159 on 12/12/2012.

[7] LHC Experimental Beampipes Committee, LEB.

[8] N. Mokhov *et al.*, CERN-LHC-PROJECT-REPORT-633.

[9] Machine protection, HL-LHC work package 7.

[10] F. Bouly *et al.*, Preliminary study of constraints, risks and failure scenarios for the high luminosity insertions at HL-LHC, in *Proc. IPAC 2014*.

[11] B. Y. Rendon, Simulations of fast crab cavity failures in the HL-LHC, contribution Presentation at the Annual HL-LHC meeting, Oct. 2013, Daresbury.

[12] L. Lari *et al.*, Simulations of beam losses on LHC collimators during beam abort failures, in *Proc. IPAC 2013*.

[13] Y. Levinsen, Machine Induced Experimental Background Conditions in the LHC, CERN-THESIS-2012-132 and LHC background study group: http://cern.ch/lbs.

[14] T. Baer, Very fast losses of the circulating LHC beam, their mitigation and machine protection, PhD Thesis, 2013.

Chapter 6

Superconducting Magnet Technology for the Upgrade

E. Todesco[1], G. Ambrosio[2], P. Ferracin[1], J. M. Rifflet[1], G. L. Sabbi[3], M. Segreti[4],
T. Nakamoto[5], R. van Weelderen[1] and Q. Xu[5]

[1]CERN, TE Department, Genève 23, CH-1211, Switzerland
[2]Fermi National Accelerator Laboratory, Batavia, IL 60510, USA
[3]Lawrence Berkeley National Laboratory, Berkeley, CA 94720, USA
[4]CEA, Saclay, 91400, France
[5]KEK, 1-1 Oho, Tsukuba, Ibaraki 305-0801, Japan

In this section we present the magnet technology for the High Luminosity LHC. After a short review of the project targets and constraints, we discuss the main guidelines used to determine the technology, the field/gradients, the operational margins, and the choice of the current density for each type of magnet. Then we discuss the peculiar aspects of each class of magnet, with special emphasis on the triplet.

1. Targets

The HL-LHC aims at gathering 3,000 fb^{-1} over ten years. As discussed in the previous section, this ambitious target can be obtained by operating with a peak luminosity leveled at 5×10^{34} cm^{-2}s^{-1}. The plan is to obtain it through higher intensity/lower emittance and a larger focusing in the Interaction Point (IP). This second part is given by the magnetic lattice; the target is to be able to reduce the beam size in the IP by a factor two, and therefore one has to double the size of the quadrupoles aperture in front of the IP (triplet).

Some of the previous proposals, done during the LHC luminosity upgrade studies [1, 2, 3], aimed at a reduction of the beam size of 30%, increasing the triplet aperture 30% (see Fig. 1 for an historical view, indicating short models which have been built). The present target of reducing the beam size in the IP by a factor of two was based on theoretical studies (see for instance [4]), and was enabled by advances in magnet technology, i.e., test results from model quadrupoles of progressively larger aperture (Fig. 1).

A critical design parameter for a superconducting quadrupole is the peak field in the coil, which is a function of the aperture times the gradient. For Nb-Ti coils

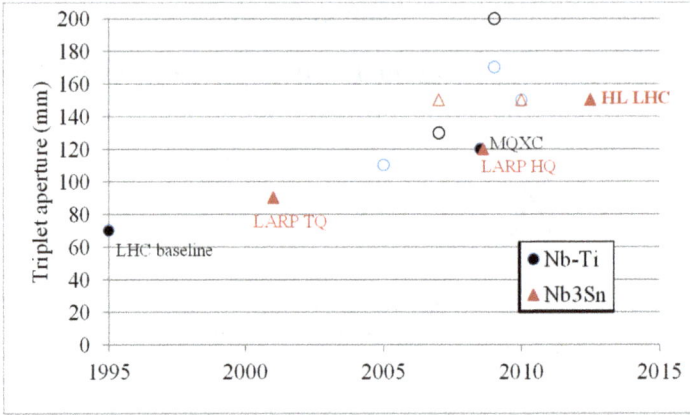

Fig. 1. Proposed aperture for the inner triplet versus time: triangles (Nb3Sn), circles (Nb-Ti), built hardware in full markers, and proposally in empty markers.

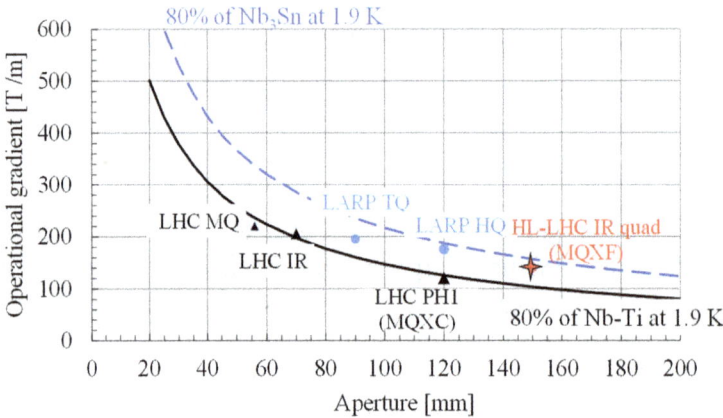

Fig. 2. Operational gradient (20% margin) versus aperture in Nb-Ti and Nb3Sn quadrupoles.

the peak field limit in operational conditions is about 8 T [5], whereas for Nb$_3$Sn this limit is ~15 T. One can prove that Nb$_3$Sn quadrupoles give 50% more gradient w.r.t. Nb-Ti for the same aperture [6] (see Fig. 2): this allows to have shorter magnets w.r.t. Nb-Ti. As explained in the previous chapter, a compact triplet means not only more space for other components, in a critical region of the tunnel, but also (and especially) additional performance: a shorter triplet means that the beam size has less longitudinal space to grow, and therefore for the same aperture one can squeeze the beam more in the IP. Moreover, a shorter triplet allows reducing the number of long range beam–beam interactions and to reduce chromatic aberrations. So, Nb$_3$Sn is the enabling technology to reach the ambitious target of the HL-LHC project.

2. Constraints

2.1. *Radiation damage and heat load*

The design of the final focus system of the upgraded LHC needs to account for the special conditions related to its proximity to the interaction points. The first important constraint for the magnetic system is the <u>radiation damage</u>, which is proportional to the integrated luminosity. Some essential components employed for magnet fabrication (epoxy resins) undergo severe degradation at 50–100 MGy. Therefore, one needs to set a safe dose limit of 10–20 MGy or switch to radiation resistant materials, as used for nuclear fusion, which can operate in the range of 100 MGy and more. For the HL-LHC we set a target for radiation damage at 30 MGy.

The second relevant constraint for the magnetic system is the <u>heat deposition on the coil</u>, which is proportional to the peak luminosity. In the stationary regime of continuous heat deposition, it induces a temperature gradient between the helium bath ($T_{bath} = 1.9$ K) and the temperature of the coil $T_{coil} = (1.9 + \Delta T)$. In the LHC triplet, the limit to the heat load is given by the requirement of having superfluid helium in the coil, i.e. a temperature lower than 2.17 K, which means $\Delta T < 0.27$ K [7]. The actual design limit is set to one third of the theoretical ΔT in order to account for uncertainties in the thermal analysis or variations in the heat load and cooling conditions. For the present inner triplet quadrupoles built with Nb-Ti conductor, this corresponds to a power deposition limit of 4 mW/cm^3, with a safety factor 3. For Nb$_3$Sn, with the same safety factor, one can withstand 12 mW/cm^3 [8].

Simulations of energy deposition in the HL-LHC show that without any shielding one has about 200 MGy peak dose and a peak heat load of 20 mW/cm^3. This regime is not acceptable for both aspects. The peak is localized in the horizontal and vertical planes. Shielding is very effective: with a 6-mm-thick tungsten shielding, one can bring these values down by a factor five, i.e. to 40 MGy and 4 mW/cm^3 [9].

Using an additional shielding in the quadrupole Q1 close to the IP (see Fig. 3), where the aperture requirement is smaller due to a smaller size of the beam, one can further reduce these values by a factor of two. Therefore, one ends up with a radiation damage similar to what is expected for the LHC (20–25 MGy) and a lower heat load (2 mW/cm^3), see Fig. 4.

The absorbers installed in the magnet bore address two of the most significant challenges of the LHC luminosity upgrade, namely the radiation damage and the heat load. To maintain the required space for the beam the final aperture of the quadrupoles has been fixed to 150 mm, i.e. slightly more than 140 mm (the double of present triplet) that was initially required by beam dynamics.

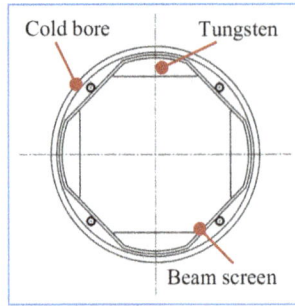

Fig. 3. Cold bore, beam screen and tungsten shielding in Q1.

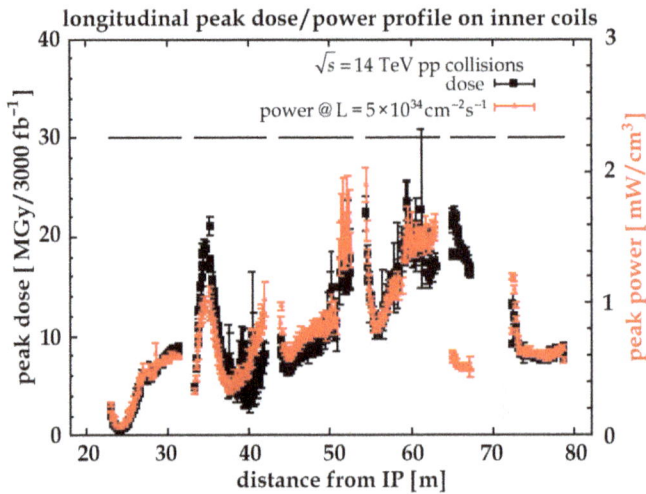

Fig. 4. Heat deposition in the coil (left scale), and radiation damage (right scale) for the 150 mm aperture triplet.

Two additional requirements point in the direction of a thick shielding as the only viable choice for the project. The total heat load on the triplet and separation dipole is 1.5 kW over 55 m, i.e. 30 W/m. The massive shielding allows to intercept about 800 W in the beam screen + shielding and remove it at intermediate temperature with higher efficiency. The remaining 700 W load needs to be removed from the cold mass at 1.9 K. This requires two heat exchangers of 70 mm diameter, barely fitting into the magnet cross-section [10]. A larger heat load would require larger heat exchangers, and larger magnet diameter, which is already at the limit of the constraints imposed by the tunnel diameter.

The second aspect is the degradation of copper Residual Resistivity Ratio (RRR) due to the radiation dose. This parameter is defined as the ratio between the resistivity at room temperature and at 1.9 K, related to the purity of copper. RRR must be > 150 to guarantee the conductor stability and a proper protection in case of quench. Recent studies pointed out that with 200 MGy the RRR is reduced by one order of magnitude [11]. Therefore, a dose of 200 MGy would also endanger the magnet operation and its protection. This degradation is partially wiped out by a warm-up to room temperature, so one could have problems in case of very long runs without warm-up. However, with the 6-mm-thick shielding, the copper RRR degradation becomes negligible.

2.2. *Field quality*

When beams are brought in collision, the beta functions in the triplet and in the separation dipole become very large, reaching peak values of ~ 20 km, i.e. five times larger than the nominal LHC values. In these conditions, the beams become very sensitive to magnetic field errors: for this reason the field quality constraints are very tight. On the other hand, at injection the interaction region gives a small contribution to the total budget of field imperfection of the accelerator and therefore the field quality targets can be significantly relaxed. For instance the b_6 systematic component in the quadrupole at injection can be as large as 25 units (i.e. 0.25% of the main field at two/third reference radius), but must be smaller than 0.5 units at high field (see previous chapter).

The field quality optimization should therefore concentrate on high field conditions. Considering that the final energy of the LHC beams could be in the range between 6.5 TeV and at 7 TeV, the field quality must also be maintained over the corresponding operational range.

A large set of correctors magnets (up to order 6) is foreseen in the layout to be able to correct field errors and/or add nonlinearities to counter beam instabilities; in fact since the beam size is very large in the correctors, they are very effective to correct any nonlinear unwanted component of the whole LHC.

2.3. *Fringe field and magnet size*

We roughly double the magnet apertures w.r.t. the LHC baseline, but the size of the cold mass is limited by the maximum cryostat size. Today in the LHC we have a cryostat with a 980 mm diameter that is not far from the limit imposed by the tunnel transverse size.

We propose to marginally increase the cold mass size from 570 to 630 mm to partly compensate for the aperture increase, with a weight increase of less than 20%. A larger increase would be difficult since some clearance is needed between the cryostat and the magnet.

In these conditions it is unavoidable to have a large magnetic field outside the cryostat: the transverse fringe field reaches ~ 50 mT on the cryostat surface. There is no specification of the allowed field in the LHC tunnel; this value depends on the specific instrumentation *in situ* (vacuum valves, beam position monitors, beam loss monitors, quench protection equipment,…) and in some cases one can envisage a displacement or shielding of the instrument (which is less invasive than shielding the magnet). An alternative solution is an active magnetic shielding, but at the price of an increased complexity of interconnections and number of components. In this phase of the design study we set a target of 50 mT maximum field on the cryostat and we do not envisage active shielding.

3. Main Design Choices

3.1. *Foreword: Loadline, critical surface and margin*

A superconducting magnet has most of the field produced by transport current, plus a second order contribution given by the iron magnetization: therefore in a first approximation the field is proportional to the current density in the coil: the relation peak field in the coil B_p versus current density j is called the loadline.

A superconducting coil can tolerate up to a given combination of field, current density and operational temperature: this is a property of the superconductor called the critical surface (see Fig. 5). Materials that can tolerate larger values of field and current density have a better performance, allowing to reach larger fields or to make more compact coils. When the loadline crosses the critical surface, one has the maximum theoretical reachable field. It is called short sample limit since the critical surface is usually measured for a short sample of conductor (see Fig. 6).

A critical choice for magnet design is the width of its winding. The peak field is proportional to the current density and to the width of the coil, so with large coils widths, the loadline in the *B-j* graph has a lower slope and one can reach higher fields (see Fig. 6). However, a magnet with larger coil is less effective, less compact, needs more superconductor, and has lower current density w.r.t. one with smaller coil width. For larger and larger coils an asymptotic field is reached, the gain becoming more and more marginal: one needs to find the optimal coil width [12].

There are two more aspects that add to what may seem a pure cost and size problem: firstly, larger current densities imply larger mechanical stress induced by the electromagnetic forces (in fact they scale with the square of the current density). Therefore, very compact magnets can lead to forces and stresses that damage the superconductor or the insulation. The second aspect is protection: in

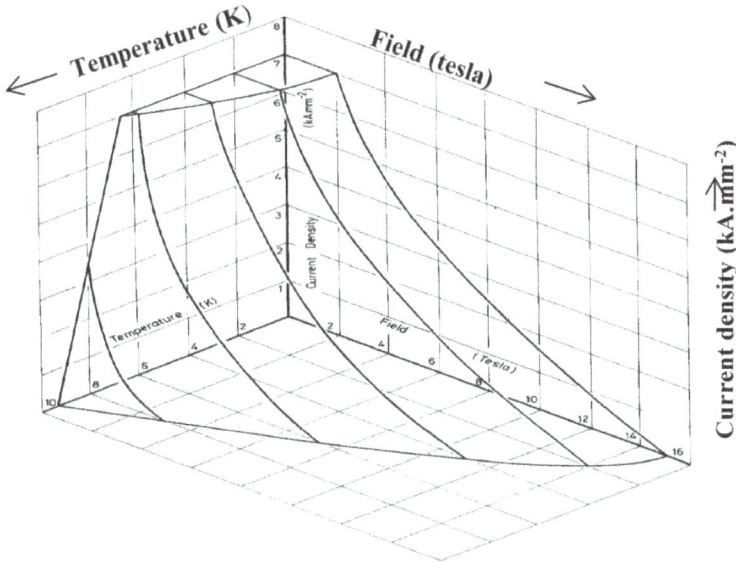

Fig. 5. Sketch of the critical surface of Nb-Ti.

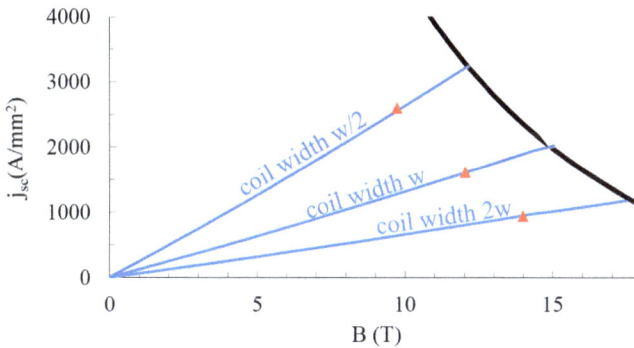

Fig. 6. Cross-section of the critical current versus field for Nb$_3$Sn at 1.9 K, and loadlines for a coil with width w, $w/2$ and $2w$; red dots indicate operational points with 20% margin.

case of a transition from the superconductive to the resistive state, the energy of the magnetic field has to be dissipated in the coil. A too large energy density brings the coil to an unsafe temperature (usually considered to be above 400 K) or temperature gradient that damages it. So both stress and protection aspects point to avoid current densities in the coil much larger than 500 A/mm^2.

Finally, the design needs to account for production and operation margins. Accelerators magnets usually operate at 50%–80% of the short sample limit, according to the magnet type and technology. Since the operational targets are usually established before magnet prototyping and production, their selection needs to take into account cost, risk and performance considerations. In the followings, we will carry out the main choices for the HL-LHC magnets: technology, coil width and operational margin.

3.2. *Technology, operational temperature, margin*

In a final focus system, performance is given by large aperture and short length in the region from the interaction point up to the separation dipole. This leads to use in the triplet the Nb_3Sn technology at 1.9 K, which allows doubling the aperture of the present Nb-Ti triplet with a moderate increase of the magnet length (see Fig. 7). In order to maximize performance, a challenging operational point of 80% on the loadline, was selected.

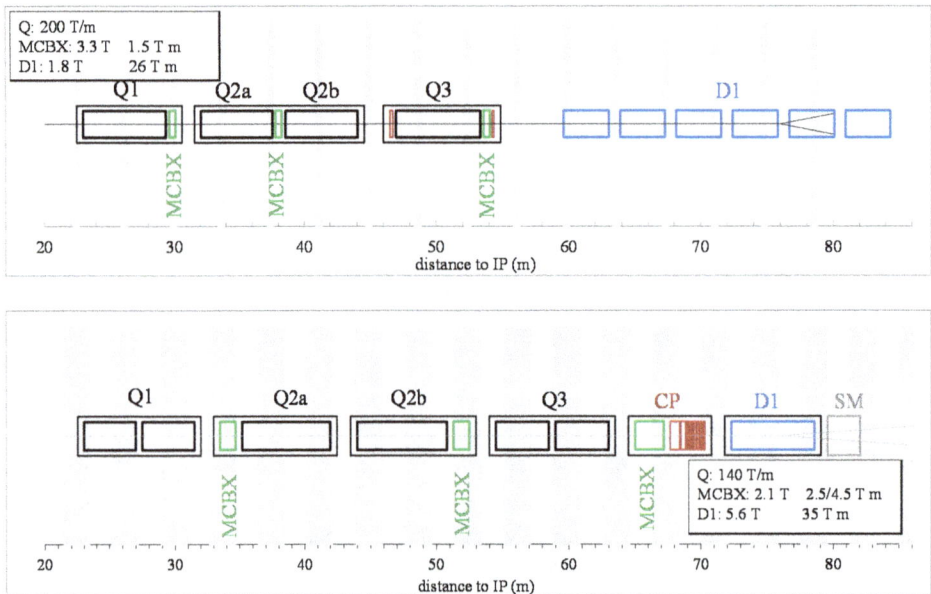

Fig. 7. Lay out of the LHC (upper part) and of HL-LHC (lower part) interaction region from first quadrupole (Q1) to separation dipole (D1).

For the separation/recombination dipole D1 (single aperture), which is presently a resistive magnet (see Fig. 7), we opt for a superconducting magnet with 5.6 T operational field, given by the Nb-Ti technology, with a more conservative operational point of 75% on the loadline [13]. The corresponding reduction in the length of D1 in the upgraded IR more than compensates for the additional space needed by the triplet; in fact, the end of D1 in the HL-LHC layout is 4 meters closer to the interaction point as compared to the LHC. The option of a Nb₃Sn magnet, considered in the past [14] has been discarded, as the gain of a few meters (3 m, with an 11 T dipole) is not considered critical in this location and has no effect on performance.

The field in the separation/recombination dipole D2 (double aperture) is imposed by field quality constraints. Here the main issue is to reduce the saturation effect. For this reason we chose an operational field of 4.2 T given by the Nb-Ti technology, with a comfortable margin (30%).

The Nb-Ti technology is also chosen for the two-in-one large aperture quadrupole Q4 [15]. With respect to the present baseline, the operational temperature is lowered from 4.5 K to 1.9 K, allowing to better exploit the Nb-Ti performance, with 20% operational margin.

3.3. *Coil width and stress*

For large aperture magnets, stress can become a limiting factor. Values of the order of 200 MPa can damage insulation for Nb-Ti magnets, or degrade conductor performances for the Nb₃Sn magnets. Therefore one has to carefully check in the conceptual design phase that the field and current density values correspond to reasonable values of stress.

For the Nb₃Sn case, in Fig. 8 we plot the operational gradient and the stress versus the coil width. We see that the gain starts to saturate at 30–40 mm, corresponding to a 150 MPa stress, which is large but still within the limits. Going from 35 to 50 mm, i.e. a 50% increase in the superconductor quantity, brings only a 10% increase in gradient. Therefore we chose a ∼ 35 mm width coil width,

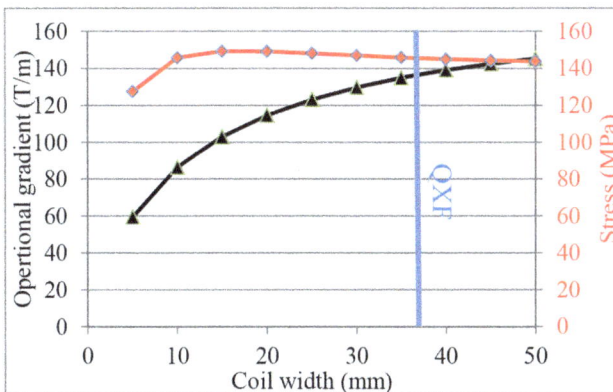

Fig. 8. Operational gradient with 20% margin (black triangles) and stress (red square) versus coil width for 150 mm aperture quadrupole in Nb₃Sn.

providing an operational gradient of ~ 140 T/m. This choice is also related to other factors as previous experience with 90 mm and 120 mm aperture quadrupoles, strand and cable geometry, and quench protection: these aspects will be discussed in the next section.

The present LHC has separation/recombination dipole magnets with a RHIC coil, one layer of 10-mm-wide Nb-Ti cable. We opt for a larger cable, i.e. the LHC main dipole 15-mm-width cable, which has the advantage of having a larger margin and is available in the CERN reserve stock. Choosing a 25% margin on the loadline, we get about 5.6 T and therefore 6.3 m for getting to the 35 T·m requirements. The stress is still manageable at 80 MPa (see Fig. 9). A larger coil width would only marginally reduce the length: two layers with the LHC cable, yielding an effective coil with of 30 mm, would provide 6.5 T (see Fig. 9), i.e. one meter shorter magnets.

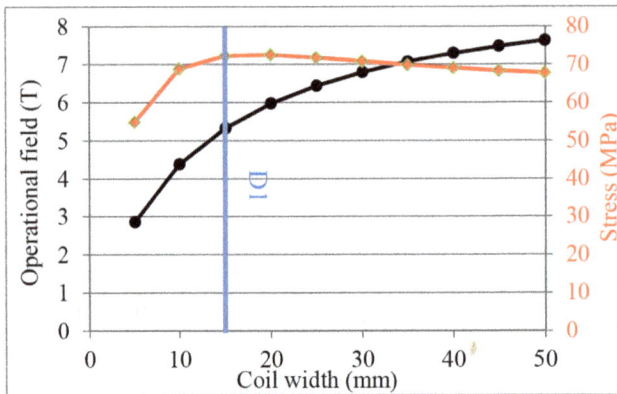

Fig. 9. Operational field with 25% margin (black triangles) and stress (red square) versus coil width for 150 mm aperture dipole in Nb-Ti.

For the D2 we have a preliminary selection of a 15-mm-width cable, a 105 mm aperture, and an operational field of 4 T with a large margin. The main problem of this magnet is a considerable magnetic cross-talk between the apertures.

For the large aperture two-in-one quadrupole Q4, we have a fixed space between the beams, and since we increase the aperture from 70 (present LHC) to 90 mm we have to limit the coil width. We explored possibilities with existing cables developed for the LHC magnets, namely a one layer with 15-mm-width cable, or two layers with 8-mm-width cable. In both cases one has a gradient around 120 T/m (20% loadline margin) with pretty low stress of 50 MPa (see Fig. 10). Doubling the cable width would create a large cross-talk between the apertures, with a modest reduction of the magnet length.

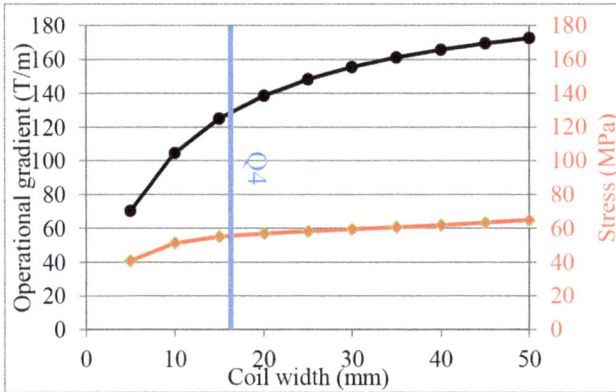

Fig. 10. Operational gradient with 20% margin (black triangles) and stress (red square) versus coil width for 90 mm aperture quadrupole in Nb-Ti.

Table 1. Parameters of the HL-LHC main magnets. Tentative values for orbit corrector and D2 are in italics.

		Triplet Q1,Q3/Q2a,b	Orbit corrector MCBX	Separ. dipole D1	Recomb. dipole D2	Large ap. 2-in-1 Q Q4
Aperture	(mm)	150	150	150	105	90
Field	(T)		2.1	5.6	4.0	
Gradient	(T/m)	140				120
Mag. Length	(m)	2×4.0/6.8	1.2/2.2	6.25	8.75	3.5
Int field	(T m)		2.5/4.5	35	35	
Int gradient	(T)	1120/938				420
Peak field	(T)	12.1	3.5	6.5	4.7	5.9
Current	(kA)	17.5	2.4	11.8	11.7	16.1
j overall	(A/mm^2)	482	497	452	448	617
Loadline margin	(%)	20%	40%	25%	40%	20%
Stored energy	(MJ/m)	1.32	0.09	0.338	0.16	0.204
Saturation	(%)	9%		12%	9%	
Material		Nb$_3$Sn	Nb-Ti	Nb-Ti	Nb-Ti	Nb-Ti
N. layers		2	1+1	1	1	1
N. turns/pole		50	74	44	58	14
Cable length/pole	(m)	2×450/700	220/400	600	1100	110
Cable width	(mm)	18.15	4.37	15.10	15.10	15.10
Cable thick. in.	(mm)	1.438	0.819	1.362	1.362	1.362
Cable thick. ou.	(mm)	1.612	0.871	1.598	1.598	1.598
Ins. thick rad	(mm)	0.15	0.105	0.13	0.13	0.13
Ins. thick azi	(mm)	0.15	0.105	0.11	0.11	0.11
No. strands		40	18	36	36	36
Strand diam	(mm)	0.85	0.48	0.825	0.825	0.825
Cu/NonCu		1.20	1.75	1.95	1.95	1.95

A summary of the technology, operational field/gradients, peak fields, and loadline margin of the main magnets of the upgrade is given in Table 1.

3.4. *Cryostats and interconnections*

The maximum length of a cryostat that can be lowered in the tunnel is 15 m, corresponding to the main dipole case. Having quadrupoles with lengths ranging from 7 m to 8 m, plus a series of orbit correctors, we are forced to have one cryostat per quadrupole (in LHC Q2a and Q2b share the same cryostat, see Fig. 7). The US LARP collaboration, in charge of the Q1 and Q3 development, has opted for a solution based on having two 4-m-long quadrupole closely connected to form Q1 and Q3 units. This reduces the risk associated to the length, even though it increases costs due to double number of coils and magnet assemblies, and requires doubling manufacturing lines. The Q2 units, under development at CERN, are designed with one 6.8-m-long quadrupole magnet (magnetic length).

We assumed to have 1 m between the cryostats for interconnection. A fine tuning of this length will be carried out during the next stages of the project. Finally, we assumed to have one corrector package and a separation dipole, each one in its cryostat. One could probably merge the two units into a single cryostat, but the gain in performance would be marginal. Also in this case the optimization between performance and risk will take place during the next phases of the project.

3.5. *Cooling*

The cooling of the triplet is provided through heat exchangers. Since the total load on the cold mass is about 15 W/m, one has to use two heat exchangers of 70 mm diameter. The alternative options of one heat exchanger of 110 mm diameter would simplify the interconnections but is not viable since it is not compatible with the magnet mechanical structure. The ideal position for a hole in the yoke of a quadrupole is at 45°, i.e. in the low field region and where less material is needed for structural reasons. An 70 mm heat exchanger is large but still fits the cold mass iron yoke. The short orbit correctors have to share the heat exchanger, i.e. the hole must be in the same positions.

Different options have been considered for the cooling of the corrector package plus the separation dipole, the most effective being another heat exchanger (of 50 mm diameter), this time at 90° from the coil midplane of the dipole.

Fig. 11.   LQ, the first "long" Nb₃Sn magnet built by LARP collaboration.

4.   The Triplet Quadrupoles Q1-Q3

4.1.   *Historical development*

The development of Nb₃Sn quadrupoles for the LHC luminosity upgrade was initiated in 2004 by the US LHC Accelerator Research Program (LARP), a collaboration of US National Laboratories and CERN (see Fig. 11). At that time the target was to reach a β^* of 25 cm and a 30% increase of the aperture, from 70 to 90 mm was considered an adequate choice both in terms of machine requirements and technological challenges. After some preliminary tests using racetrack coils, the 1-m-long Technological Quadrupole (TQ) series were developed to address key manufacturing and design issues for $\cos 2\theta$ coils [16]. Two mechanical structures were tested, one based on stainless steel collars and the other on a Al shell pre-loaded using water-pressurized bladders and interference keys [16]. After the test of several models, the bladder and key structure demonstrated a better capability of controlling stress and was selected for the length scale-up from 1 m to 3.4 m (Long Quadrupole (LQ) series), with a first successful test in 2009 (see Fig. 11).

Meanwhile, several works were pointing at the possibility of using apertures larger than 90 mm to increase the upgrade performance [5]. In order to study the feasibility of larger apertures, and demonstrate the capability to incorporate field quality and alignment requirements, in 2008 LARP started the development of the 120-mm-aperture High-field Quadrupole HQ [17]. A successful HQ test at CERN in early 2012 supported a decision to further increase to 150 mm aperture for the triplet quadrupoles (QXF) [19]. The most advanced solutions used in TQ, LQ and HQ are now being applied to the larger aperture quadrupole. So MQXF is essentially a scaling of the design of HQ [20]. The guideline is to keep all features that have been shown to work properly in the LARP magnets.

4.2. Strand and cable

As the aperture in QXF is 25% larger than in HQ, a corresponding increase of the coil width is desirable. In order to minimize deviations from established LARP designs, a two layer coil layout is maintained and the increase in coil width is obtained with an increase in cable width, requiring a larger strand and/or more strands per cable, the option of having one additional layer being excluded to avoid complexity in the coil fabrication. The number of strands is limited by cable mechanical instabilities which affect the winding process, and/or damage to the superconducting strands during the cabling operation. For MQXF, it has been decided to limit the number of strands to 40, which is also the upper limit of the CERN winding machine. A key-stoned cable with 40 strands is already rather difficult to be optimized. TQ cable had 27 strands and 10 mm width, and HQ had 35 strands with 15 mm width. The number of strands and the cable width fixes the strand diameter to 0.85 mm. This is a marginal increase w.r.t. the HQ case, which had 0.8 mm. In all cases we tried to minimize the changes w.r.t. the HQ cases to rely on established design solutions and avoid significant delays to overcome new issues.

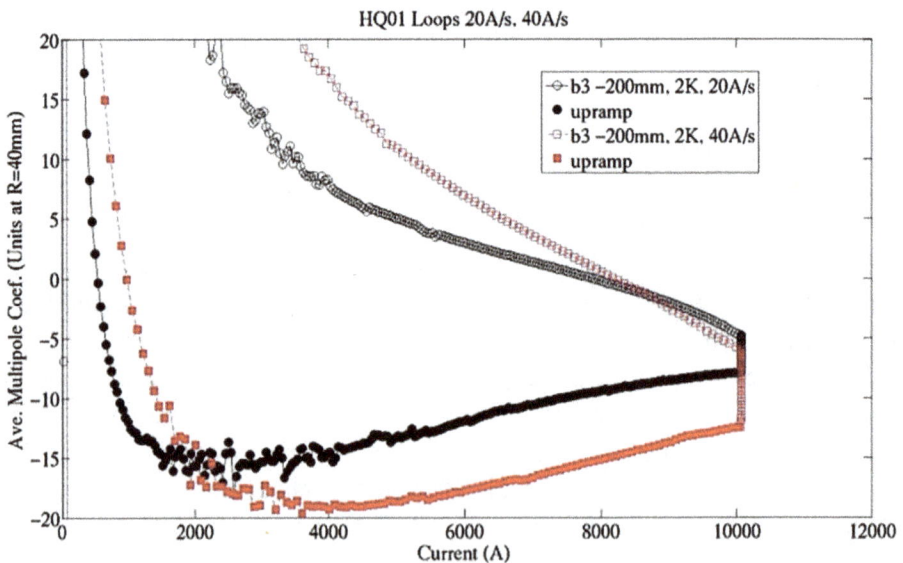

Fig. 12.   Dependence of b_3 along the ramp for different ramp rates: case of cable without core (HQ01e).

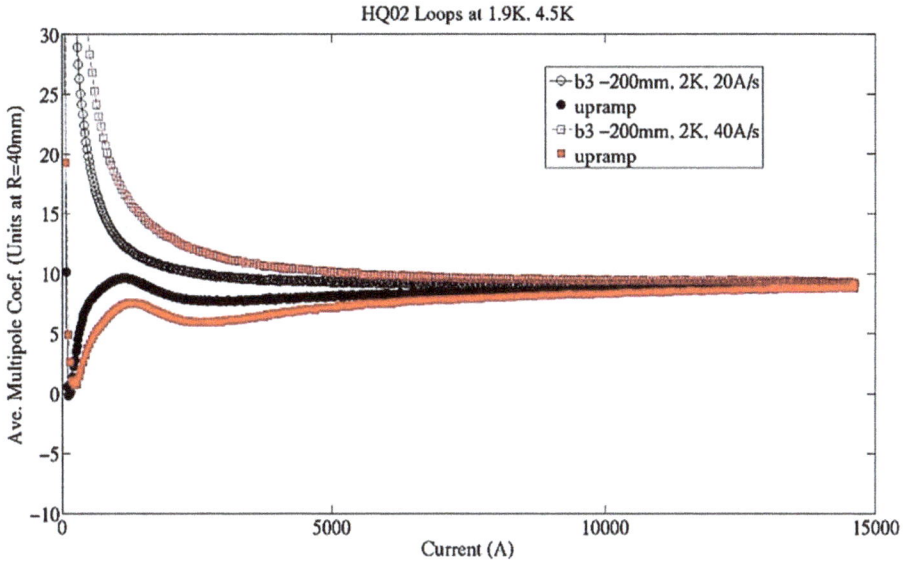

Fig. 13. Dependence of b_3 along the ramp for different ramp rates: case of cable with core (HQ02a).

The other novel aspect is the use of a stainless steel core (25 μm thick) in the cable to increase the inter-strand resistance. Previous LARP quadrupoles, built without cored cables, showed a clear indication of a very low inter-strand resistance (of the order of 0.1–0.5 $\mu\Omega$) [21], producing (i) a severe degradation of quench performance with increasing ramp rate, affecting the capability to perform a fast discharge without quench and (ii) a degradation of field quality, visible as non-allowed components with large dependence on ramp rate, and decay of several units even at high field, with times of the order of a few seconds (see Fig. 12). The second short model HQ02, built with cored cable, proved to cure these issues with an increase of the effective inter-strand resistance by more than one order of magnitude (see Fig. 13).

4.3. *Coil*

The coil is a double layer, four block coil. Two wedges provide the required flexibility to tune the field quality to optimal values. The basic layout of the conductor blocks (Fig. 14) is similar to what has been used in HQ. In particular similar pole angles are chosen for both layers. This approach has been shown to minimize the peak coil stresses. A novel method allowing an exhaustive study of the optimization of the cross-section [22] has shown that the selected option provides a short sample gradient which is very close (less than 1%) to the maximum gradient attainable with this type of cable (see Fig. 15). In operational conditions the peak field in the coil is 12.2 T, corresponding to a ratio between peak field and gradient times aperture of about 1.15.

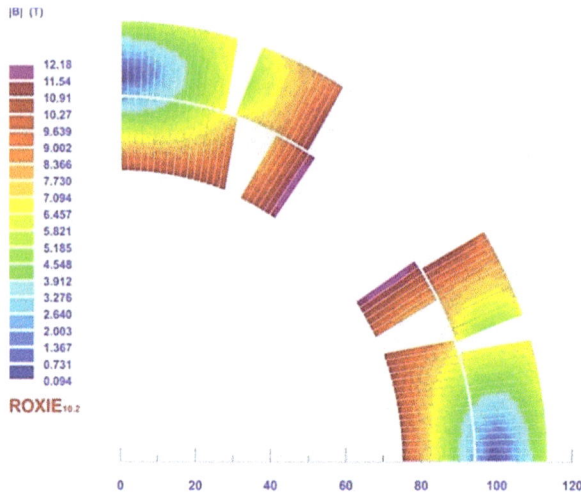

Fig. 14.   MQXF coil cross-section (one quarter shown), and field in operational conditions.

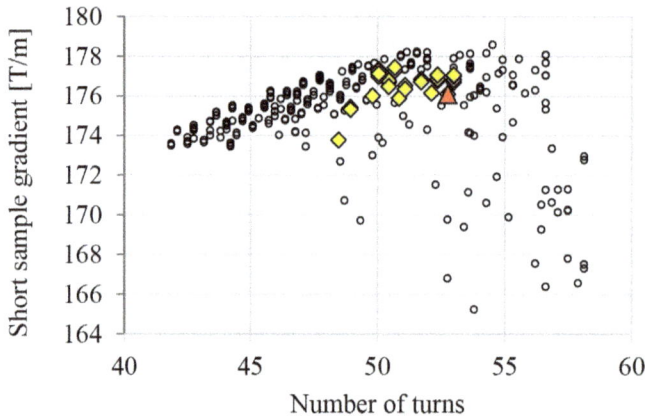

Fig. 15.   Short sample gradient provided by 300 different cross-section, satisfying field quality (chosen cross-section: triangle).

4.4. *Mechanical structure*

The Lorentz forces are contained by an aluminium shell (see Fig. 16). During the assembly at room temperature, a prestress of the order of 50 MPa is applied to the coil through the insertion of keys in the slots opened by bladders. During the cool down, the Al cylinder stress increases to about 150 MPa. This procedure has been used in several models, proving to be an efficient and accurate way to control the stress in the magnet. As the magnet is energized, the pre-load provided by the mechanical structure is replaced by the internal loads generated inside the coils by the electro-magnetic forces (see Fig. 8). Full alignment is maintained at all steps of coil fabrication, magnet assembly and powering. An additional stainless steel vessel is needed for He containment — unfortunately, stainless steel cannot provide the adequate stress increase since its thermal contraction factor is too small, and aluminium cannot provide He containment since it cannot be welded, so two cylinders are needed.

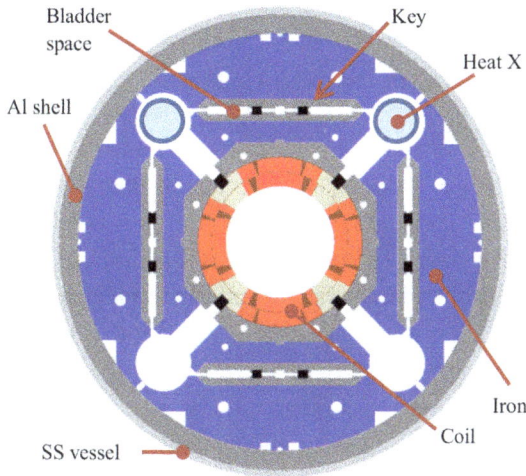

Fig. 16.   MQXF cross-section, including heat exchangers.

4.5. *Protection*

The inductance of the QXF magnets is 8–10 mH/m, the lowest value being at nominal current and the highest in the linear regime of non-saturated iron, i.e. at injection. Since the baseline is to have two 8-m-long magnets powered in series, we have an inductance of ~ 0.13 H. The current is 17.5 kA, so the dump resistor is limited to 50 mΩ to avoid having voltages that exceed 900 V at the beginning of the current dump. This means that the time constant of the circuit is of the order of 2 s, and the dump extracts a negligible fraction of the energy stored in the magnetic field. As in the LHC dipoles, the only solution is to use the thermal inertia of the magnet coil to dissipate the energy of the magnetic field. Pending further analysis and verification, a design constraint to remain below ~ 350 K in all points of the coils was adopted.

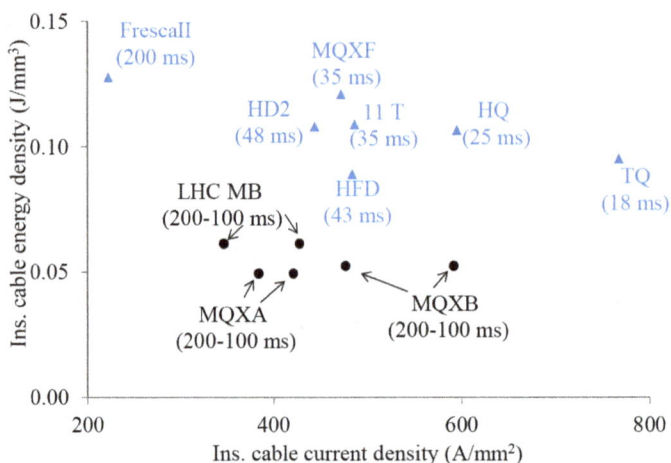

Fig. 17.   Snapshot of the protection issues in Nb-Ti and Nb₃Sn magnets: energy density on the coil versus current density, and time margin (circles: Nb-Ti, triangles: Nb₃Sn).

Both Nb-Ti and Nb₃Sn windings have a similar enthalpy from 2 K to 300 K of the order of 0.6 J/mm^3. So the first physical quantity to check is the energy density, i.e. the stored energy divided by the volume of the coil. Note that due to the time scale involved in these phenomena (a fraction of second), the structure components as collars and yoke are too far to share the burden of the heat dissipation — that's why we consider the energy density only over the coil volume. For typical Nb-Ti magnets this value is around 0.05 J/mm^3 (see Fig. 17). In our case, as in many other Nb₃Sn magnets, we are at twice this value, so still well within the enthalpy limit but with half margin.

The key point is to prevent excessive energy dissipation at the initial quench location, which can lead to coil damage due to high local temperature and stress, (hot spot). This requires distributing the energy over the whole magnet, by ensuring rapid transition of the entire winding to the normal conducting state in the fastest possible time. This is done as in most accelerator magnets through quench heaters, i.e. strips of stainless steel which are powered as soon as the quench is detected, and whose heat is transferred via conduction to the coil, pushing it above the critical temperature. The temperature margin in Nb₃Sn is in the range of 5–10 K, the lowest value being reached in the high field zone where the margin on the loadline is 20%. Integrating the specific heats, one finds that has to give 30 μJ/mm^3 in the high field zone. Note that this value is a factor 10 larger w.r.t. to the Nb-Ti case.

A simple way to compare the protection challenge is to compute the time budget (time margin) for the protection system available to quench all magnet, setting 300 K as the maximum temperature reached by the coil [23]. An advantage of this quantity is that it depends only on the magnet design, and not on the quench features (high field or low field, propagation, etc.) and on the protection system. On the other hand, to make the estimate of the warmest point reached in the magnet (so-called hotspot temperature) one needs other hypothesis on the quench location, efficiency of heaters, propagation, etc.

The time margin is of the order of 100 ms for Nb-Ti magnets (see Fig. 17). In general one needs a few ms to build enough resistance to have a measurable voltage (voltage thresholds are usually set at 100 mV). Then a validation window of 10 ms is used to avoid having false signals. Then the switch of the circuit disconnecting the power converter and dumping the current on the external resistor or on a diode is opened (2 ms). At the same time the heaters are fired. Typical times between the heater firing and the quench of the coil induced by the heaters is 10–20 ms [24] (see Fig. 18).

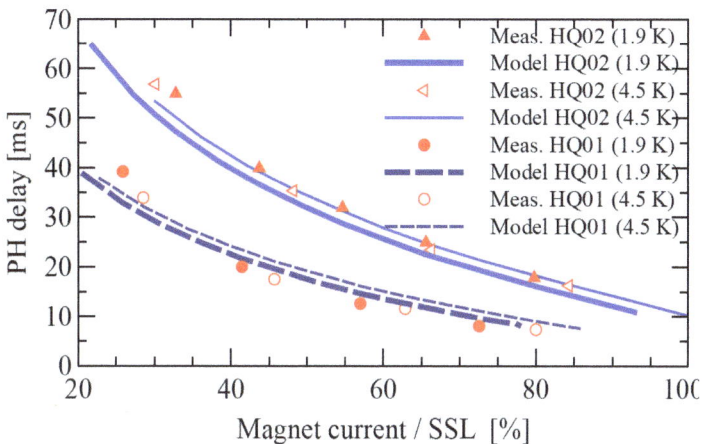

Fig. 18. Quech heater delay versus operational current, in HQ01 and HQ02 featuring 25 and 75 μm thickness of polymide between heatrs and coil.

Therefore, 100 ms seems a comfortable margin, and 50 ms seems a minimal value necessary to have a safe protection system. This time budget is too small in TQ and HQ magnets (15–25 ms). QXF has a larger margin since the current density has been lowered, but it is still on the edge: the present best estimates give about 35 ms.

If one considers heaters on the outer part of the outer layer, protection has to rely on the propagation of the heat from outer to inner layer, which takes of the order of 20 ms. To increase the thin margin, LARP quadrupoles made use of quench heaters on the inner part of the inner layer; these heaters were effective, but in some cases they showed partial detachment after successive quenches.

Moreover they act as a thermal barrier in a region which is critical for heat transfer, so for the triplet one would need a special geometry with gaps. Studies to find an effective solution are ongoing. The present baseline is to have heaters both on the outer part of the outer layer and on 50% of the inner part of the inner layer.

Another possibility is to use quench heaters in the interlayer, but they should be heat treated with the coil, so they would need a completely different technology. A third option is to find ways of making heaters more effective, and/or creating thermal bridges from the outer to the inner layer without endangering the insulation.

4.6. Field quality and shimming

When the beams are squeezed in the interaction point, the optical functions in the triplet are very large and the beam dynamics becomes very sensitive to any field imperfection in the triplet. Field quality of the triplet must satisfy tight constraints. The main challenges are (i) a reproducibility of the transfer function of less than one unit (ii) control the low order harmonics within one unit. On the other hand, the nonlinearities coming from the large iron saturation (about 10%, as in HQ, see Fig. 19) can be compensated through an adequate powering of the magnets, provided that the effect is reproducible. Results from the LARP program show that this level of reproducibility is obtained, and that there is a good understanding of the quadrupole main component behavior as a function of the current and of the ramping direction (see Fig. 19).

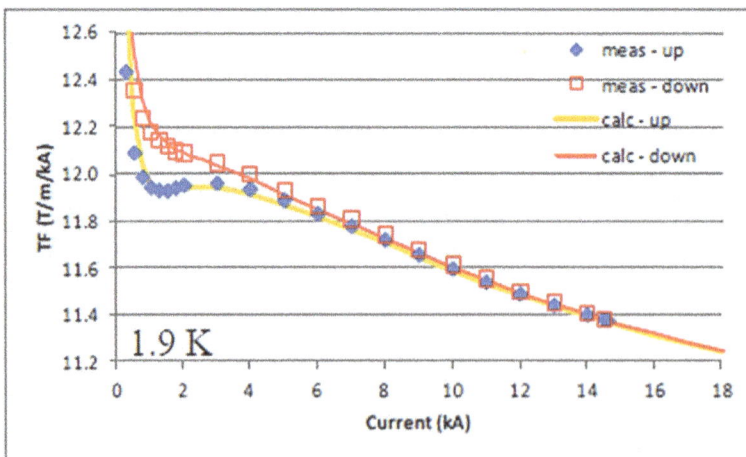

Fig. 19. Ratio between gradient and current for ramping up and down of the quadrupole model HQ.

The low order harmonics are related to the asymmetries of the components and of the assembly. Here, results have not yet shown that the achieved quality is compatible with the beam targets, since in some case several units of non-allowed low-order harmonics (a_3, b_3, a_4, b_4) have been found (see Figs. 12, 13). In principle there is no reason suggesting that Nb_3Sn has a significantly worse field quality for these harmonics (related to asymmetries, i.e. to coil geometry) w.r.t. Nb-Ti. If these effects would persist, a magnetic shimming [25] can be used to compensate for a few large harmonics (typically two at the same time). The technique is based on inserting magnetic rods in holes in the collars and magnetic bars in the spaces used by the bladders. By placing magnetic shims in an asymmetric way, one can compensate up to several units of low order harmonics [26].

Usually a lot of emphasis is put on the first allowed harmonics b_6. In fact, this harmonic is not the most critical for the beam, as it is a high order. Moreover, from the point of view of the magnet builder, it is pretty easy to control b_6 through the cross-section geometry. At injection one has about 20 units given by the magnetizations, which are within the beam dynamics targets.

5. Correctors

5.1. *Orbit correctors*

The beam dynamics requirements for the orbit correctors are 2.5 T·m both in horizontal and vertical plane for the correctors installed next to Q1 and Q2, and 4.5 T·m for the correctors installed next to Q3. This increase in integrated strength is obtained with a corresponding increase in length. The magnets have to operate at any combination of horizontal and vertical field, i.e. a square in the (B_x, B_y) plane. In the LHC we have nested magnets providing 3 T in each plane, with 70 mm aperture. The main challenge of the nested magnet is the management of the large torque (10,000 N·m per meter length of the magnet) coming from the Lorentz forces. In the LHC, this is kept by impregnation — the two layers are "glued" together, and not with a mechanical lock.

For the HL-LHC we consider a nested magnet with an operational field of 2.1 T, giving 1.2 m and 2.1 m magnetic length respectively. With a Rutherford cable composed of 18 strands of 0.45 mm diameter, with a width of 4.5 mm, one can build one-layer coils of a classical $\cos\theta$ layout, reaching 2.1 T with 40% margin. This case ([27], see Fig. 20) has been considered in S-LHC preparatory phase program, set up in the frame of the previous project LHC upgrade phase I, now superseded by HL-LHC. Please note that the peak field is 3.5 T, close to twice the nominal field. This is due to the presence of two perpendicular fields (giving a

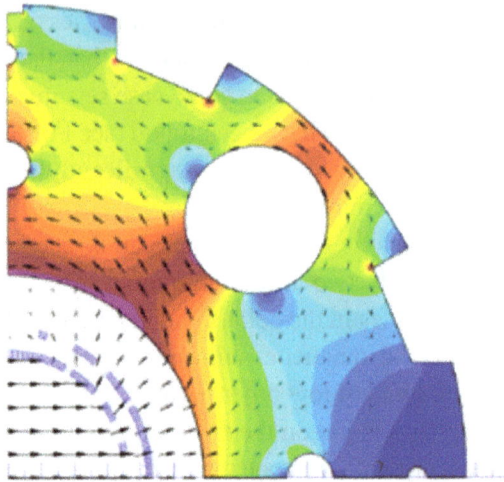

Fig. 20.   Cross-section of the orbit correctors developed for SLHC studies (140 mm case [27]).

factor $\sqrt{2}$) plus the ratio coil peak field/bore field, which is ~ 1.3. Large ratios peak field/bore field are unavoidable in dipoles where the coil width is thin with respect to the aperture.

A double layer coils, both for B_x and B_y, would enable reaching ~ 3 T with ~ 40% margin; this option is also being considered, as it also would allow to make additional room available to the triplet and for the interconnections.

The nested option is challenging from the point of view of the mechanical structure, and to ensure reliability we require that the torque has to be controlled through a mechanical locking. A non-nested option would possibly allow to raise the field, but it would lose the space needed for interconnections and coil ends. Estimating at 100 mm the length of the coil ends per side, and 300 mm for interconnections, the non-nested option gives a 1.5 m longer layout. CIEMAT has taken the responsibility of the magnet design and of the construction of the 1.2-m-long prototype.

5.2.   Linear and nonlinear correctors

The correction of the triplet imperfections and misalignment requires a large set of correctors. The first requirement is a skew quadrupole, with about 1 T·m integrated force, to correct a tilt in the triplet. Then we have sextupole, octupole, decapole and dodecapole. Requirements for the normal and skew terms are the same, with the exception of the normal b_6, four times larger than a_6 since this is an allowed multipole of the quadrupole and therefore has a larger systematic and random component. The beam dynamics requirements, given in Table 2, are based on the field error tables of the triplet and of the separation dipole, with an additional safety factor 2 up to order four, and 1.5 for orders five and six. INFN-LASA is in charge of the correctors design and of the construction of prototypes.

In the LHC we have nested correctors, with up to five magnets nested. This solution saves space, but makes operation more complex. For a non-nested solution a key point is to have very short heads, otherwise all the space is lost in heads and interconnections. In the framework of the S-LHC studies, a superferric technology [28] was used to build some prototypes with 140 mm aperture (see Figs. 21 and 22). The magnet has the same cross-section as a resistive magnet, with coils serving to magnetize the iron poles and yoke. In this case, (i) the field quality is given by the shape of the iron poles and not by the precise location of the coils, and (ii) the field is limited at ~ 1.5–2 T due to iron saturation.

Table 2.   Parameters of the HL-LHC correctors.

Name	Multipole	Coil length (m)	Force (T m)	Peak field (T)
MCQSX3	a_2	0.716	0.9830	2.20
MCTX3	b_6	0.339	0.0860	2.10
MCTSX3	a_6	0.087	0.0168	2.10
MCDX3	b_5	0.079	0.0254	2.00
MCDSX3	a_5	0.079	0.0254	2.00
MCOX3	b_4	0.137	0.0458	1.25
MCOSX3	a_4	0.137	0.0458	1.25
MCSX3	b_3	0.121	0.0625	1.25
MCSSX3	a_3	0.121	0.0625	1.25

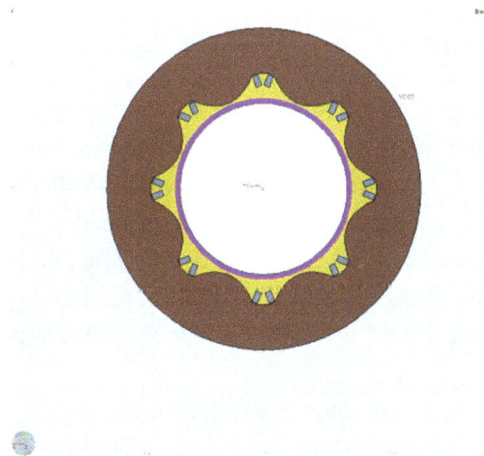

Fig. 21.   Superferric correctors cross-section [28].

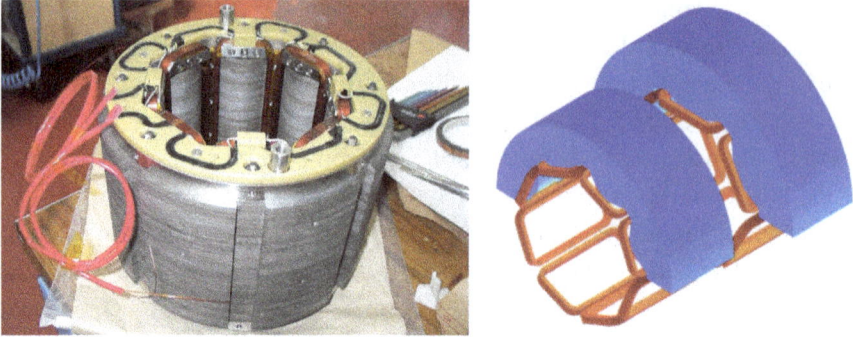

Fig. 22. Superferric correctors [28].

One advantage is that coils are not directly exposed to the aperture, so the magnet is resistant to radiation and additional shielding can be put to protect the coils. The second advantage is that the heads can be made extremely short, with small diameter cable and sharp bends (see Fig. 22), so what is lost in the non-nested option is partially recovered by the shorter heads. It has been also checked that the longitudinal interference between different correctors is negligible even with short interconnection of 80 mm (see Fig. 22, right). The last advantage is that operational current is ~ 100 A, since the conductor is a small single wire. This also simplifies the numerous current leads needed to power this large set of correctors.

This magnet may become a good application for MgB_2 conductor, a novel material characterized by potentially low cost, and current densities vs field properties similar to Nb-Ti. It would be the first application of this conductor to accelerator physics.

6.   The Separation Dipole D1

The LHC separation dipole is a 20-m-long resistive magnet, made of 6 modules of 3.4 m length, providing 26 T·m (see Fig. 7). The new specification of integrated field is 35 T·m. The replacement of the resistive units with a single Nb-Ti magnet allows recovering the additional space which is needed by the longer triplet. Keeping the same aperture of the triplet, and the same shielding, one can verify that the collision debris induce a heat load and a radiation dose within the project targets. So the replacement of a resistive magnet with a superconductive one is justified.

The main challenges in the magnet design are the large aperture, fringe fields and field quality. The large aperture gives large stored energies and forces, so a proper mechanical structure must be developed. With such a large aperture the fringe field becomes an issue: with a 5 T operational field in 150 mm aperture, one needs ~ 200 mm of iron to avoid fringe fields. In case of 15 mm coil width and 15 mm spacers, the magnet size reaches $150 + (15 + 15 + 200) \times 2 = 610$ in mm diameter, i.e. about the same size of the triplet quadrupole cold mass. This suggest to (i) do not push field to very large values, restricting the study to one-layer coil (ii) have a mechanical structure where forces are taken by the yoke and collars are simple spacers: in this way, more iron is available for shielding.

The baseline is to set the working point at 75% of the loadline, with a Nb-Ti 15-mm-width cable as in the LHC main dipole, providing 5.6 T operational field (see Fig. 23). In this way a 6.3-m-long magnet provides the required 35 T·m. The iron is largely saturated at nominal field, with a 12% decrease of ratio field/current w.r.t the linear case. Such a large saturation has a relevant impact on field quality, which becomes the main challenge. A careful iron shaping can reduce this effect, following the example of what has been done for the RHIC dipoles [29]. The impact on b_3 can be reduced from the initial values of several tens of units (for a circular iron without holes) to a few units along the operational range (see Fig. 24). Optimization is done at high field. Since there is some uncertainty about the actual energy of the LHC, two cases with 6.5 TeV and 7 TeV have been considered. When choosing the final iron cross-section the energy will be established.

The mechanical structure is similar to the MQXA [30], with support given by the iron yoke locked by keys. This structure has the advantage of reducing the collar size, leaving more space to iron and reducing the fringe field.

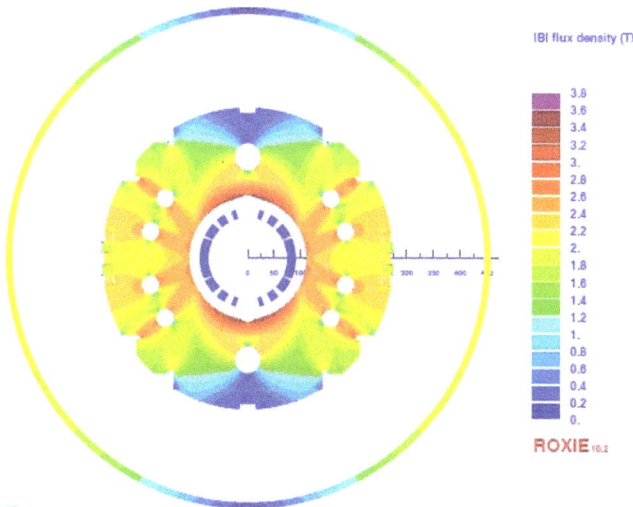

Fig. 23.   Cross-section of the separation dipole [13].

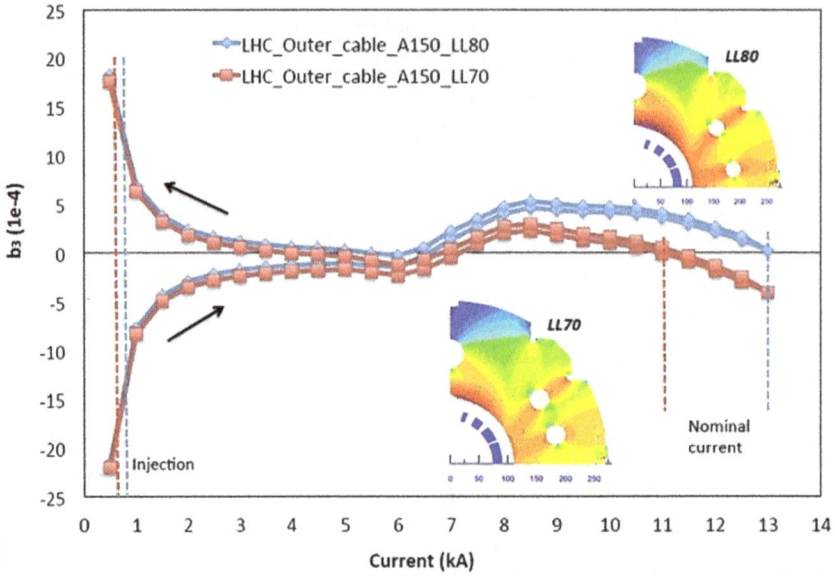

Fig. 24.   Cross-section of the separation dipole [13].

7. The Recombination Dipole D2

The recombination dipole needs the same integrated force of 35 T·m to bring the beams back to parallel trajectories, with the nominal spacing of 192 mm. In the LHC this is done by a two-in-one 10-m-long superconducting magnet with ~ 3 T operational field, and 80 mm aperture. Due to the larger beam size one needs to increase this aperture to about 105 mm. In these conditions, since the beam spacing is unchanged, even with a 15-mm thin coil and 15 mm spacing for collars, only a few cm are left between the two apertures, which have the field pointing the same direction. In these conditions, the main design challenge is to decouple the magnetic field in the two apertures and ensure good field quality.

For these reasons, we consider an operational field of 4 T (much lower than D1), giving a magnet length of 9 m. Even with this conservative design choice, using iron yoke as a shield between two apertures leads to unacceptably large saturation effects. Therefore, a different approach was proposed in a study performed by the LARP collaboration [31]: the iron yoke is designed primarily for low saturation effects, and the resulting large but current-independent cross-talk between the apertures is corrected with an asymmetric arrangement of the conductor blocks.

With this approach, it is possible to reach 4 T at 1.9 K with a comfortable 35% margin, and satisfying the field quality requirements. The fringe field on the cryostat surface is an issue, but can be cured by an oval iron as in the RHIC design. The mechanical structure will also be challenging, since little space is available for collars, a free-standing collar solution (as it adopted for the Q4, see next section) becomes unlikely. So the iron has to support the coil, as in the separation dipole, but in a two-in-one magnet. The design will be performed by INFN-Genova.

8. The Large Aperture Two-in-One Quadrupole

The two-in-one large aperture quadrupole MQY is a 70-mm-aperture magnet with Nb-Ti cable providing 160 T/m at 4.2 K. For the upgrade, the aperture has to be increased to 90 mm. The integrated gradient requirement given by beam dynamics lowers from the LHC baseline of 540 T down to 420 T.

   As in the recombination dipole, increasing the aperture with the constraint of the fixed beam separation makes the aperture cross-talk larger, and field quality becomes critical. The case is easier w.r.t the recombination dipole since (i) the magnet is a quadrupole and not a dipole and (ii) one has 15 mm less aperture (see Fig. 25).

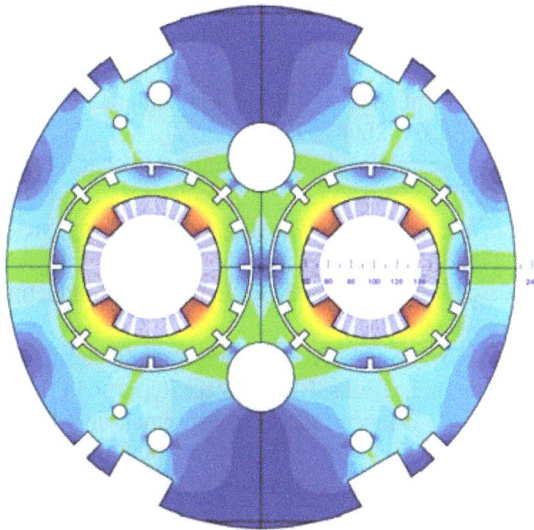

Fig. 25.   Cross-section of the large aperture two-in-one quadrupole Q4 [15].

   The study has been carried out by CEA-Saclay. Since the integrated gradient requirement is not so large, we decided to have also in this case only 15 mm of coil width. We assume an operational temperature of 1.9 K, as for D2 and crab cavities. Two options were studied, namely a two layer with 8-mm-width cable, and a single layer with 15-mm-width cable. They both provide 120 T/m with 20% margin on the loadline, and very similar field quality. This would give a 3.5-m-long magnet. With these numbers, it is clear that having a second layer of 15 mm width cable would only marginally increase the gradient (to 160 T/m, see Fig. 10), with a negligible decrease of magnet length, but with much worse field quality.

A 15-mm-thick coil, and the reduced level of forces w.r.t. the recombination dipole, leaves the possibility of having free-standing collars as mechanical structure, with the iron yoke playing only a role for alignment. Field quality can be well optimized.

The baseline is to use the 15-mm-width LHC main dipole cable (outer layer). Since the magnet is short, this can be done with spare cables lengths from the LHC production which could not be used for winding main dipoles. This requires 16 kA operational current, but has the advantage that a dump resistor is enough for the protection, i.e. no quench heaters are needed.

References

[1] O. Brüning *et al.*, LHC luminosity upgrade: a feasibility study, LHC Project Report 626 (2002) 98 p.

[2] J. P. Koutchouk *et al.*, A solution for phase-one upgrade of the LHC low-beta quadrupoles based on Nb-Ti, LHC Project Report 1000 (2007) 23 p.

[3] R. Ostojic *et al.*, Conceptual design of the LHC interaction region upgrade: phase I, LHC Project Report 1163 (2008) 42 p.

[4] J. P. Koutchouk, Investigations of the parameter space for the LHC luminosity upgrade, EPAC (2006) 556–558.

[5] L. Rossi, State of the art superconducting accelerator magnets, *IEEE Trans. Appl. Supercond.* **12**, 219–227 (2002).

[6] L. Rossi, E. Todesco, Electromagnetic design of superconducting quadrupoles, *Phys. Rev. STAB* **9**, 102401 (2006).

[7] N. Mokhov *et al.*, Protecting LHC IP1/IP5 component against radiation resulting from colliding beam interactions, LHC Project Report 633 (2003) 55 p.

[8] V. V. Kashikhin *et al.*, Quench margin measurement in Nb_3Sn quadrupole magnet, *IEEE Trans. Appl. Supercond.* **19**, 2454–2457 (2009).

[9] L. Esposito *et al.*, Fluka energy deposition studies for the HL-LHC, in *IPAC 2013*, 1379–1381.

[10] R. Van Weelderen, Cooling aspects for Nb_3Sn inner triplet quadrupoles and D1, talk on 8th May 2012 given at LARP-Hilumi meeting at FNAL, available at www.cern.ch/hilumi/wp3.

[11] R. Flukiger, T. Spina, The behaviour of copper in view of radiation damage in the LHC luminosity upgrade, CERN Yellow Report 2013-006, 76–82 (2013).

[12] A. Tollestrup *et al.*, The development of superconducting magnets for use in particle accelerators: from Tevatron to the LHC, *Rev. Sci. Accel. Tech.* **1**, 185–210 (2008).

[13] Q. Xu *et al.*, Design optimization of the new D1 dipole for HL-LHC upgrade, *IEEE Trans. Appl. Supercond.* **24**, 4000104 (2014).

[14] A. Den Ouden *et al.*, Progress in the development of an 88-mm bore 10 T Nb₃Sn dipole magnet, *IEEE Trans. Appl. Supercond.* **11**, 2268–2271 (2011).

[15] M. Segreti, J. M. Rifflet, Studies on large aperture Q4, reports I, II and III, (2012–2013), available at http://www.cern.ch/hilumi/wp3.

[16] H. Felice *et al.*, Test results of TQS03: a LARP shell-based Nb3Sn quadrupole using 108/127 conductor, *J. Phys. Conference Series* **234**, 032010 (2010).

[17] F. Borgnolutti *et al.*, Fabrication of a second generation of Nb₃Sn coils for the LARP HQ02 quadrupole magnet, *IEEE Trans. Appl. Supercond.* **24**, 4003005 (2014).

[18] G. Ambrosio *et al.*, Test results and analysis of LQS03 third long Nb₃Sn quadrupole by LARP, *IEEE Trans. Appl. Supercond.* **23**, 4002204 (2013).

[19] E. Todesco *et al.*, A first baseline for the magnets in the high luminosity LHC insertion regions, *IEEE Trans. Appl. Supercond.* **24**, 4003305 (2014).

[20] P. Ferracin *et al.*, Magnet design of the 150 mm aperture low-beta quadrupoles for the high luminosity LHC, *IEEE Trans. Appl. Supercond.* **24**, 4002306 (2014).

[21] X. Wang *et al.*, Multipoles induced by interstrand coupling currents in LARP Nb₃Sn quadrupoles, *IEEE Trans. Appl. Supercond.* **24**, 4002607 (2014).

[22] F. Borgnolutti, Magnetic design optimization of a 150 mm aperture Nb₃Sn low-beta quadrupole for the HiLumi LHC, *IEEE Trans. Appl. Supercond.* **24**, 4000405 (2014).

[23] E. Todesco, Quench limits in the next generation of magnets, CERN Yellow Report 2013-006 (2013) 10–16.

[24] T. Salmi *et al.*, Modeling heat transfer from quench protection heaters to superconducting cables in Nb₃Sn magnets, CERN Yellow Report 2013-006 (2013) 30–37.

[25] R. Gupta, Tuning shims for high field quality in superconducting magnets, *IEEE Trans. Magn.* **32**, 2069–2073 (1996).

[26] P. Hagen, Study of magnetic shimming in triplet magnets, Milestone Report 36 of HiLumi project, available at http://www.cern.ch/hilumi/wp3.

[27] R. Ostojic *et al.*, Conceptual design of the LHC interaction region upgrade: phase I, LHC Project Report 1163 (2008) 14.

[28] F. Toral *et al.*, Development of radiation resistant superconducting corrector magnets for the LHC upgrade, *IEEE Trans. Appl. Supercond.* **23**, 4101204 (2013).

[29] M. Anerella *et al.*, The RHIC magnet system, *Nucl. Instrum. Meths. A* **499**, 280–315 (2003).

[30] Y. Ajima *et al.*, The MQXA quadrupoles for the LHC low-beta insertions, *Nucl. Instrum. Meths. A* **550**, 499–513 (2005).

[31] R. Gupta, G. L. Sabbi, X. Wang, talks given at HiLumi Daresbury meeting (2013), available at https://indico.cern.ch/conferenceDisplay.py?confId=257368.

Chapter 7

Crab Cavity Development

R. Calaga[1], E. Jensen[1], G. Burt[2] and A. Ratti[3]

[1]CERN, BE Department, Genève 23, CH-1211, Switzerland
[2]University of Lancaster, STFC, Lancaster, UK
[3]Lawrence Berkeley National Laboratory, Berkeley, CA 94720, USA

The HL-LHC upgrade will use deflecting (or crab) cavities to compensate for geometric luminosity loss at low β^* and non-zero crossing angle. A local scheme with crab cavity pairs across the IPs is used employing compact crab cavities at 400 MHz. Design of the cavities, the cryomodules and the RF system is well advanced. The LHC crab cavities will be validated initially with proton beam in the SPS.

1. Crab Cavities

For higher luminosity operation, proton beams are squeezed to very small β^* at IP1 and IP5 (well below the nominal 55 cm). Controlling the effect of the large number of parasitic collisions requires a non-zero crossing angle. A non-zero crossing angle in combination with small β^* however implies a geometric reduction of the luminosity $R_\varphi = (1+\varphi^2)^{-1/2}$ due to non-perfect overlap of the colliding bunches, illustrated in Fig. 1, left.

Here we use the Piwinski parameter $\varphi \equiv \frac{\theta_c \, \sigma_z}{2 \, \sigma_x^*}$, which is half the crossing angle $\theta_c/2$ normalized to the bunch aspect ratio (width/length). Crab Cavities are able to deflect the head and the tail of a bunch sideways in opposite directions such that their tilt at collision exactly compensates for the crossing angle (Fig. 1, right).

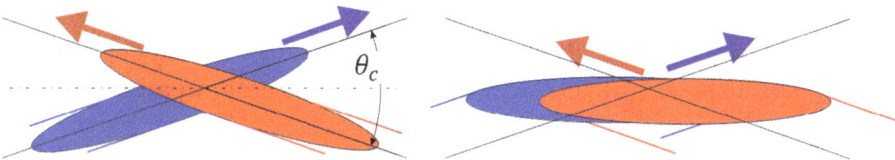

Fig. 1. Schematic of bunches colliding with an inefficient overlap due to non-zero crossing angle (left) and the geometry of crab crossing (right).

R. Calaga et al.

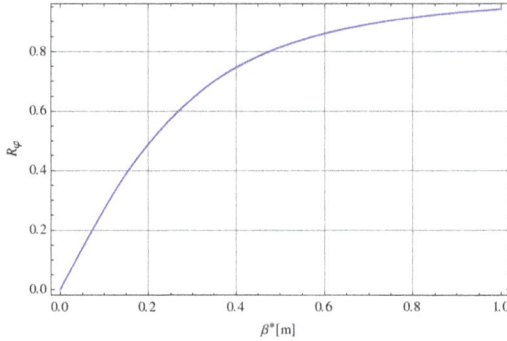

Fig. 2.   Luminosity reduction factor R_φ as function of β^* for typical LHC parameters ($\Delta = 9.5\sigma$, $\varepsilon_n = 3.75\,\mu m$).

The dependence of R_φ on β^* for typical LHC parameters is plotted in Fig. 2. Consequently the potential gain with crab cavities is approximately a factor two in luminosity for $\beta^* = 20\,cm$, even larger with smaller β^*.

2. Global and Local Schemes

Two schemes were considered for crab crossing in the LHC, the global scheme (cf. Fig. 3, left) and the local scheme (Fig. 3, right). The global scheme would require a single crab cavity per beam, installed, e.g. near Point 4, where presently all LHC RF systems are installed.[a] The transverse kick introduced by this cavity, different for the head and the tail of each bunch, is equivalent to a closed orbit distortion, i.e. head and tail would follow their individual closed orbit around the ring, their tilt wobbling around the unperturbed closed orbit of the bunch center. It is clear that this scheme introduces severe constraints on the betatron phase advance between the location of the crab cavities and the IPs. It is also inconsistent with the different crossing angles implemented in IP1 (vertical crossing) and IP5 (horizontal crossing). Furthermore, the collimator settings would have to allow for the wobbling bunches. The advantage however is that the existing dogleg region near Point 4 with more available space and an increased beam separation of 42 cm would allow for larger cavities and more RF infrastructure.

The local scheme on the other hand introduces a localized perturbation upstream of the IP where crabbing is required and compensates for it downstream, such that through the rest of the ring the bunches remain unperturbed. This scheme requires 2 to 4 cavities per beam and per IP, so at least 8 cavities (and up to 16) if

---

[a]More than one cavity may be required if the required kick voltage cannot be obtained with a single cavity.

only the high luminosity regions (IP1, IP5) are considered. This scheme does not have the optics constraints of the global scheme between the two IP's and allows for the different crossing planes in IP1 and IP5. On the other hand, it requires more cavities and in particular cavities that are compact enough for the nominal beam pipe distance of 194 mm.

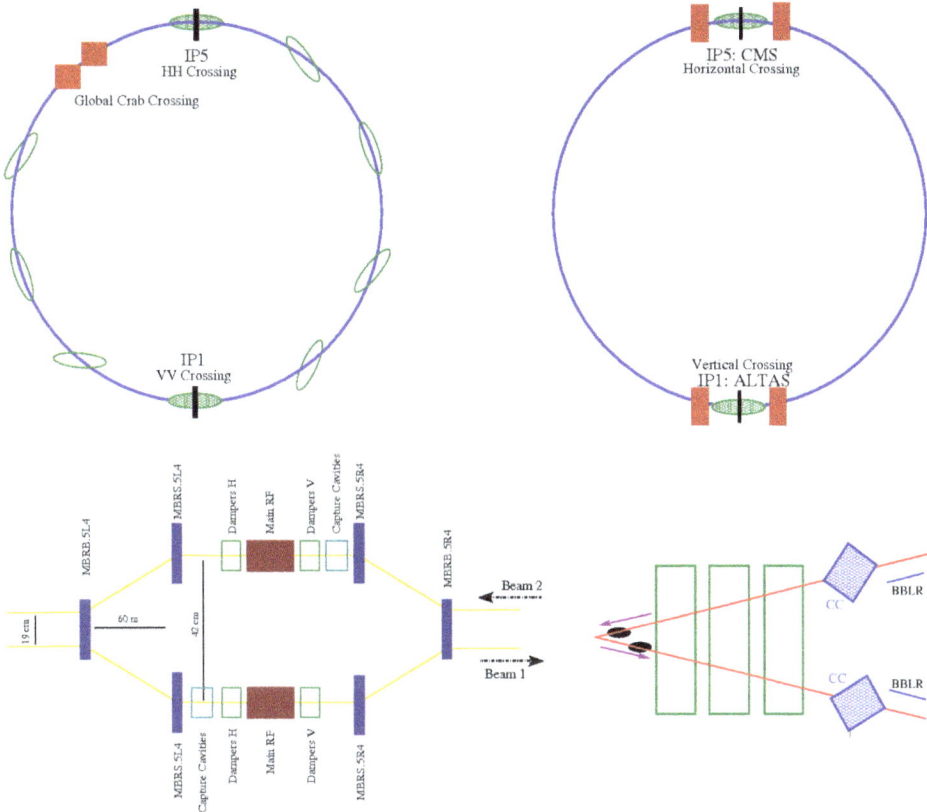

Fig. 3. Schematic for a global scheme (left) at IR4 and local scheme (right) at IP1 and IP5.

For its significantly better performance reach and its compatibility with machine protection, the local scheme has been chosen for the LHC luminosity upgrade and the resulting challenging spatial constraints and technology choices are discussed in the following section.

3. Technology Choice and Spatial Constraints

The upgrade lattice for a 7 TeV proton beam calls for superconducting crab cavities at a frequency of 400.79 MHz. The frequency choice is primarily driven by the long proton bunches, but it is also convenient to use the same frequency as for the accelerating RF system. Four modules per experiment at the two high luminosity experiments (ATLAS and CMS) will be used to perform the rotation locally without perturbing the rest of the LHC machine. Each module consists of nominally four cavities and is expected to provide the required nominal kick voltage of 12 MV per beam per side of each collision point [1]. Seen from the IP, the cavities are placed outside the recombination dipole D2, where the beams are completely separated and in their individual beam pipes spaced by 194 mm, and the β-functions in the crossing plane are sufficiently large to minimize the required voltage. Due to remaining optical constraints, the ideal betatron phase advance of $\frac{\pi}{2}$ between IP and crab cavities may not be realized exactly; the orbit bump from the crossing angle is closed prior to the entry into the crab cavities to minimize beam loading effects with trajectory offsets.

The tightest constraint results from narrow beam pipe spacing in the transverse plane. Measuring from the electric center of the cavity (where the integral $\int_{-\infty}^{\infty} E_z e^{j\frac{\omega}{c}z} dz$ of the operating mode vanishes), the beam pipe at both ends of the cavity must leave a disk of radius 42 mm clear; this will allow the transverse alignment of the cavity without reducing the aperture for the beam. To allow passage of the second beam pipe (distance center to center 194 mm) it is required that the cavity transverse size does not extend beyond 145 mm from the same electric centre. Since both vertical and horizontal crossing are used, these tight constraints have to be respected in the crossing plane and the plane orthogonal to it, as indicated in Fig. 4.

In the longitudinal plane, the constraints are primarily dictated by the proximity to the neighboring elements, namely the D2 recombination dipole and the Q4 matching section quadrupole. A total of 13.3 m is reserved for two pairs of four cavities per IP [2]. The β-functions around IP5 are sketched in Fig. 5.

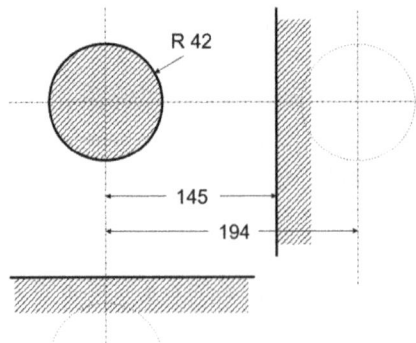

Fig. 4.   Beam pipe separation and the maximum allowed cavity envelope in the LHC for crab cavities.

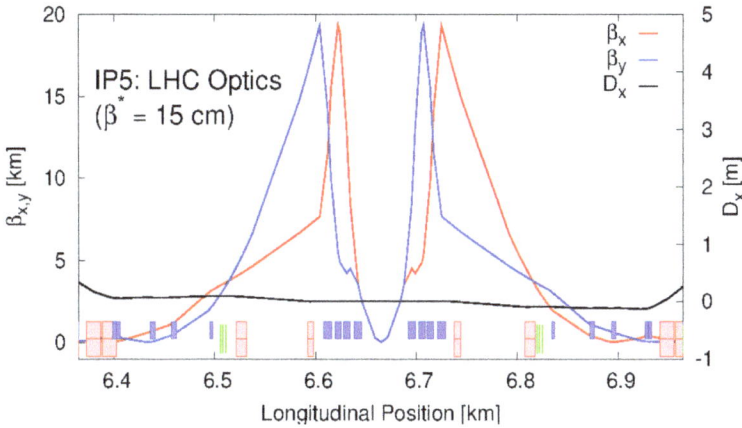

Fig. 5. Optical function in the interaction region 5 with lattice elements dipoles (red), quadrupoles (blue) and crab cavities (green).

4. Compact Cavity Design Options

Until 2006, compact crab cavities were considered "exotic" — they seemed to violate the rule of thumb for the transverse size of a superconducting cavity for the ratio radius to wavelength of 0.45 for the accelerating mode, 0.6 for the dipole mode. For a 400 MHz cavity crab cavity, this would imply a cavity equator radius of 450 mm — too large even for the dogleg area near Point 4. For this reason, the baseline frequency for an LHC crab cavity before 2008 was an 800 MHz cavity [3]. However, for this high frequency, the tails of nominal length bunches ($\sigma_z = 75$ mm, $4\sigma_z \sim 0.4\lambda$) would already suffer from the curvature of the sinusoidal RF wave. The challenge for "compact" crab cavities at 400 MHz was to have a "radius" of 145 mm as indicated in Fig. 4, requiring a ratio of radius to wavelength of below 0.193. Already for this reason, compact crab cavities are "unconventional" and have to be considered a real novelty in the panorama of new accelerator technologies.

Studies on compact crab cavities were intensified on both sides of the Atlantic and in April 2009 one of these unconventional designs, the 4-rod cavity (originally proposed by JLAB [4]), became task 10.3 in the FP7 project EuCARD [5] and was further developed in a collaborative effort led by University of Lancaster. In November 2011, the LHC crab cavities became a key part as Work Package 4 of the project HL-LHC, started as FP7 Design Study HiLumi-LHC. After only three years of intense research on compact crab cavities, four valid proposals had emerged, namely the 4-rod cavity, the ridged waveguide deflector, the parallel-bar cavity and the quarter-wave resonator (cf. Fig. 6, left to right).

Shortly after, the parallel-bar cavity and the ridged waveguide deflector designs merged to become the RF dipole (RFD), and the quarter-wave resonator was symmetrized and became the double quarter-wave resonator (DQWR); prototypes of the three remaining designs were subsequently fabricated by industry (see Fig. 7).

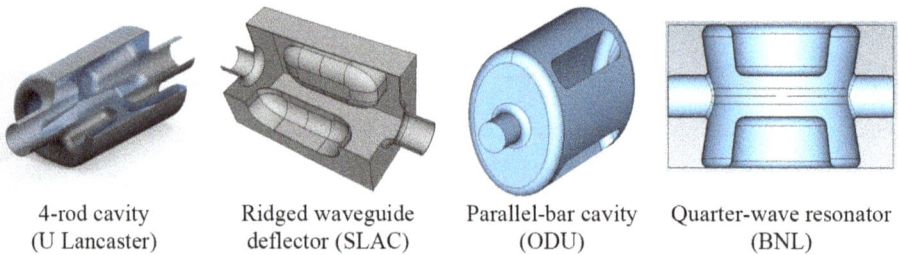

| 4-rod cavity | Ridged waveguide | Parallel-bar cavity | Quarter-wave resonator |
| (U Lancaster) | deflector (SLAC) | (ODU) | (BNL) |

Fig. 6.   Four candidate designs for compact crab cavities in 2011 [6].

| 4-rod cavity | RF dipole | DQWR |
| (U Lancaster) | (SLAC/ODU) | (BNL) |

Fig. 7.   Nb prototypes fabricated by Niowave Inc. of the three remaining compact crab cavities.

5. Present Status of Prototype Cavities

These new topologies make it possible to integrate the cryomodules in the present LHC interaction region and simultaneously be used for the alternating crossing schemes in the two IPs. As a first step of RF design validation, the prototypes of all three cavities underwent comprehensive vertical tests to validate their performance (field levels, quench limit, ramping behavior, microphonics and multipacting).

The 4-rod cavity was first tested in November 2012 at CERN of up to a deflecting voltage of 1.5 MV (nominal 3.4 MV) and further surface processing was necessary before proceeding further which is suspected as a moderate multipactor zone cavity [7]. A vacuum leak due to irregular features on the knife edge of the NbTi flanges posed additional limitations. Further tests after the repair of the flanges and additional surface treatments resulted in kick voltage of approximately 3 MV before quenching. The RF dipole cavity was tested in April 2013 at Jefferson Laboratory to a maximum of 7 MV kick voltage. The Q_0 of the cavity at the nominal field of 3.4 MV was approximately $3 \cdot 10^9$ corresponding to a surface resistance of 34 nΩ cavity [8]. At the quench voltage of 7 MV, the peak surface electric and magnetic fields are 75 MV/m and 131 mT, which are close to or above the state of the art in the field of superconducting RF. The RF dipole also reached well beyond the nominal voltage at 4.2 K and was limited by the available input power. The double quarter-wave cavity was tested in November 2013 at Brookhaven National Lab to a maximum of 4.6 MV kick voltage. The Q_0 of the cavity at the nominal field was approximately $2 \cdot 10^9$ with a surface resistance of 28 nΩ at low field [9]. The observed cavity quench at approximately 5 MV was due to localized heating in one of the HOM ports where the surface magnetic field is the highest.

6. RF Multipoles, Coupler Kicks and Limits

The crab cavity designs presently considered are such that they lack axial symmetry. Therefore, they can potentially exhibit all higher order components of the main deflecting field. Assuming an azimuthal variation of $\sim \cos(n\varphi)$ of the dipolar field, the transverse kick for a particle traversing the cavity can be expressed as sum of its multipolar components (using the notation and formalism derived in [10],

$$\Delta p_\perp(r,\varphi) = \frac{1}{c}\int_0^L F_\perp dz = \sum_{n=1}^{\infty} \Delta p_\perp^{(n)}(r,\varphi),$$

where $\Delta p_\perp^{(n)}$ are the multipolar components of the RF kick expressed in terms of the transverse electromagnetic fields integrated over the cavity length. For ultra-relativistic particles it is useful to express the RF multipoles in same units as standard magnetic multipoles with an essential difference being that RF multipoles are complex in nature:

$$b_n = \frac{1}{ec}\int F_\perp^{(n)} dz = \frac{nj}{\omega}\int E_{acc}^{(n)} dz \ [\text{Tm}^{2-n}].$$

Due to certain symmetries inherent to each design, only odd multipoles have a non-zero component. Therefore, the first important multipole is the sextupolar component b_3 of the deflecting field. Long term tracking simulations with the optical functions of the HL-LHC indicate that the b_3-component should be limited to approximately $1\,\mathrm{Tm}^{-1}$ to limit the degradation of the dynamic aperture by less than 1σ for orbit offsets of 1.5 mm [11]. For all three cavities, the imaginary part of the kick of all multipoles is negligible within the accuracy of the calculation. Hence their contribution in the crabbing phase is generally small. This orbit stability is compatible with the beam loading specifications [12].

7. Frequency Tuning System

A number of procedures from the fabrication, final surface treatment and cavity cool-down will determine the final shape and the frequency of the cavity. During the cavity filling with RF fields, the Lorentz forces on the cavity will further perturb the frequency by (0.5–1) MHz in some of the cavities. A "slow" mechanical tuning system is required to alter the cavity shape to compensate for the frequency changes and bring it in resonance. The frequency can fluctuate during the operation due to external forces, which has to be compensated dynamically by an ancillary tuning system. Different tuning mechanisms have been proposed and adopted for each cavity design. A modified Saclay II type tuner with longitudinal force on the cavity to elongate or contract the cavity is designed [13]. A JLAB scissor jack mechanism to contract or elongate the cavity for the RF dipole is proposed due to the symmetry of the cavity along the beam axis [14]. Exploiting the coaxial symmetry of the double quarter-wave in the plane of deflection, a coaxial tuner combined with the Helium vessel is being designed [15]. The 3D models of the three designs are shown in Fig. 8.

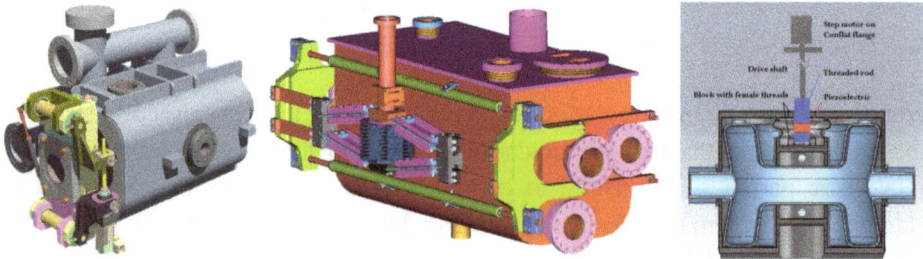

Fig. 8. Tuning Concepts adopted for the 4-rod cavity (left), the RF dipole (middle) and the double quarter-wave cavity (right).

To make the cavity transparent to the beam if crabbing is not needed, two options are considered: either the cavities are independently detuned by $Q \cdot f_{\mathrm{rev}}$ from resonance, where Q is the betatron tune to suppress coupled bunch instabilities in the crabbing mode or counterphasing of a group of cavities is used.

A tuning resolution of a quarter of the final cavity bandwidth is required due to the available RF power. Therefore, a dual tuning system with a large span in the MHz range and an ancillary fine tuning system with a resolution in the 100's of Hz range is yet to be realized.

8. RF System and Controls

8.1. *Beam loading and RF power requirement*

For a beam centered in the crab cavities, there is no beam loading. Therefore, the minimum required power is dictated by the losses in the superconducting cavity (negligible) and the required power to maintain the cavity stably on resonance. This power decreases monotonically with Q_L. However, due to unavoidable offsets and drifts in a circulating beam (for example, injection oscillations), a non-zero beam loading proportional to the shunt impedance and the circulating beam current is induced. A sufficient bandwidth and the corresponding power are required to compensate for the unavoidable orbit offsets. Figure 9 shows the required forward power as a function of Q_L for a beam that is centered and off-centered by 1 mm.

The power has a broad minimum of approximately 25 kW for Q_L between $3 \cdot 10^5$ and $1.5 \cdot 10^6$. The cavity field can be kept constant by adjusting the forward power for the corresponding displacement. Selection of an optimal value in the broad minimum is a compromise between the tuning precision feasible and the minimization of the field fluctuations from the amplifier electronics [16]. To

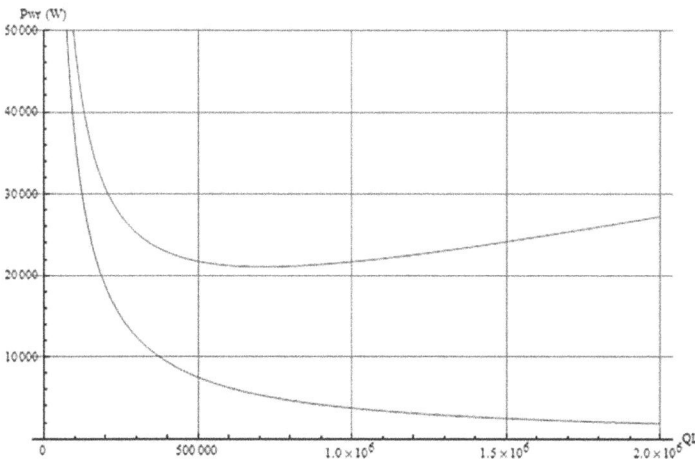

Fig. 9. Forward power vs. cavity Q_L for centered (red) and 1 mm offset (blue) beams. Assumed $R/Q = 300\,\Omega$, 3 MV RF, 1.1 A DC [12].

minimize the power needed to compensate for fast tune variations it is also desirable to keep the cavity bandwidth larger than the frequency of the mechanical modes which favors a lower Q_L.

8.2. *Power amplifier and input coupler*

A LEP type 400 MHz (40–80) kW tetrode is presently being tested as a possible option for the power amplifier for their efficiency and stability [12]. The bandwidth is approximately 1 MHz. Two such tetrodes can be combined to provide a maximum of 80 kW which would provide sufficient margin for RF and beam manipulations. It is perhaps imaginable that solid state amplifiers replace the tetrodes in the power range of interest on the longer term. The input coupler design corresponding to this power level resulted in a choice of inner diameter of 27 mm and a corresponding coaxial outer tube diameter of 62 mm. Due to the relatively low average power, air cooling of the inner antenna can be considered. The cavity will interface with the cryomodule assembly via a double wall tube, which will serve as a common platform for all three cavities. A coaxial ceramic will provide the air to vacuum interface with appropriate bellows between the double wall tube and the cryomodule. The window assembly will be equipped with vacuum gauge, electron monitor and arc detection devices. These types of windows have been

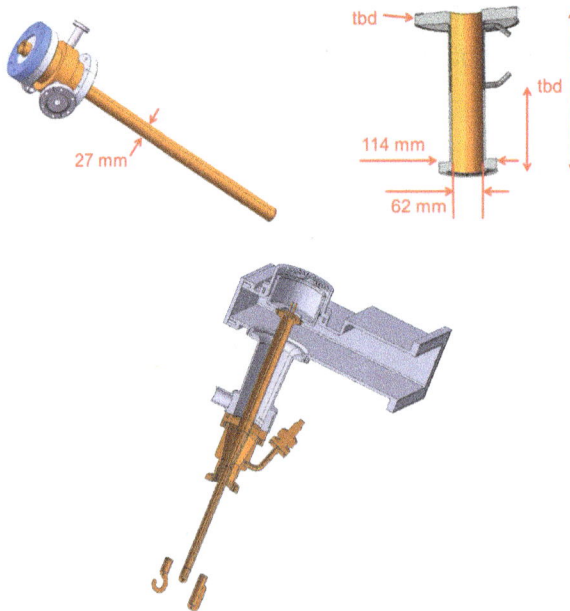

Fig. 10. Crab cavity input antenna (top left), ceramic windows and the double wall tube assembly (top right) and the complete power coupler assembly (Courtesy E. Montesinos).

widely used in the SPS cavities reliably at high power for many years [17]. Schematic of the input antenna and the ceramic assembly with the double wall tube is shown in Fig. 10.

8.3. *RF controls and machine protection*

Limitations from the round turn loop delay for cavity control should be taken into account for the fast feedback to cope with effects from fast RF failures. The amplifier driven by a feedback system feeds a compensating current to cancel the beam current. The cavity impedance is then effectively reduced by the feedback gain. Therefore, the limiting factor in the RF chain is the round-turn group delay. Therefore, a short distance between the cavity and the power amplifiers is preferred [16]. Above a certain feedback gain, the loop delay will drive the feedback into electrical oscillations. The minimum effective impedance is

$$R_{\min} \sim \frac{R}{Q} \omega_0 T ,$$

where ω_0 is the RF frequency, R/Q the classic cavity parameter and T the group delay of the feedback loop. Therefore, a radiation free cavern close to the crab cavity location in the LHC tunnel is required to keep the RF feedback delay to less than about 1.5 μs. This allows a significant reduction of the cavity impedance seen by the beam.

A phase modulation of the accelerating cavities is required to minimize the transient beam loading effects. If the crab cavities are operated from the fixed RF frequency references, it will result in a 60 ps maximum displacement of a bunch center from the zero phase in the crabbing field. For the longitudinal displacement of the luminous region this may be acceptable given the 1 ns bunch length; the resulting transverse offset of the bunch centroid in the IP however (see below under "RF noise and stability") will require that the LLRF synchronizes bunch by bunch correctly taking the actual phase modulation into account. The effect of phase modulation and the correction methods can be measured in the planned crab cavity test in the SPS.

A rapid change of the field in one cavity (for example a fast quench or a power supply trip), the LHC Beam Dump System (LBDS) will act to extract the beam in a minimum time of three turns (270 μs). Two kinds of interlocks are foreseen: slow (on BPMs) and fast (on RF). The RF controls should minimize the effect on the beam within the three turns to avoid abrupt displacements which can potentially damage the machine elements. Therefore, independent power systems of each cavity with a short delay cavity controller are proposed [16]. Figure 11 shows the proposed LLRF architecture. A central controller between the two systems across the IP makes the required corrections to adjust the cavity set points as necessary.

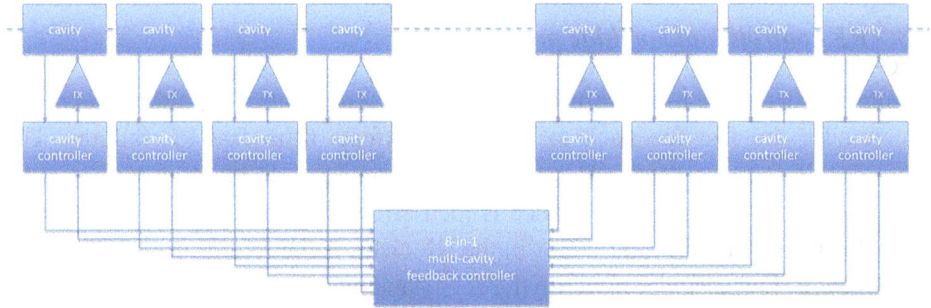

Fig. 11.   Proposed LLRF architecture for one ring at one IP. Cavity controller: strong RF feedback ($<1.5\,\mu$s loop delay) regulating individual cavities; 8-in-1 multi-cavity feedback controller: global feedback regulating the relative crabbing-uncrabbing actions. Loop delay about $5\,\mu$s [16].

The BPM interlock post-mortem, i.e. the last recorded trajectories could be used to study the effect on beam during a cavity failure. Operationally, it is preferred to have a low $Q_{ext}(\sim 10^5)$, as the cavity frequency is less sensitive to perturbations. However, it is assumed that machine protection may benefit from a high $Q_{ext}(\geq 5\cdot 10^5)$ to help avoid fast reaction on the frequency and phase changes of cavity. Consequently, the cavity will be more sensitive to external perturbations.

8.4.   RF noise and stability

Cavity voltage amplitude jitter introduces a residual crossing angle at the IP proportional to the error as shown in Fig. 12, left. It is sufficient that this residual crossing angle is much smaller ($<1\%$) than the geometric angle leading to a tolerance of electronics [18]:

$$\frac{\Delta V}{V} \ll \frac{1}{\varphi},$$

where φ is the Piwinski parameter. A phase error in the RF wave causes an offset of the bunch rotation axis translating into a transverse offset at the IP (cf. Fig. 12, right). The offset at the IP is given by

$$\Delta x_{\mathrm{IP}} = \frac{c\cdot\theta_c}{\omega_{\mathrm{RF}}}\delta\varphi_{\mathrm{RF}}$$

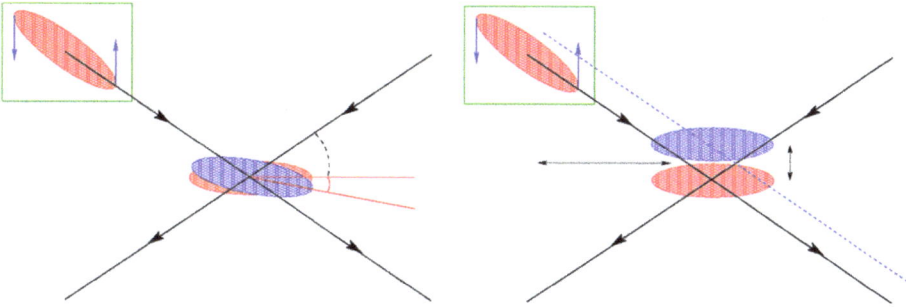

Fig. 12. Schematic of cavity voltage amplitude error leading to a residual crossing angle (left) and a phase error leading to an offset at the IP (right).

where θ_c is the full crossing angle (cf. Fig. 1) and φ_{RF} is the crab cavity phase with respect to the synchronous particle. For the HL-LHC parameters, the voltage error ratio should be kept to below 0.1%. The challenging aspect is to control the phase jitter across the IP to below $5 \cdot 10^{-3}$ degrees to minimize transverse emittance growth. This corresponds to a transverse displacement of 5% of the beam size at the IP [19].

The amplitude and phase control must be achieved also during filling and ramping with "zero" field in the cavities. Smooth transition between no-crabbing and crabbing must be realized. A single reference generated in a surface building above the accelerating cavities is sent over phase-compensated links to respective crab cavities at IP1 (ATLAS) and IP5 (CMS). An alternative would be to re-generate the bunch phase from a local pick-ups [12].

8.5. *Impedance budget and higher order mode damping*

On resonance, the impedance of the fundamental deflecting mode is canceled between the positive and negative sideband frequencies, which are symmetric around ω_{RF}. When the cavity is not operational the impedance of the fundamental deflecting modes has to be damped by appropriate feedback.

For higher order modes (HOMs), both narrow band and broadband impedance should be minimized throughout the entire energy cycle as LHC will accelerate and store beams of currents of 1.1 A (DC). Tolerances are set from impedance thresholds estimated from [20].

The longitudinal impedance has approximately a quadratic behavior in the region of interest with the minimum threshold value at approximately (300–600) MHz. The total maximum allowed impedance from each HOM, summing over all cavities in one beam, assuming that the HOM falls exactly on a beam harmonic, is set at $< 200 \ k\Omega$, so if all 16 cavities have identical HOM frequencies, the longitudinal impedance must not exceed 12.5 kΩ per cavity. For frequencies higher than 600 MHz, the threshold is higher ($\propto f^{5/3}$), but the same threshold was imposed. Modes with frequencies above 2 GHz are expected to be Landau-damped due to natural frequency spread and synchrotron oscillations.

In the transverse plane, the impedance threshold is set by the bunch-by-bunch feedback system with a damping time of $\tau_D = 5$ ms [12]. Assuming the pessimistic case that the HOM frequency coincides with the beam harmonic, the maximum impedance is set to be < 4.8 MΩ/m. Again, assuming 16 cavities per beam, the maximum allowed impedance per cavity is 0.3 MΩ/m. Analogous to the longitudinal modes, frequencies above 2 GHz are expected to be Landau-damped due to natural frequency spread, chromaticity and Landau octupoles. It should be noted that there are nominally only eight cavities per each transverse plane, so the threshold per cavity is higher, but the 0.3 MΩ/m is given assuming that the crossing plane between the experiments could become the same as a worst case scenario.

Due to the very tight impedance thresholds, the distribution of the HOM frequencies due to manufacturing errors can help relax the tolerances. The beam power deposited in the longitudinal HOMs can become significant when the frequencies coincide with bunch harmonics. The HOM couplers were dimensioned to accept a maximum of 1 kW to be able to cope with HL-LHC beams [21].

8.6. *Cavity transparency and operation*

The crab cavities must cope with the various modes of the collider cycle: filling, ramping and physics. During filling of the 2748 bunches into the LHC, ramping or operation without crab cavities, the cavities are detuned (+1.5 kHz). With a positive non-integer tune ($Q_h = 64.3$, ω_b/ω_{rev} above an integer), the cavity should be tuned above the RF frequency to make the mode $l = -64$ stabilizing (see Fig. 13). Although RF feedback is not mandatory for stability with a detuned cavity, however, for accurate knowledge and control of the cavity tune and field, active feedback is preferred. Active feedback will also keep the beam induced voltage zero if the beam is off-centered. As the kick is provided by a set of four cavities, counter-phasing is preferred to make the cavity invisible to the beam.

On flat top, detuning (if used) is reduced while keeping the cavity voltage small using counter-phasing. The RF feedback keeps the cavity impedance small (beam stability) and compensates for the beam loading. Subsequently the crab cavities are rephased synchronously to obtain the desired kick voltage. Any leveling scheme is possible. With a circulator between TX and cavity, the TX response is not affected by the cavity tune. This is very favorable for the proposed active compensation scheme, with a cavity being gently moved from parked position to on-tune.

In physics, with the crabbing on, the active RF feedback will provide precise control of the cavity field. The RF feedback reduces the peak cavity impedance and transforms the high Q resonator into an effective impedance that covers several revolution frequency lines (Fig. 13). The actual cavity tune has no big impact on stability anymore; the growth rates and damping rates are much reduced.

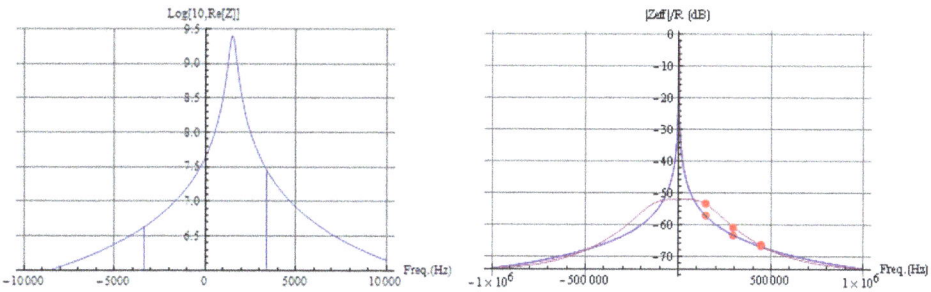

Fig. 13.   Left: Real part of the deflecting mode impedance as a function of detuning from 400 MHz. Right: Effective impedance seen by the beam with the RF feedback on (red) and off (blue).

9. Integration into SPS and LHC

The first proof of principle system is foreseen to be tested in the SPS prior to the realization of the full LHC crab cavity system. The primary aim of these tests is to validate the technology with proton beams and establish a robust operational control of a multi-cavity system for the different modes of operation.

9.1. *SPS-BA4 test setup*

The SPS ring is equipped with a special bypass (Y-chamber) with mechanical bellows that can be displaced horizontally (see Fig. 14). This allows for a test module to be placed out of the beam during regular operation of the SPS and only moved in during the dedicated machine development. This setup is essential both due to aperture limitations of the crab cavities and the risk associated with leaving the cavities in the beam line with different modes of operation in the SPS.

A 2-cavity cryomodule is envisioned for installation in the 2016–17 end of the year technical stop. This cryomodule will consist of all the main elements that need to be validated with the LHC type beams prior to a full installation in the LHC interaction regions. A cryogenic box in the BA4 region is presently being prepared to deliver 2 K Helium for the test operation of the crab cavities [12].

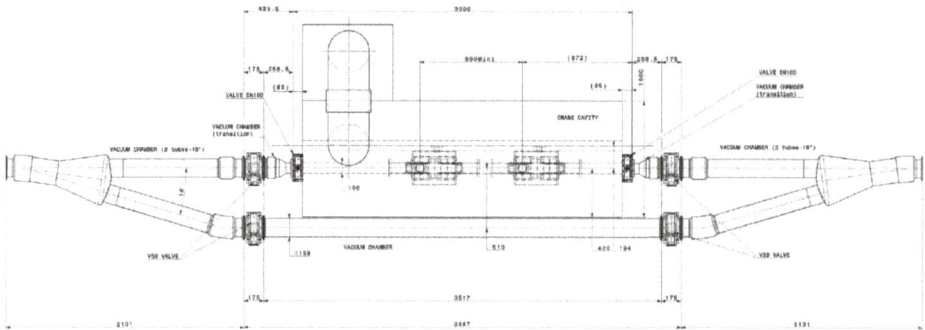

Fig. 14. SPS-BA4 bypass for the installation of a 2-cavity crab cavity module for the first beam tests.

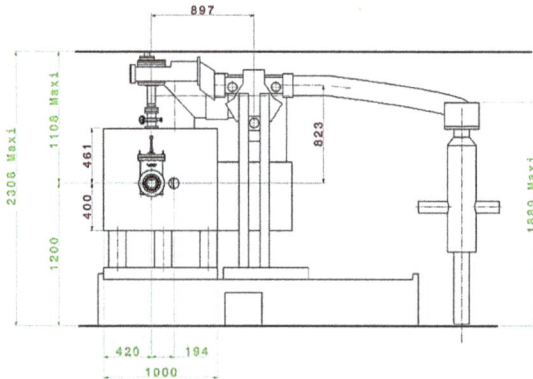

Fig. 15. Cryomodule and RF system layout in the BA4 cavern (left) and a 400 MHz Tetrode amplifier under test (right).

Two transmission lines (coaxial or waveguide) could feed the RF power from the tetrode amplifiers to the cavities (see Fig. 15). Placement of the amplifiers on a movable table together with the cryomodule, transmission lines and circulators is considered. A possible 3D integration of the cryomodule and the RF assembly in the BA4 region is shown in Fig. 16.

9.2. LHC integration constraints

Due to a complete change of the interaction region, the integration of the crab cavities in the LHC is combined with the rest of the magnetic elements and undertaken by Work Package 15 (Integration and Installation) with input from WP4 (Crab Cavity) and WP2 (Accelerator Physics and Performance). The RF system demands an independent control of each of the eight cavities per IP side

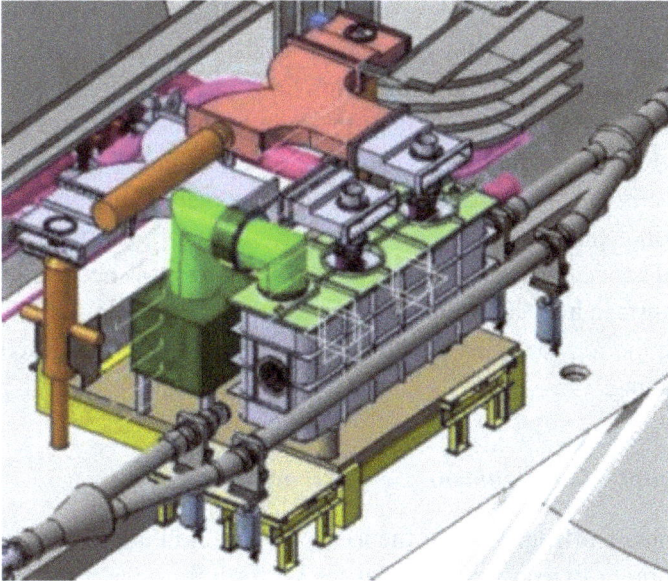

Fig. 16.   Possible 3D integration of the cryomodule, RF assembly and the cryogenics in the BA4 region of the SPS.

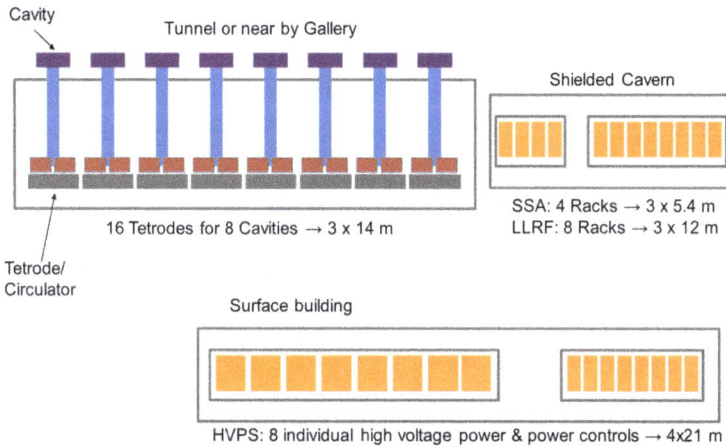

Fig. 17.   Schematic of the RF system layout in the LHC tunnel with respective number of electronics racks required close to the cavities and on the surface building.

with the shortest delay loops between the RF transmitter and the cavity (see Fig. 17). To minimize RF group delay a short distance between amplifiers and cavities is required. Short delay is already in place for the ACS main RF system in Point 4. A service gallery parallel to the tunnel would allow for sufficient shielding to sensitive RF electronics and possibly easy access to the RF equipment.

Near Points 1 and 5, the closest area to house any equipment is in the RR caverns at about 80 m from the crab cavities. They were considered to place RF electronics. However, the transmission lines are of significant transverse dimension and passing eight such lines through the tunnel cross-section is a major challenge. More importantly, high radiation levels in the RR caverns practically exclude placement of sensitive RF electronic equipment in them.

Considering these reasons, it is highly desirable to either extend the experimental service gallery towards the crab cavity regions near Points 1 and 5 or to consider surface installations above them for RF equipment. A study is ongoing to determine the feasibility of the civil engineering with minimal perturbation to the LHC running.

9.3. *Positioning and alignment*

The positioning and alignment of the 4-cavity system in the LHC is an important challenge in order to respect the tolerances set forth from the beam dynamics and RF. These can be generally classified into four groups and effects are briefly described:

(1) Transverse rotation of the individual cavities with respect to each other or the cryostat introduces a parasitic crossing angle in the non-crossing plane, thereby counteracting the compensation scheme. In addition, it can also cause a non-closure of the crab bump in the crossing plane. Using a similar analogy to voltage modulation of the cavity, a transverse rotation of approximately $0.3°$ per cavity can be tolerated.

(2) Tilt of the cavities with respect to the longitudinal cryomodule axis in the crossing plane introduces a beam loading in part of the cryomodule proportional to the tilt. Analogous to the 1 mm tolerance from the power requirements, the tilt with respect to the longitudinal axis should be less than $0.06°$. The tilt in the non-crossing plane is nominally more relaxed, but due to the alternating crossing scheme, the same tolerance is imposed in both planes.

(3) Transverse displacement of cavities with respect to each other inside a cryostat. This is analogous to (2) and an intra-cavity alignment of approximately 0.7 mm is set primarily to minimize the beam loading and multipolar effects.

(4) Longitudinal displacement of cavities with respect to each other inside a cryostat from their nominal position is less critical as the deviation can be compensated by adjustments of the individual cavity set point voltages to account for changes in the optical functions. The tolerance is set by respecting a change less than 0.1% of the nominal voltage, which corresponds to approximately (1–2) cm.

The SPS tests will play a key role in establishing the feasibility of such tight tolerances and the use of remote alignment techniques, both passive and active to achieve the requirements. In the SPS, this includes the movement of the support table into and out of the beam by approximately 55 cm while respecting the internal alignment tolerances.

References

[1] R. Calaga, Comments on crab cavity voltage (unpublished), https://espace.cern.ch/HiLumi/WP4/Shared%20Documents/Reports/crabvolt.pdf, 2012.

[2] R. de Maria *et al.*, Upgrade optics with crab cavities, presented at the 4th Crab Cavity Workshop (LHC-CC10), CERN, Geneva, 2010, http://indico.cern.ch/event/100672.

[3] CARE-HHH Mini workshop – LHC Crab Cavity Validation, CERN, 2008, http://indico.cern.ch/conferenceDisplay.py?confId=38210.

[4] C. Leemann and C. G. Yao, A highly effective deflecting structure, *Linac1990*, Albuquerque, NM.

[5] http://eucard.web.cern.ch/eucard/index.html.

[6] LARP CM16, Montauk, NY, https://indico.fnal.gov/conferenceDisplay.py?confId=4041.

[7] R. Calaga *et al.*, First Test Results of the 4-rod Crab Cavity, in *Proc. IPAC2013*, Shanghai, 2013.

[8] S. Da Silva *et al.*, Compact Superconducting RF-dipole Cavity Designs for Deflecting and Crabbing Applications", in *Proc. IPAC2013*, Shanghai, 2013.

[9] B. Xiao *et al.*, presented at the 6th Crab Cavity Workshop, LHC-CC13, CERN, 2013, https://indico.cern.ch/event/269322.

[10] A. Grudiev *et al.*, Study of Multipolar RF Kicks from the Main Deflecting Mode in Compact Crab Cavities for LHC, in *Proc. IPAC12*, New Orleans, 2012.

[11] J. Barranco *et al.*, RF multipoles: modelling and impact on the beam, presented at the 2nd HiLumi Collaboration Meeting, Frascati, 2012, https://indico.cern.ch/event/183635.

[12] P. Baudrenghein *et al.*, Functional Specifications of the LHC Prototype Crab Cavity System, CERN-ACC-NOTE-2013-003, CERN, 2013.

[13] T. Jones *et al.*, Helium Vessel/Tuner, UK 4Rod Cavity, presented at the Crab Cavity Engineering Meeting, FNAL, 2012, https://indico.cern.ch/event/136807.

[14] H. Park *et al.*, Helium Vessel/Tuner, ODU-SLAC RF Dipole, presented at the Crab Cavity Engineering Meeting, FNAL, 2012, https://indico.cern.ch/event/136807.

[15] B. Xiao *et al.*, Helium Vessel/Tuner, BNL Quarter Wave, presented at the Crab Cavity Engineering Meeting, FNAL, 2012, https://indico.cern.ch/event/136807.

[16] P. Baudrenghien, LLRF for Crab Cavities, 2nd HiLumi LHC/LARP Meeting, Frascati, 2012, https://indico.cern.ch/event/183635.

[17] E. Montesinos, Power coupler interfaces and RF system, presented at the Crab Cavity Engineering Meeting, FNAL, 2012, https://indico.cern.ch/event/136807.

[18] K. Oide and K. Yokoya, *Phys. Rev. A* **40**, 315 (1989).

[19] R. Calaga *et al.*, in *Proc. LHC Performance Workshop 2010*, Chamonix, 2010.

[20] A. Burov, Impedance budgets for crab cavities, presented at 5th Crab Cavity Workshop (LHC-CC11), Nov. 14–15, 2011; E. Shaposhnikova, Impedance effects during injection, energy ramp & store, presented at 4th Crab Cavity Workshop LHC-CC10, Dec 10th, 2010.

[21] R. Calaga and B. Salvant, Comments on crab cavity HOM power, 2013 (unpublished).

Chapter 8

Powering the High-Luminosity Triplets

A. Ballarino and J. P. Burnet

CERN, TE Department, Genève 23, CH-1211, Switzerland

The powering of the magnets in the LHC High-Luminosity Triplets requires production and transfer of more than 150 kA of DC current. High precision power converters will be adopted, and novel High Temperature Superconducting (HTS) current leads and MgB_2 based transfer lines will provide the electrical link between the power converters and the magnets. This chapter gives an overview of the systems conceived in the framework of the LHC High-Luminosity upgrade for feeding the superconducting magnet circuits. The focus is on requirements, challenges and novel developments.

1. Introduction

The powering of the High-Luminosity Triplets requires production and supply of about 160 kA DC current for feeding twenty electrical circuits. The triplet system comprises three quadrupole optical elements made of four Nb_3Sn magnets (Q1, Q2a, Q2b and Q3), a Nb-Ti separation dipole (D1) and several Nb-Ti corrector magnets. The low-beta quadrupoles are fed via two circuits with trims and all other magnets are individually powered.

Two main features distinguish the powering system of the High-Luminosity Triplets from that of the present LHC triplet configuration:

(1) A significant higher current has to be generated and transported (about 160 kA to be compared with the present need of 40 kA), because of the high operating current of the low-beta quadrupoles and of the superconducting separation dipoles — the latter replaces the currently resistive low-field dipole modules;

(2) The cold powering relies on the use of hundreds of meters long high temperature superconducting (HTS) lines (hereafter called "links") providing the electrical connection between the current leads and the magnets.

In the LHC, power converters and current leads are both located in underground areas, the first either in galleries parallel to the main tunnel (UA zones) or directly in the tunnel (RR and UJ alcoves and below the dipole magnets) and the second in cryostats, which are at each end of the LHC sectors and in line with the superconducting magnets. Conventional copper cables connect the power converters to the current leads.

The cold powering system for the High-Luminosity upgrade incorporates novel superconducting lines that enable feeding the magnets from power converters and current leads located at a remote distance, i.e. either in surface buildings, located above ground near the LHC access shafts, or underground in radiation free access areas [1].

The use of superconducting links brings a number of benefits to the system that can be summarized as follows:

(1) Location of the power converters in areas with no radiation to avoid the problem associated with the SEU (Single Event Upset), i.e. events that result stochastically from single interactions between energetic ionizing particles and electronic components and that can affect the performance of the power converters;

(2) Easier access of personnel for maintenance, tests and interventions on equipment like current leads and power converters in radiation free areas, in accordance with the CERN principle of radiation protection that optimizes doses to personnel exposed to radiation by keeping them As Low As Reasonably Achievable (ALARA);

(3) Removal of the current leads and associated cryostats from the accelerator ring, thus making space available for other accelerator components. The LHC Points 1 and 5, where the High-Luminosity insertions will be located, are among the most critical areas of the LHC machine, and removal of some large components will make easier integration of new equipment.

According to the present baseline, for the powering of the High-Luminosity Triplets at P1 and P5, power converters are planned to be located in surface buildings and connected to the magnet circuits via superconducting links containing tens of cables transferring all together about 160 kA. One link per triplet is needed. Each superconducting link is about 300 m long and has to span a vertical distance of about 80 m.

A solution similar to that of the High-Luminosity Inner Triplets is being studied also for the powering of the magnets in the LHC matching sections and in arc at LHC P1 and P5. Each of these systems requires two additional links per point. The links provide the electrical connection from the surface to the underground areas.

A solution for remote powering is proposed also for LHC P7 [2, 3]. In this case, power converters and current leads are located in an underground radiation-free gallery, which serves as access to the LHC ring (TZ76). At P7, two superconducting links are needed. Each of them is about 500 m long and transfers a total DC current of about 30 kA.

The superconducting links for the LHC upgrade are under development at CERN. They represent a challenging and unprecedented development of compact and high-current superconducting transfer lines meeting all requirements needed for use as integral part of a cold powering system for the LHC accelerator.

A program has been launched at CERN for developing radiation hard electronics for power converters that could be operated in the LHC tunnel in zones exposed to radiation. This program, which started in 2010, aims at making available power converters that can withstand doses of up to about 100 Gy/year when exposed to neutron (1 MeV) and hadron (20 MeV) fluences of respectively $1.5 \cdot 10^{11}$ n/(cm^2·year) and $5 \cdot 10^{10}$ p/(cm^2·year).

2. Powering the High-Luminosity Triplets

The present LHC Inner Triplets have a complex powering scheme based on the use of nested circuits [4, 5]. Stability in current reaches the required performance of one per million of maximum current. However, during operation of the machine it was found that the MTTR (Mean Time To Repair) in case of faults in the power converters is much longer than for the other LHC circuits [6]. This is a drawback of the complexity of the powering layout which makes debugging the circuit more difficult to handle.

The powering scheme proposed for the High-Luminosity Inner Triplets [7] tries to simplify the electrical circuits with the purpose of improving availability of the machine while providing full flexibility for beam optics. The quadrupoles of the High-Luminosity Triplet (Q1, Q2 and Q3) are powered via two main circuits, each one equipped with a trim power converter (see Fig. 1). The two Q2a units are powered in series with a 200 A trim converter on Q2a. Q1 and Q3 are powered in series with a 2 kA trim converter on Q3. The separation dipole (D1) and the corrector magnets (the nonlinear correctors and the dipole correctors) are individually powered.

The current rating of the powering equipment is reported in Table 1. The new power converters will be based on the same principle of the present LHC modular switch-mode power converters [8]. All the converters will have redundant modules to maximize the availability of the machine. The 2.4 kA units, working in 4-quadrant, require a dedicated development. In the LHC, 4-quadrant power converters were required only for corrector circuits operated at currents of up to 600 A.

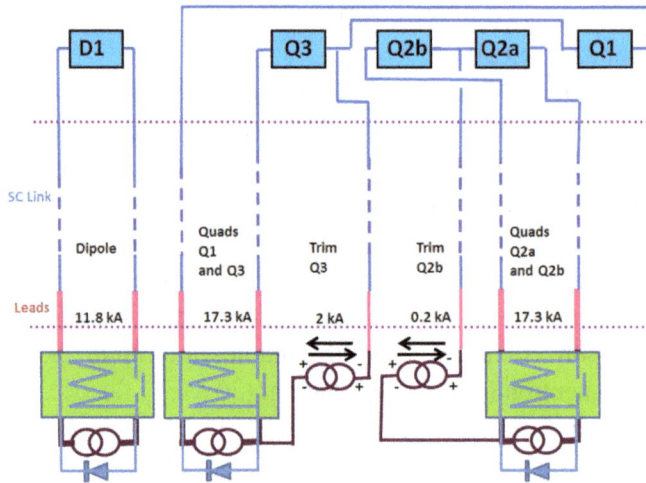

Fig. 1.   High-Luminosity Inner Triplet powering scheme.

Table 1. Current rating of magnets [9] and of powering equipment and number of power converters (PC), superconducting (SC) cables and current leads per triplet.

Circuit	Magnet current (kA)	PC rating (kA)	Cold powering system rating (kA)	Number of PC/SC cables/leads
Quadrupole	17.3	20	20	2/4/4
Trim on Q3	±2	±2.4	3	1/1/1
Trim on Q2b	±0.3	±0.4	0.4	1/1/1
Corrector dipole	2.4	3	3	6/12/12
Nonlinear corrector	±0.1A	±0.12	0.12	9/18/18
Separation dipole	11.8	14	20	1/2/2

3. Cold Powering System

The cold powering system [10] consists of HTS current leads integrated in a distribution feed-box cryostat, of a superconducting link, which is a long cryostat enclosure containing an assembly of HTS cables, and of two interconnection boxes connecting the link at its colder side to the magnet cryostat in the tunnel and at its warmer side to the distribution feed-box cryostat. Electrical connections between superconducting cables take place in the interconnection boxes, which house also the cryogenic instrumentation needed for control and operation. The cold powering system relies on cooling with helium gas supplied from the tunnel [11]. The gas warms up while absorbing the static load of the link cryostat and it is recuperated at the surface after cooling of the link and of the current leads. The superconducting cables in the link are designed for operation at a maximum temperature of 25 K — with nominal operation at about 20 K. The availability of helium gas at about 5 K in the cryogenic system enables the use of MgB_2 superconductor.

Novel cables made from different types of high temperature superconducting materials are being developed at CERN for use in the LHC superconducting links [2, 10]. For the links powering the High-Luminosity Triplets, an extensive program aiming at the development of both MgB_2 round wires and cables was carried out during the last three years. Thanks to a development program between CERN and the company Columbus Superconductors, MgB_2 round wires with the mechanical properties required for cabling were produced and extensively characterized [12, 13]. The first ever made high-current superconducting (20 kA at 24 K) cables assembled from reacted MgB_2 round wires were then assembled and successfully measured at CERN in a system-like configuration [10, 14]. A dedicated test station was built for measuring up to 20 m long superconducting cables operated in helium gas at any temperature in the range from 5 K to 70 K.

In contrast with superconducting transmission lines developed for electrical power distribution, where one or a maximum of three cables are contained in the same cryogenic envelope, the links for the LHC contain tens of cables rated at different DC currents ranging from a minimum of 120 A up to a maximum of 20 kA. For the powering of the High-Luminosity Triplets, each of the four links to be integrated at LHC P1 and P5 contains six cables rated at 20 kA, fourteen cables rated at 3 kA, four cables rated at 0.4 kA, and eighteen cables rated at 0.12 kA (see Fig. 2). The total current transferred by the assembly of these forty-two cables is 165 kA DC.

The cable assemblies are incorporated in semi-flexible cryostats of the CRYOFLEX® type. The present baseline, which is to be confirmed through on-going integration studies, envisages integration in the LHC tunnel of the cryostat with the cables already pulled in at the surface. To limit the risks associated with high-current resistive joints operated in helium gas environment, the cables are planned to be assembled in one single unit length with no splices between cables inside the link. This is possible thanks to the availability of unit lengths of MgB_2 wire in excess of 1 km. The cryostat consists of four corrugated pipes and it includes an actively cooled thermal shield.

The forty-two multi cable assembly has an external diameter of about 65 mm and a mass of about 11 kg/m. The external diameter of the CRYOFLEX® semi-flexible cryostat containing the cable assembly is 220 mm. This link will cover 80 m of vertical distance for transferring the current from the surface down to the LHC underground areas (see Fig. 3).

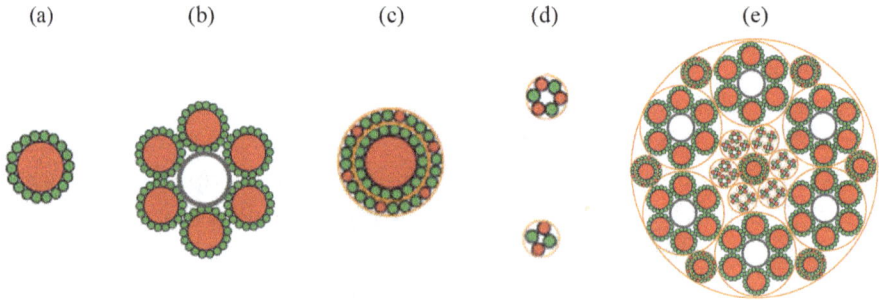

(a)   (b)   (c)   (d)   (e)

Fig. 2. Cables made with MgB₂ round wire. (a) One of the six sub-units making a 20 kA cable, $\Phi \sim 6.5$ mm; (b) 20 kA cable, $\Phi \sim 19.5$ mm; (c) concentric 2×3 kA cable, $\Phi \sim 8.5$ mm; (d) 0.4 kA cable (top) and 0.12 kA cable (bottom), $\Phi \sim 3$ mm; (e) 165 kA cable assembly (6×20 kA, 7×3 kA, 4×0.4 kA, 18×0.12 kA), $\Phi \sim 65$ mm. MgB₂ wire is in green, copper stabilizer is in red.

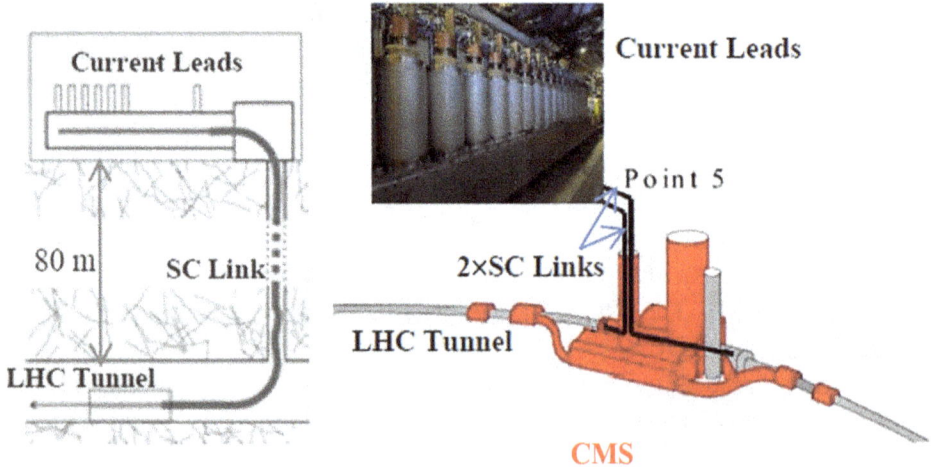

Fig. 3. From left: schematic layout at LHC P1 and P5, with superconducting links spanning across a vertical distance of 80 m and connecting the current leads, located in cryostat in surface buildings, to the LHC magnets in the tunnel; integration at LHC P5, with two superconducting links powering the insertion magnets developed in the framework of the LHC High-Luminosity upgrade.

4. Conclusions

A new powering scheme for the High-Luminosity Triplets is proposed with the goal of maximizing the availability of the machine and freeing precious space in the underground areas. The powering system will require the transfer of about 160 kA DC current for feeding twenty electrical circuits. Novel superconducting transfer lines are under development to enable the feeding of the magnets from power converters and current leads located in surface buildings. The type of LHC power converters with high current stability will be extended to cover the upgrade requirements. The powering layout proposed for the High-Luminosity Triplets will be reviewed in a more advanced phase of the project, when additional input from the magnet system is available.

A powering system is being studied for the High-Luminosity matching section magnets in IR1 and IR5, with superconducting links to remove onto the surface the cryogenic feedboxes and the power converters. The technical challenges are similar to those of the system feeding the Inner Triplets.

References

[1] A. Ballarino, *IEEE Trans. Appl. Supercond.* **21**, 980–984 (2011).
[2] A. Ballarino, J. Fleiter, J. Hurte, M. Sitko, G. Willering, *Physics Procedia* **36**, 855–858 (2012).
[3] Y. Yang, E. Young, W. Bailey, C. Beduz and A. Ballarino, *Physics Procedia* **36**, 1337–1342 (2012).
[4] F. Bordry and H. Thiesen, LHC inner triplet powering strategy, in *Proceedings of the 2001 Particle Accelerator Conference*, Chicago, 2001.
[5] F. Bordry, D. Nisbet, H. Thiesen and J. Thomsen, Powering and control strategy for the main quadrupole magnets of the LHC inner triplet system, presented at the *13th European Conference on Power Electronics and Applications* (EPE'2009), Barcelona, CERN/ATS 2010-022.
[6] H. Thiesen, M. Cerqueira-Bastos, G. Hudson, Q. King, V. Montabonnet, D. Nisbet and S. Page, High precision current control for the LHC main power converters, *Proceedings of IPAC'10*, Kyoto, Japan, 2010.
[7] A. Ballarino, Magnet powering for HL-LHC, 4th HL-PLC meeting, 2013, http://indico.cern.ch/event/239311/.
[8] F. Bordry, V. Montabonnet and H. Thiesen, Soft switching (ZVZCS) high current, low voltage modular power converter (13 kA, 16 V), EPE'01.
[9] E. Todesco, H. Allain, G. Ambrosio, G. Arduini, F. Cerutti, R. De Maria, L. Esposito, S. Fartoukh, P. Ferracin, H. Felice, R. Gupta, R. Kersevan, N. Mokhov, T. Nakamoto, I. Rakno, J. M. Rifflet, L. Rossi, G. L. Sabbi, M. Segreti, F. Toral, Q. Xu, P. Wanderer and R. Van Weelderen, *IEEE Trans. Appl. Supercond.* **24**, 4003305 (2014).

[10] A. Ballarino, Development of superconducting links for the LHC machine, *Supercond. Sci. Technol.* **27**, 044024 (2014).

[11] A. Ballarino, Preliminary report on cooling options for the cold powering system, CERN-ACC-2013-010, Deliverable Report, FP7 Large Hadron Collider Design Study, 2013, http://cds.cern.ch/record/1557215/files/CERN-ACC-2013-010.pdf.

[12] B. Bordini, A. Ballarino, D. Richter, V. Cubeda, G. Grasso, R. Piccardo and M. Tropeano, Electro-mechanical characterization of the MgB2 wire developed for the LHC Superconducting Link Project, presented at *EUCAS 2013, Proc.SuST*.

[13] R. Piccardo, F. Barberis, M. Capurro, P. Cirillo, E. Zattera, A. Ballarino, B. Bordini, V. Cubeda, D. Nardelli, M. Tropeano and G. Grasso, Room temperature stress-strain curve and its correlation to the superconducting transport properties of MgB2 ex-situ wires developed for the LHC Superconducting Link Project, presented at *EUCAS 2013*.

[14] S. Giannelli, A. Ballarino, B. Bordini, J. Hurte and A. Jacquemod, First measurements of MgB2 cables operated in helium gas at up to 35 K, CERN Internal Note, EDMS N. 1476839.

Chapter 9

Cryogenics for HL-LHC

L. Tavian, K. Brodzinski, S. Claudet, G. Ferlin,
U. Wagner and R. van Weelderen

CERN, TE Department, Genève 23, CH-1211, Switzerland

The discovery of a Higgs boson at CERN in 2012 is the start of a major program of work to measure this particle's properties with the highest possible precision for testing the validity of the Standard Model and to search for further new physics at the energy frontier. The LHC is in a unique position to pursue this program. Europe's top priority is the exploitation of the full potential of the LHC, including the high-luminosity upgrade of the machine and detectors with an objective to collect ten times more data than in the initial design, by around 2030. To reach this objective, the LHC cryogenic system must be upgraded to withstand higher beam current and higher luminosity at top energy while keeping the same operation availability by improving the collimation system and the protection of electronics sensitive to radiation. This chapter will present the conceptual design of the cryogenic system upgrade with recent updates in performance requirements, the corresponding layout and architecture of the system as well as the main technical challenges which have to be met in the coming years.

1. Introduction

The upgrade of the cryogenics for HL-LHC will consist of:

- The design and installation of two new cryogenic plants at P1 and P5 for the high luminosity insertions. This upgrade will be based on a new sectorization scheme aiming at separating the cooling of magnets in this insertion regions form magnets of the arcs as well as on a new cryogenic architecture based on electrical feed boxes located at ground level and vertical superconducting links.
- The design and installation of a new cryogenic plant at P4 for SRF cryo-modules.
- The design of new cryogenic circuits at P7 for HTS links and displaced current feed boxes.

- Cryogenic design support for cryo-collimators and 11 T dipoles at P2, P7 and, if needed, also at P3, P5 and P7.

Figure 1 shows the overall LHC cryogenic layout including the upgraded infra-structure.

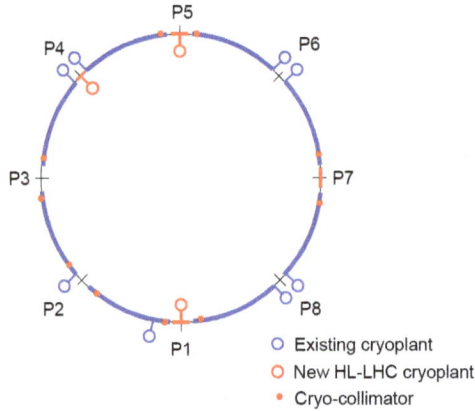

Fig. 1. Overall LHC cryogenic layout including the upgraded infrastructure.

Table 1. LHC upgraded beam parameters for 25-ns bunch spacing.

Parameter		Nominal	Upgrade
Beam energy, E	[TeV]	7	7
Bunch population, N_b	[protons/bunch]	$1.15 \cdot 10^{11}$	$2.2 \cdot 10^{11}$
Number of bunches per beam, n_b	[–]	2808	2808
Luminosity, L	[cm^{-2}s^{-1}]	10^{34}	$5 \cdot 10^{34}$
Bunch length	[ns]	1	1

2. LHC Machine Upgrades

2.1. Upgraded beam parameters and constraints

The main parameters impacting on the cryogenic system are given in Table 1. With respect to the nominal beam parameters, the beam bunch population will be double and the luminosity in the detectors of the high-luminosity insertions at Points 1 and 5 will be multiplied by a factor of 5.

These upgraded beam parameters will add new constraints on to the cryogenic system:

- The collimation scheme must be upgraded by adding collimators in particular in the continuous cryostat close to Points 2 and 7, and may be also at P3, P1 and P5. The corresponding integration space must be created by developing shorter but stronger 11-T cryo-dipoles. As the new collimators will work at room temperature, cryogenic bypasses are required to guarantee the continuity of the cryogenic and electrical distribution. Figure 2 shows the nominal and upgraded layouts of the continuous cryostat. Halo control for HL-LHC may also require the installation of hollow electron lenses at Point 4, making use of a superconducting solenoid. While not yet in the HL-LHC baseline, this device may be the best option for controlling particle diffusion speed within and depopulating the halo of the high-power hadron beams, avoiding uncontrolled sudden losses during critical operations, like squeezing for putting beam in collision. Figure 3 shows the nominal and upgraded layouts of the Point 4 insertion region, considering the installation of e-lens and of a new SC RF system.
- The increase of the level of radiation to electronics in the tunnel will require relocating power convertors and related current feed boxes in an access gallery at Point 7 and at ground level at Points 1 and 5. New superconducting links will be required to connect the displaced current feed boxes to the magnets. Figures 4 and 5 shows the nominal and upgraded layouts of the insertion regions at Points 1, 5 and 7.
- To better control the bunch longitudinal profile, reduce heating and improve the pile-up density, new cryo-modules of 800-MHz RF cavities will be added to the existing 400-MHz cryo-modules at Point 4 creating a high-harmonic RF system (see Fig. 3). Actually, discussions are under way if a better scheme

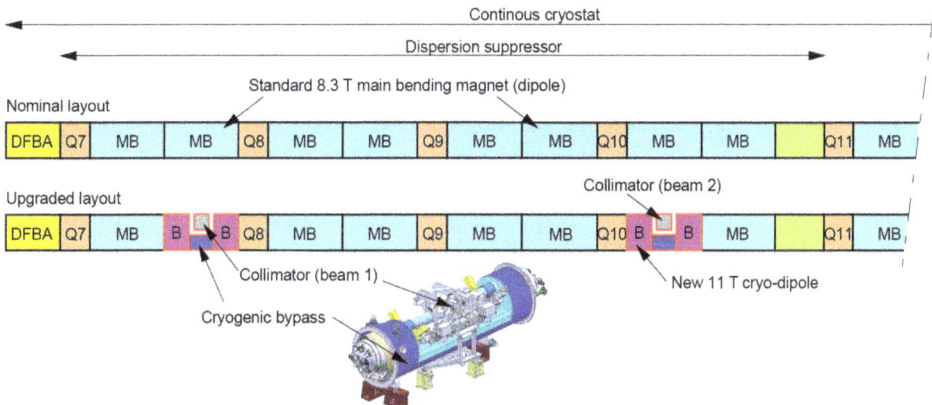

Fig. 2. Upgraded layout of the continuous cryostat (at Points 1, 2, 5 and 7).

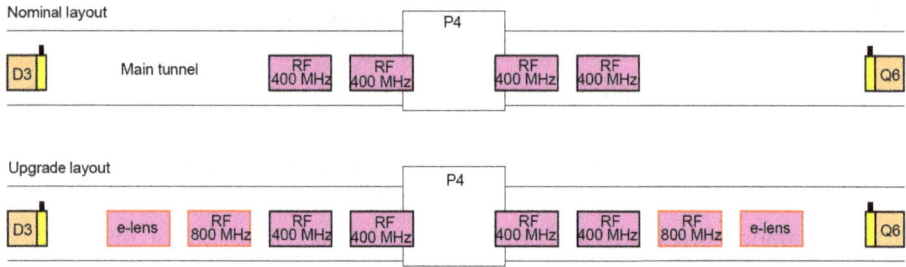

Fig. 3. Upgraded layout of Point 4 insertion region.

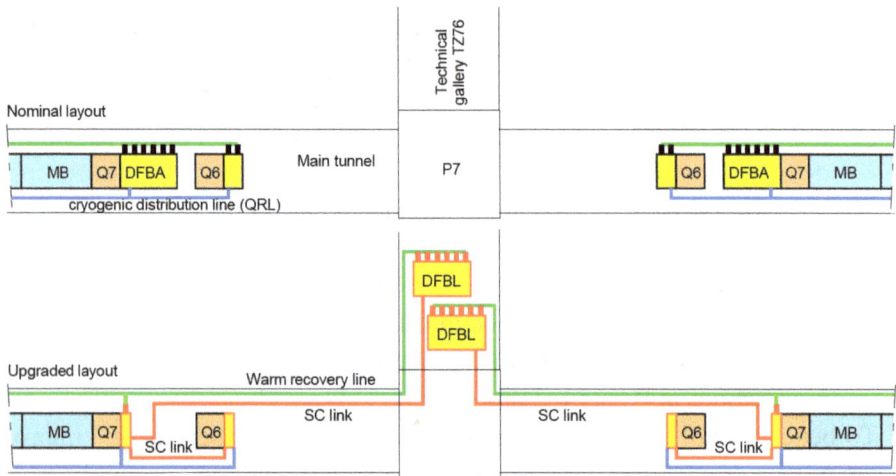

Fig. 4. Upgraded layout of Point 7 insertion region.

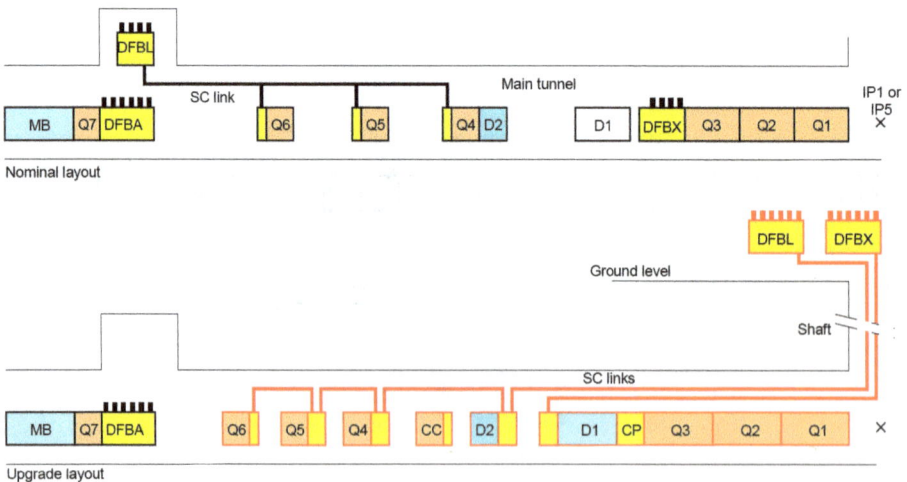

Fig. 5. Upgraded layout of Point 1/5 insertion region (half insertion).

would be installing a new 200-MHz SCRF system, rather than the 800 MHz. However, from the cryogenic point of view the requests are similar, so we will consider in the following text the 800-MHz system that is in an advanced phase of study.

- To improve the luminosity performance, by recovering the geometric luminosity reduction factor (loss of overlap between colliding bunches and increased effective cross-section at the IP) and also possibly to allow the levelling of the luminosity for limiting pile up at the maximum acceptable value for the high-luminosity detectors, cryo-modules of crab cavities (CC) will be added at Points 1 and 5 (see Fig. 5).
- Finally, the matching and final focusing of the beams will require completely new insertions cryo-magnets at Points 1 and 5 (see Fig. 5).

3. Temperature Level and Heat Loads

In Table 2 the static heat inleaks are reported, for the different temperature levels. For new equipment, the thermal performance of supporting systems, radiative insulation and thermal shields is considered identical to the one of existing LHC equipment.

Table 3 gives the dynamic heat loads on the HL-LHC machine. The main concern is the electron-cloud impingement on the beam screens which can only be reduced by an efficient beam scrubbing (dipole off) of the beam screens which is today not demonstrated. Without efficient beam scrubbing (dipole on), the e-cloud activity will remain high (more than 4 W per meter and per beam) in the arcs and dispersion suppressors (DS). This heat deposition corresponds to about

Table 2. Static heat inleaks of HL-LHC machine (w/o contingency).

			Nominal	Upgrade
4.6–20 K	Beam screen circuit (arc + DS)	[mW/m]	140	
	Beam screen circuit (IT)	[mW/m]	125	
	Beam screen circuit (MS)	[mW/m]	578	
1.9 K	Cold mass (arc + DS)	[mW/m]	170	
	Cold mass (IT)	[mW/m]	1250	
	Crab-cavities	[W per module]	0	25
4.5 K	Cold mass (MS)	[mW/m]	3556	
	400 MHz RF module	[W per module]	200	
	800 MHz RF module	[W per module]	0	120
	Electron-lens	[W per module]	0	12
20–300 K	Current lead	[g/s per kA]	0.035	

Table 3. Dynamic heat loads on HL-LHC machine (w/o contingency).

			Nominal	Upgrade
4.6–20 K	Synchrontron radiation (arc + DS)	[mW/m per beam]	165	310
	Image current (arc + DS + MS)	[mW/m per beam]	145	522
	Image current (IT low-luminosity)	[mW/m]	475	1698
	Image current (IT high-luminosity)	[mW/m]	166	596
	E-clouds (arc + DS) (dipole off)	[mW/m per beam]	271	41
	E-clouds (arc + DS) (dipole on)	[mW/m per beam]	4264	4097
	E-clouds (IT high luminosity)	[W per IT]	200	600
	E-clouds (IT low-luminosity)	[W per IT]	200	500
	E-clouds (MS)	[mW/m per beam]	2550	383
	Secondaries (IT beam screen P1 and P5)	[W per IT]	0	650
1.9 K	Beam gas scattering	[mW/m per beam]	24	45
	Resistive heating in splices	[mW/m]	56	56
	Secondaries (IT cold mass P1 and P5)	[W per IT]	155	630
	Secondaries (DS cold mass P1 and P5)	[W per DS]	37	185
	Qrf crab-cavities	[W per module]	0	24
4.5 K	Qrf 400 MHz	[W per module]	101	366
	Qrf 800 MHz	[W per module]	0	183
	E-lens	[W per module]	0	2
20–300 K	Current lead	[g/s per kA]	0.035	

twice the local cooling limitation given by the hydraulic impedance of the beam screen cooling circuits. In addition, the corresponding integrated power over a sector (more than 25 kW) is not compatible with the installed capacity of the sector cryogenic plants. For e-cloud deposition in the arcs and dispersion suppressors, efficient (dipole off) or inefficient (dipole on) beam scrubbing is considered.

The beam screens of the new inner triplets at P1 and P5 will be equipped with tungsten shielding which will be able to stop about half of the secondaries particles escaping the high-luminosity detectors. Despite of this shielding the 1.9 K load, i.e. the energy that collision debris deposit onto the magnet coil and cold mass, increases by four times with respect to the nominal LHC case. This W-shielding reduces the overall refrigeration cost and increases the lifetime of the inner-triplet coils.

4. Impact on Existing Sector Cryogenic Plants

With new cryogenic plants dedicated to the cooling of cryogenic equipment in the P1, P4 and P5 insertions, the cooling duty of the existing sector cryogenic plants will be reduced and more equally distributed. Figure 6 shows the required cooling capacities for the different temperature levels and compares them to the nominal cooling requirements and to the installed capacities. The low-load sectors equipped with upgraded ex-LEP cryogenic plants have lower installed

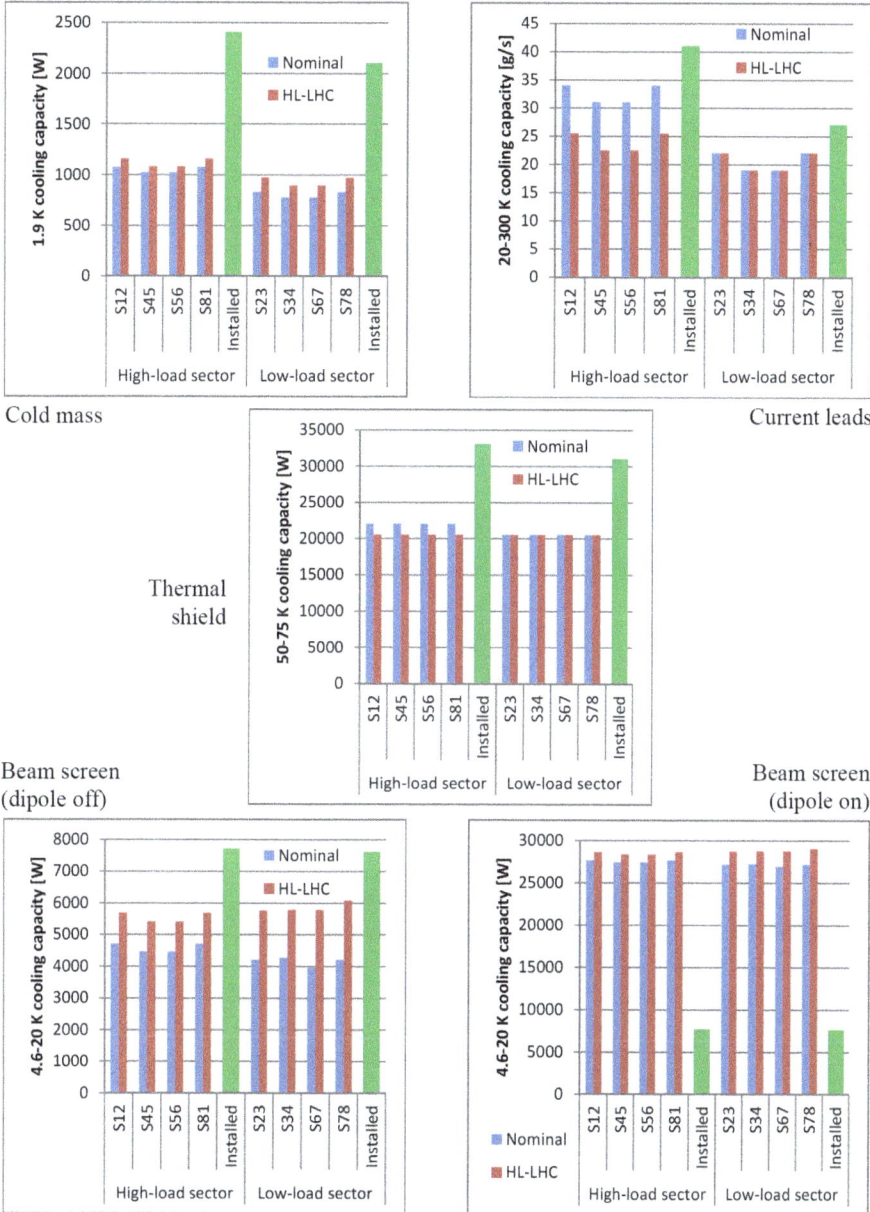

Fig. 6. Cooling capacity requirement of sector cryogenic plants.

capacity than the four cryogenic plants especially ordered for LHC for the high-load sectors. For HL-LHC, sufficient capacity margin still exist providing that the beam scrubbing of dipole beam-screens is efficient (dipole off).

5. New Cryogenics for P4 Insertion

Figure 7 shows the cryogenic architecture of the P4 insertion consisting of:

- a warm compressor station (WCS) located in a noise-insulated surface building and connected to a helium buffer storage;
- a lower cold box (LCB) located in the UX45 cavern and connected to a cryogenic distribution valve box (DVB) as well located in the UX45 cavern;
- main cryogenic distribution lines connecting the cryo-modules to the distribution valve box;
- auxiliary cryogenic distribution lines interconnecting the new infrastructure with the existing QRL service modules (SM) and allowing redundancy cooling with adjacent-sector cryogenic plants;
- a warm-helium recovery line network.

Concerning the installed capacity ($Q_{installed}$) of the new cryogenic plant, some uncertainty (f_u) and overcapacity (f_o) margins have to be introduced in the following equation:

- $Q_{installed} = f_o \times (Q_{static} \times f_u + Q_{dynamic})$ with $f_o = 1.5$ and $f_u = 1.5$.

Table 4 gives the installed capacity of the P4 cryogenic plant required at the different temperature levels. The P4 cryogenic plant will require an equivalent capacity of about 6 kW at 4.5 K.

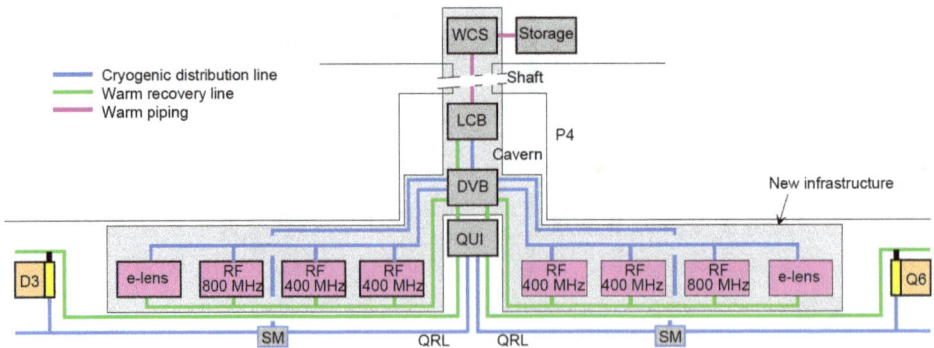

Fig. 7. Upgraded cryogenic architecture at P4.

Table 4. Installed capacity requirements of the new cryogenic plant at P4.

Temperature level		Static	Dynamic	Installed	Equivalent installed capacity @ 4.5 K [kW]	
4.5 K	[W]	1144	1736	5223	5.6	5.8
50–75 K	[W]	1000	0	2250	0.2	

6. New Cryogenics for High-Luminosity Insertions at P1 and P5

Figure 8 shows the cryogenic architecture of the P1 and P5 high-luminosity insertions consisting of:

- a warm compressor station (WCS) located in a noise-insulated surface building and connected to a helium buffer storage;
- an upper cold box (UCB) located in a ground level building;
- a quench buffer (QV) located at ground level;
- one or two cold compressor boxes (CCB) in underground cavern;
- two main cryogenic distribution lines (one per half-insertion);
- two interconnection valve boxes with existing QRL cryogenic line allowing redundancy with the cryogenic plants of adjacent sectors.

Table 5 gives the installed capacity of the P1 and P5 cryogenic plants required at the different temperature levels and using the same uncertainty and overcapacity margins than for P4. The cryogenic plants will require an equivalent capacity of about 18 kW at 4.5 K, including 3 kW at 1.8 K.

At P1 and P5, the superconducting magnets of the ATLAS and CMS detectors are cooled with dedicated cryogenic plants. A possible redundancy with detector cryogenic plants could be interesting in case of major breakdown of the detector cryogenic plants. The corresponding power requirements are about 1.5 kW at 4.5 K for CMS and 3 kW at 4.5 K for ATLAS.

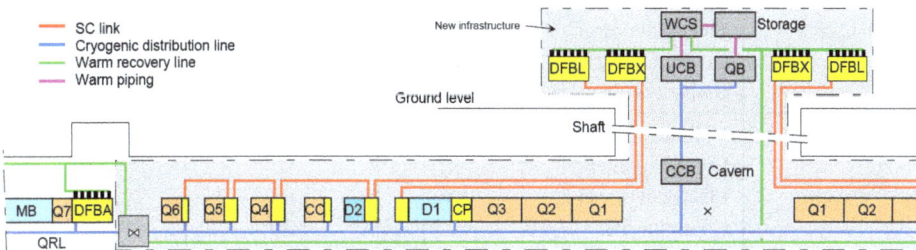

Fig. 8. Upgraded cryogenic architecture at P1 and P5.

Table 5. Installed capacity requirements of the new cryogenic plants at P1 and P5.

Temperature level		Static	Dynamic	Installed	Equivalent installed capacity @ 4.5 K [kW]	
1.9 K	[W]	433	1380	3045	12	
4.5 K	[W]	196	8	452	0.5	
4.6–20 K	[W]	154	2668	4348	2.4	18
50–75 K	[W]	4900	0	7350	0.5	
20–300 K	[g/s]	16	16	59	2.6	

The cooling capacity of 3 kW at 1.8 K is higher than the 2.4-kW installed capacity of a LHC sector which corresponds to the present state-of-the-art for the cold compressor size. Consequently:

- larger cold compressors have to be studied and developed, or,
- parallel cold compressor trains have to be implemented (one 1.5 kW train per half insertion), or,
- duplication of the first stage of cold compression to keep the machine within the available size.

Table 6. Building and general service requirement.

Cryogenic system				P1 and P5	P4
Warm compressor building	Surface	[m²]		700	400
	Crane	[t]		20	20
	Electrical power	[MW]		4.6	2.0
	Cooling water	[m³/h]		540	227
	Compressed air	[Nm³/h]		30	20
	Ventilation	[kW]		250	100
	Type	[-]		Noise-insulated (~108 dBA)	
Surface "SD" building	Surface	[mxm]		30x10	N/A
	Height	[m]		12	N/A
	Crane	[t]		5	N/A
	Electrical power	[kW]		50	N/A
	Cooling water	[m³/h]		15	N/A
	Compressed air	[Nm³/h]		90	N/A
Cavern	Volume	[m³]		200	300
	Local handling	[t]		2	2
	Electrical power	[kW]		100	20
	Cooling water	[m³/h]		20	20
	Compressed air	[Nm³/h]		40	30

7. Building and General Service Requirement

Table 6 gives the buildings and general services at P1, P4 and P5 required by the cryogenic infrastructure. At P4, the required surface and volume for the warm compression station and for the cold box are respectively available in the existing SUH4 building and in the UX45 cavern.

8. Conclusion

The HL-LHC project will require a large cryogenic upgrade with new cryogenics challenges. New cooling circuits must be designed to extract up to 13 W/m on 1.9 K cold-mass superconducting cables while keeping sufficient margin with respect to resistive transition limit, as well as up to 23 W/m on inner-triplet beam-screens with possibly a different operating range (40–60 K) and with a large dynamic range which will require specific cryogenic-plant adaptation studies. New cooling method of HTS links, current feed boxes and crab cavities must be developed and validated. The resistive transition containment and helium recovery via cold buffering must be designed for efficient operation. Concerning cryogenic plants, larger 1.8 K refrigeration capacities beyond the present state-of-the-art must be developed including large capacity (1500/3000 W) sub-cooling heat exchangers. This upgrade will be implemented within to main phases. During the second long shutdown of LHC in 2018–2019 calendar years, the upgrade at Point 7 and Point 4 will be implemented. The remaining part will be implemented during the third long shutdown of LHC in 2023–2025 calendar years.

References

[1] V. Baglin, Ph. Lebrun, L. Tavian and R. van Weelderen, 2012, Cryogenic beam screens for high-energy particle accelerators, in *Proceedings of ICEC24*, Fukuoka, Japan (2012), pp. 629-634.

[2] R. Calaga, Crab cavities for the LHC upgrade, in *Proceedings of the Chamonix 2012 Workshop on LHC Performance*, Chamonix, France (2012).

[3] P. Granieri, L. Hincapié, R. van Weelderen, Heat transfer through the electrical insulation of Nb_3Sn cables, in *Proceedings of MT23*, Boston, USA (2013).

[4] P. Granieri and R. van Weelderen, Deduction of steady-state cable quench limits for various electrical insulation schemes with application to LHC and HL-LHC Magnets, in *Proceedings of MT23*, Boston, USA (2013).

[5] L. Rossi and O. Brüning, High luminosity large hadron collider: A description for the European Strategy Preparatory Group, CERN-ATS-2012-236 (2012).

[6] E. Todesco *et al.*, A first baseline for the magnets in the high-luminosity LHC insertion regions, in *Proceedings of MT23*, Boston, USA (2013).

[7] E. Todesco *et al.*, Design studies for the low-beta quadrupoles for the LHC luminosity upgrade, in *Proceedings of ASC 2012*, Portland, Oregon, USA (2012).

Chapter 10

The "Environmental" Challenges:
Impact of Radiation on Machine Components

M. Brugger, F. Cerutti and L. S. Esposito

CERN, TE Department, Genève 23, CH-1211, Switzerland

The High Luminosity LHC upgrade poses demanding requirements in terms of energy deposition, in particular around the high luminosity experiments where the Inner Triplet elements and the separation dipole will be exposed to unprecedented levels of radiation, challenging their reliability and lifetime. Dedicated Monte Carlo studies have been conducted in order to characterize the debris-machine interaction and define a suitable shielding.

Moreover, also the areas adjacent to the LHC tunnel, where the installation of electronics equipment is envisaged, will be significantly impacted. In this context, cumulative damage (Total Ionizing Dose and/or Non-Ionizing Energy Loss) and stochastic effects have to be taken into account in an appropriate Radiation Hardness Assurance strategy, including the specification of the radiation environment, required test and qualification procedures, and corresponding radiation monitoring.

1. Collision Debris

Proton–proton inelastic collisions taking place in the LHC inside its four big detectors generate a large number of secondary particles, on average about 100 (120) per collision with 3.5 (7) TeV beams, but with very substantial fluctuations over different events. Moving from the interaction point (IP), this multiform population evolves, even before touching the surrounding material, because of the decay of unstable particles (in particular neutral pions decaying into photon pairs). Figure 1 illustrates the composition of the debris at 5 mm from the point of a 14 TeV center-of-mass collision, featuring a $\sim 30\%$ increase in number of particles and a clear prevalence of photons (almost one half) and charged pions ($\sim 35\%$).

Most of these particles are intercepted by the detector and release their energy within the experimental cavern. However, the most energetic ones, emitted at small angles with respect to the beam direction, travel farther in the vacuum and reach the accelerator elements, causing a significant impact on the magnets along the

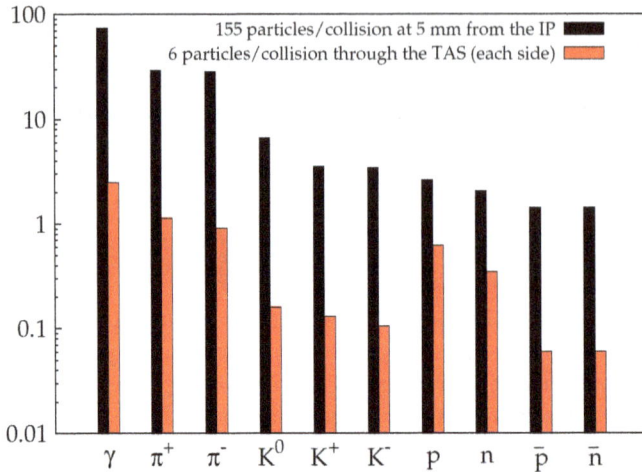

Fig. 1.   Number of debris particles per collision at 5 mm from the interaction point (black histogram) and at the exit of each TAS aperture (red).

Insertion Regions (IRs), in particular the final focus quadrupoles and the separation dipole. Figure 1 shows also the breakdown of the debris component going through the (future) 60 mm aperture of the TAS (Target Absorber Secondaries) absorber, a protection element consisting of a copper core 1.8 m long located at 20 m from the IP and representing the interface between the detector and the accelerator. The TAS absorbers are installed only at each side of the high-luminosity IRs, namely IR1 and IR5, hosting the ATLAS and CMS detectors respectively, since their protection role, in fact limited to the first quadrupole, is not needed for luminosities up to 0.2×10^{34} cm^{-2}s^{-1} [1].

Despite the fact that the number of particles per collision leaving the TAS aperture is more that one order of magnitude lower than the total number of debris particles, they carry about 80% of the total energy, 40% per each side. At the nominal HL-LHC luminosity (5×10^{34} cm^{-2}s^{-1}), this represents about 3800 W per side that is inevitably impacting the LHC elements and is dissipated in the machine, in the nearby equipment (e.g. electronics, racks, ...) and in the tunnels walls.

It is fundamental to study how these particles are lost in order to implement all the necessary protections to shield the most sensitive parts of the LHC magnets and the most delicate components. For these purposes, Monte Carlo simulations of the particle interaction with matter play an essential role, relying on a sophisticated implementation of physics models and an accurate 3D-description of the region of interest.

A specific problem is represented by the electronics sensitivity to radiation. The above described particle debris emerging from the IP (together with an additional loss contribution from beam–gas interactions) will impact equipment being present in the areas adjacent to the LHC tunnel (UJs, RRs). Respectively installed (present or future) control systems are either fully commercial or based on so-called COTS (Commercial-Off-The-Shelf) components, both possibly affected by radiation. This includes the immediate risk of so-called Single Event Effects (SEE) and a possible direct impact on beam operation, as well as in the long-term, also cumulative dose effects (impacting the component/system lifetime) which additionally have to be considered.

For the tunnel equipment in the existing LHC, certain radiation tolerant design criteria were already taken into account prior construction. However, most of the equipment placed in adjacent and partly shielded areas was not conceived nor tested for their current radiation environment. Therefore, given the large amount of electronics being installed in these areas, during the past years a CERN wide project called R2E (Radiation To Electronics) [2] has been initiated to quantify the danger of radiation-induced failures and to mitigate the risk for nominal beams and beyond to below one failure a week. The respective mitigation process included a detailed analysis of involved radiation fields, intensities and related Monte Carlo calculations; radiation monitoring and benchmarking; the behaviour of commercial equipment/systems and their use in the LHC radiation fields; as well as radiation tests with dedicated test areas and facilities [2, 3].

In parallel, radiation induced failures were analyzed in detail in order to confirm early predictions of failure rates, as well as to study the effectiveness of implemented mitigation measures. Figure 2 shows the actual number of SEE failures measured during 2011 and 2012 operation, the achieved improvement (please note that the failure rate measured during 2011 already included mitigation measures implemented during 2009 and 2010), as well as the goal for operation after LS1 and during HL-LHC.

Aiming for annual luminosities of up to 300 fb^{-1}, it is clear that machine availability has to be maximized during HL-LHC in order to successfully achieve the physics goal. This implies that existing electronic control systems are either installed in fully safe areas, sufficiently protected by shielding or adequately radiation tolerant. The last implies existing equipment, but also any future equipment to be possibly installed in R2E critical areas to be conceived in a specific way.

In the following, we will give details about the geometrical IR model developed for Monte Carlo simulation studies (Section 2), then we will describe how and where the different debris particle species are captured (Section 3), and in Section 4 we will provide estimates of the energy deposition. Finally, Section 5 summarizes the overall R2E requirements for HL-LHC and provides a first overview of design criteria and radiation levels expected in areas of concern.

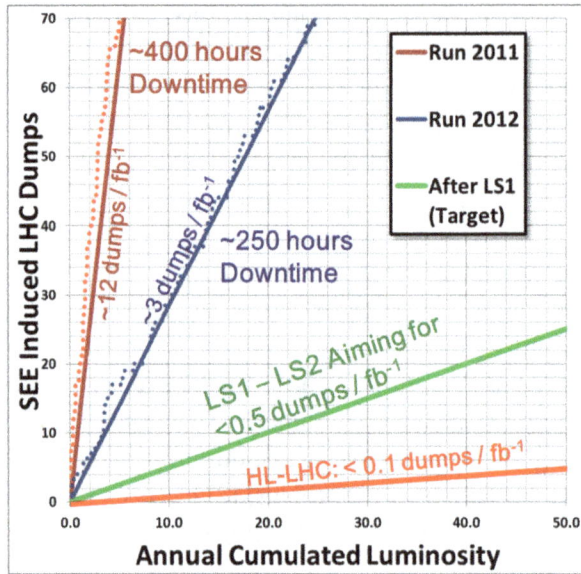

Fig. 2.   LHC beam dumps due to single-event effects against beam luminosity. Dots (2011 and 2012) refer to measurements, whereas lines show annual averages for both, past and future operation.

2. Beamline Model

Figure 3 shows the HL-LHC layout from the TAS to the separation dipole D1, as it is implemented in FLUKA [4–7]. The final focus quadrupoles are arranged in a triplet configuration, Q1-Q2-Q3, that is DFD (defocusing-focusing-defocusing) in the vertical plane for the outgoing beam. A corrector package (CP), that includes a skew quadrupole and eight high order correctors (from sextupole to dodecapole, normal and skew), is placed between the triplet and the D1. Two different technologies are in place to build the magnet coils: the novel Nb_3Sn for the final focus quadrupoles and the well established Nb-Ti for the corrector magnets and the D1.

To guarantee a sufficient protection of these magnets from radiation, an octagonal stainless steel beam screen equipped with 6 mm tungsten absorbers on the mid-planes is placed inside the cold bore all along the triplet, the CP and the D1, except in Q1 where the tungsten thickness is increased to 16 mm, compatible with the aperture requirements. The two beam screen versions are shown in the top insets of Fig. 3. The absorbers attached externally to the beam screen have a negligible thermal contact with the cold mass. Therefore, from the point of view of energy deposition, the beam screen function is two-fold:

(i) it shields the coils,
(ii) it removes a sizable part of the heat load from the 1.9 K cooling system.

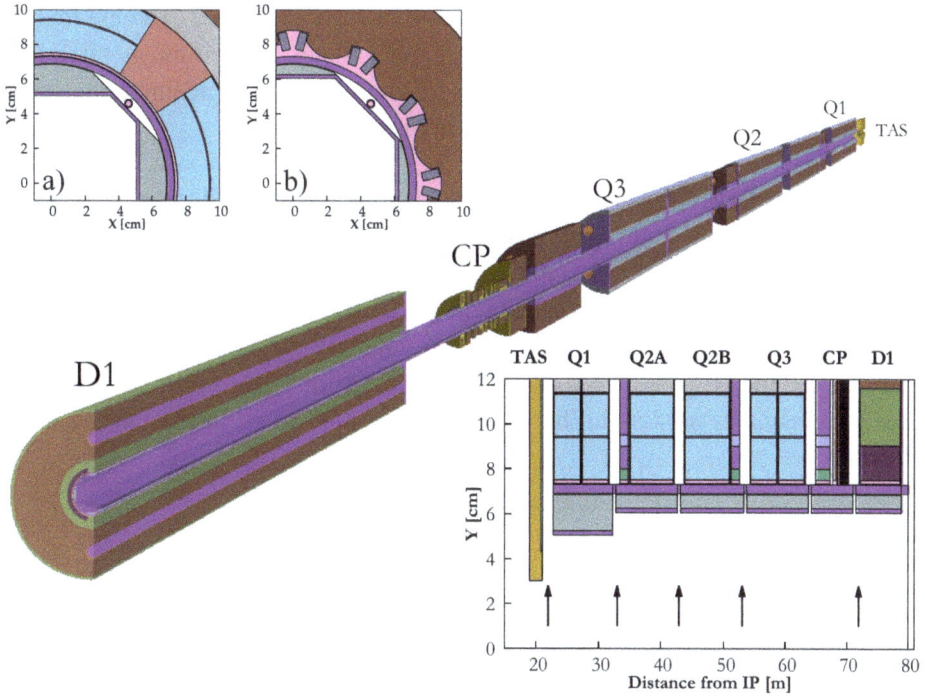

Fig. 3. Prospective view of the HL-LHC geometry from TAS to separation dipole D1 as implemented in FLUKA. Bottom inset: beam line aperture. Top inset: magnet cross-section (one quadrant) showing the beam screen with 16 mm tungsten absorbers in the Q1 (a) and with 6 mm absorbers in the normal dodecapole (b). Each color corresponds to a different material.

The present HL-LHC layout foresees six cryostats: four for the triplet quadrupoles (Q1, Q2A, Q2B and Q3), one for the CP and the last one for the D1. Over the interconnects, the distance between the magnets is 1.5–1.7 m. This represents a vulnerable point in the shielding, since an interruption of the beam screen is necessary therein. As a reasonable baseline, we assume here 500 mm interruption of the beam screen in the middle of the interconnects.

3. Radiation Capture

The particles exiting from the TAS and entering the LHC vacuum pipe can be divided in two categories:

- neutrally charged particles (mainly photons, with a smaller contribution of neutrons) that travel along a straight line. They are not affected by any field and interact as soon as they hit an obstacle along their trajectory,

- positively and negatively charged particles (mainly pions and protons) that are steered and eventually captured by the magnetic fields, as their magnetic rigidity is lower than the one of the circulating protons. They first encounter the strong magnetic fields of the final focus quadrupoles.

The evolution of the differential fluences[a] of the main particle species is shown in Fig. 4 at different longitudinal positions along the triplet.

The proton spectrum is characterized by the peak at the 7 TeV beam energy. This is due to single diffractive events[b] at the IP, imparting to the beam protons a small angular kick and energy loss, and consequently making them to be lost farther away in the Dispersion Suppressor or cleaned out by the collimation system. Lower momentum protons in the 2–3 TeV range start to be intercepted from Q3 onwards.

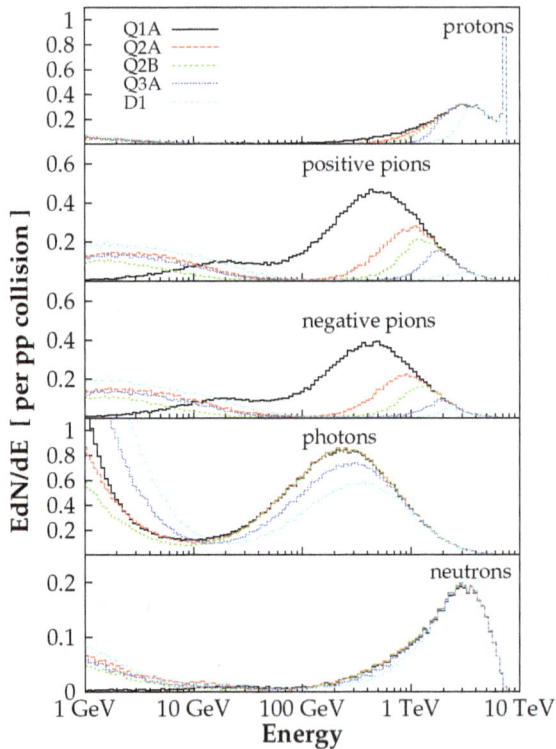

Fig. 4.   Energy spectra of different particle species in the vacuum pipe at the positions along the triplet indicated by the arrows in the bottom inset of Fig. 3.

---

[a]Track-length volume density.
[b]Protons that have undergone an elastic interaction are not simulated here. They are cleaned by the collimation system.

Negatively and positively charged pions have a softer spectrum and most of them are immediately captured by the magnetic field of the first quadrupole. The increase in the lower part of the spectrum (less then few ten GeV) comes from pion production by interactions in the upstream part of the triplet.

Neutral particles, not affected by the magnetic field, are captured only when they run into a restriction of aperture. Few hundred GeV photons start to be intercepted inside the Q2B, where the shadow provided by the TAS ends. Neutrons are concentrated in the TeV region and, due to their relatively small angle, proceed farther in the machine. To stop these neutral particles, a TAN (Target Absorber Neutral) absorber is placed in front of the recombination dipole D2, at the beginning of the Matching Section.

4. Energy Deposition

For steady state losses, like the ones due to the collision debris, we can distinguish two scale sizes for the energy deposition[c]:

- total power deposited on an assemble of elements that must be included in the budget of heat load to be evacuated,
- local (order of cm^3) energy deposition on sensitive parts of a magnet, namely the superconducting coils.

The first represents a crucial parameter for the design of the cryogenic system. The latter has to be evaluated in order to assure not to surpass the margin for the magnet quench, as well as to reach the desired lifetime (over a long term, radiation can deteriorate the cable electric properties and eventually damage their mechanical structure). For the purposes of quench risk evaluation, the reference quantity is the peak power density averaged over the entire radial dimension of the inner coil layer, that is the most exposed to the radiation field. As for radiation damage, one has to calculate the peak dose on a finer radial binning (~ 3 mm), since relevant material degradation can be localized with heat diffusion playing no role in this respect. Along the other dimensions, typical resolutions are ~ 10 cm longitudinally and 2-degree azimuthally.

Like the present machine, two collision schemes are foreseen at the high luminosity IRs: the protons are led to the interaction point with 295 μrad half-crossing angle either on the vertical or on the horizontal plane. This implies that the debris leaving the IP flies preferentially on the respective plane toward the IR elements. The combination of vertical crossing with the above mentioned optics configu-

---

[c]Definitely we are not referring to microscopic processes, that are well below the minimum millimeter scale considered throughout this chapter.

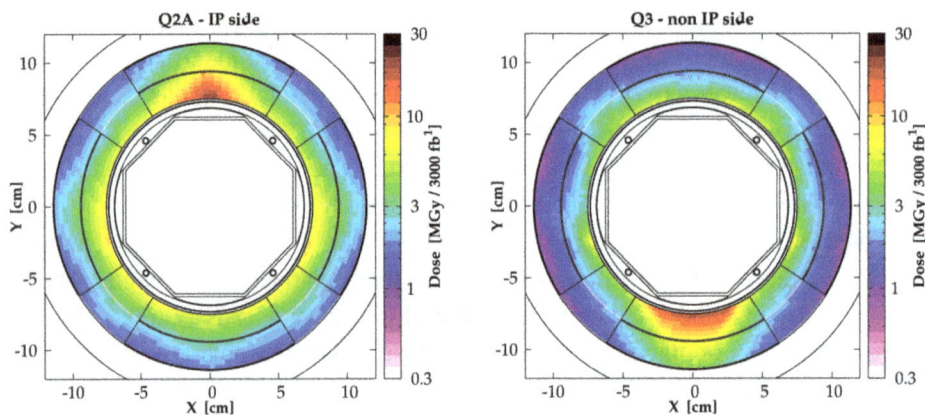

Fig. 5.   Dose distribution after 3000 fb^{-1} integrated luminosity over a transverse section of the Q2A (left) and Q3 (right) coils. Vertical crossing angle pointing upward has been assumed at the IP. Geometry details (beam screen, cold bore, collar) are also shown with no respective dose estimation.

ration of the triplet (DFD) represents the worse case from the energy deposition point of view, since vertically scattered particles are more efficiently caught by the defocusing Q1 field. Figure 5 shows a couple of characteristic dose distributions over a transverse section of Q2 (IP side) and Q3 (non IP side). The energy is mainly deposited along the magnet mid-planes (particularly on the vertical one), with strong radial and azimuthal gradients. It is worth to note the inversion of the peak position from top (in Q2) to bottom (in Q3) because of the Q2 over-focusing effect while particles travel along the triplet.

Figure 6 (red curve) shows the obtained peak dose longitudinal profile. The pattern features a local maximum at the end of Q1/beginning of Q2A driven by the Q1 defocusing field. The other maxima at the IP side of the Q2B, Q3 and CP are due to the interruption of the beam screen over the interconnects. As mentioned in Section 2, a 500 mm gap has been assumed. Reducing this gap would lower these peaks, although a minimum interruption of few ten cm cannot be avoided, because it is necessary to install the beam tube bellow between two consecutive cryostats.

In the considered configuration one does not exceed 40 MGy after 3000 fb^{-1} (at the beginning of the CP). At the nominal HL-LHC luminosity, the peak power density stays within 2 mW/cm^3 all along the final focus quadrupoles and the separation dipole. This value is safely below the expected quench limit, that is assumed to be 40 (12) mW/cm^3 for Nb$_3$Sn (Nb-Ti) coils [8,9].

Figure 6 shows also the case where 6 mm absorbers (instead of 16 mm) are used in the Q1 too, as for the rest of the magnet string. The maximum dose on Q1 becomes ~5 times higher and, moreover, there is no shadow effect on the downstream element, where a peak dose of ~50 MGy is reached.

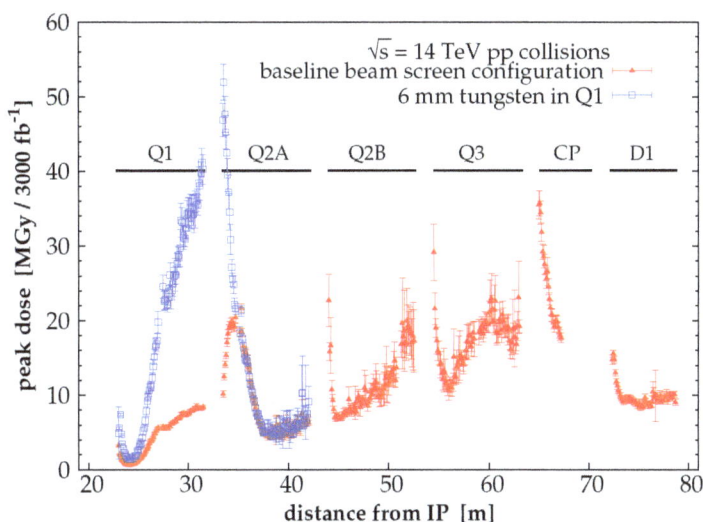

Fig. 6. Longitudinal profile of the peak dose at 3000 fb^{-1} along the final focus quadrupoles, the CP and the separation dipole in case of the baseline beam screen configuration (triangles) with 16 mm tungsten absorbers in Q1 and 6 mm in downstream elements. The case with 6 mm tungsten absorbers in Q1 is also shown (boxes). Vertical crossing at the IP has been considered. Horizontal lines indicate the total magnetic lengths of the elements hosted inside the same cryostat.

The total heat load deposited is about 1300 W at the nominal HL-LHC luminosity. This load is removed by two different cooling systems (one for the magnet cold masses and the other one for the beam screen), each of them in charge to evacuate about one half of it.

5. Radiation to Electronics

Radiation damage to electronics is often considered with space applications. However, it is important to note that the radiation environment encountered at the LHC, the high number of electronic systems and components partly exposed to radiation, as well as the actual impact of radiation induced failures strongly differ from the context of space applications. While for the latter application design, test and monitoring standards are already well defined, additional constraints, but in some cases also simplifications have to be considered for accelerator environment.

The mixed particle type and energy field encountered in the relevant LHC areas is composed of charged and neutral hadrons (protons, pions, kaons and neutrons), photons, electrons and muons ranging from thermal energies up to the GeV range. This complex field has been extensively simulated by the FLUKA Monte Carlo code and benchmarked in detail for radiation damage issues at the LHC [10, 11]. The observed radiation is due to particles generated by proton–proton (or ion–ion) collisions in the LHC experimental areas (as previously discussed in this chapter), distributed beam losses (protons, ions) around the machine, and to beam interacting with the residual gas inside the beam pipe. The proportion of the different particle species in the field depends on the distance and on the angle with respect to the original loss point, as well as on the amount (if any) of installed shielding material. In this environment, electronic components and systems exposed to a mixed radiation field will experience three different types of radiation damages: these are displacement damage, damage from the Total Ionizing Dose (TID) and the SEEs.

The first two are of cumulative nature and are measured through TID and non-ionizing energy deposition (NIEL[d], generally quantified through accumulated 1-MeV neutron equivalent fluence), where the steady accumulation of defects cause measurable effects which can ultimately lead to device failure. As for stochastic SEE failures, they form an entirely different group as they are due to the direct ionization by a single particle, able to deposit sufficient energy through ionization processes in order to disturb the operation of the device. They can only be characterized in terms of their probability to occur as a function of accumulated High Energy (> 5–20 MeV) Hadron fluence. The probability of failure will strongly depend on the device as well as on the flux and nature of the particles. In the context of HL-LHC, several tunnel areas close to the LHC tunnel, and partly not sufficiently shielded, are or are supposed to be equipped with commercial or not specifically designed electronics which are mostly affected by the risk of SEEs, whereas electronics installed in the LHC tunnel will also suffer from accumulated damage in the long-term [12].

For this purpose, during the first years of LHC operation, the radiation levels in the LHC tunnel and in the shielded areas have been measured by using the CERN RadMon system [13] dedicated to the analysis of radiation levels possibly impacting installed electronic equipment. Table 1 summarizes the level of accumulated High Energy Hadron (HEH) fluence measured during 2012 for the most critical LHC areas where electronic equipment is installed and which are relevant for the HL-LHC project, together with the expected radiation levels for nominal LHC performance (50 fb^{-1}y^{-1}). The HEH fluence measurements are based on the RadMon reading of the Single Event Upsets (SEU) of SRAM memories whose sensitivity was extensively calibrated in various facilities [14]. The results obtained during 2012 LHC proton operation show that the measurements very well compare with previously performed FLUKA calculations and observed differences can actually be attributed to changes of operational parameters not considered in the calculations [15]. In a first approximation, the measured radiation levels can also be used

---

[d]Non-Ionizing Energy Losses.

Table 1.   Predicted and measured annual HEH fluence in critical shielded areas for a cumulated ATLAS/CMS luminosity of 15 fb^{-1} during 2012 operation, extrapolated to expected nominal and HL-LHC performance (50 fb^{-1}y^{-1} for nominal and 300 fb^{-1}y^{-1} for HL-LHC performance, except for P8 where 2012 can already be considered as almost nominal and HL-LHC refers to a 5-fold increase). For HL-LHC an estimate for corresponding annual TID levels is also given.

LHC Area	Prediction (HEH/cm^2)	Measured (HEH/cm^2)	Nominal (HEH/cm^2)	HL-LHC (HEH/cm^2)	HL-LHC (Dose/Gy)
UJ14/16	1.4×10^8	1.6×10^8	5×10^8	3×10^9	6
RR13/17	2.0×10^8	2.5×10^8	8×10^8	5×10^9	10
UJ56	1.6×10^8	1.5×10^8	5×10^8	3×10^9	6
RR53/57	2.0×10^8	2.5×10^8	8×10^8	5×10^9	10
UJ76	2.1×10^7	6.0×10^7	2×10^8	1×10^9	2
RR73/77	2.9×10^7	5.0×10^7	2×10^8	1×10^9	2
UX85B	4.3×10^8	3.5×10^8	4×10^8	2×10^9	4
US85	1.3×10^8	8.8×10^7	9×10^7	4×10^8	1

to extrapolate towards HL-LHC by purely scaling with annual luminosity (see last two columns of Table 1), however keeping in mind that operational and layout parameters (beam energy, crossing angle, TAN design, absorbers, etc.) can have a not negligible impact on the final values, especially for the RRs close to the Matching Section. The resulting values clearly indicate that any equipment installed in the LHC tunnel will not only suffer SEE failures, but will also be heavily impacted by the TID effects thus limiting the respective lifetime.

To provide one specific example, based on available FLUKA calculations for the present LHC at nominal parameters and applying a simplified scaling with only cumulative luminosity, Fig. 7 shows the distribution of high-energy hadrons for LHC-P1 including the adjacent UJ, UL and RR areas (please note that the radiation levels in the UJ refer to the layout with only limited shielding, already improved along the R2E mitigation measures prior and during LS1). Any control equipment (commercial or based on commercial components) to be installed in these areas, clearly has to be proven to be sufficiently radiation tolerant. For comparison, as mentioned earlier, during the last years of operation we already had there a number of radiation induced failures on commercial equipment for radiation levels corresponding to 10^8–10^9 cm^{-2}y^{-1} (which is about 1000–10000 more than what one would get at surface due to cosmic radiation).

For the current R2E project, this allowed us deducing an acceptable limit of 10^7 cm^{-2}y^{-1} annual radiation level, leading to the definition of so-called protected areas (in terms of overall risk of radiation induced failures). Therefore, for HL-LHC any installation of non-tested (and not specifically designed) electronic equipment in the UJs, part of the ULs and RRs is clearly to be avoided or subject

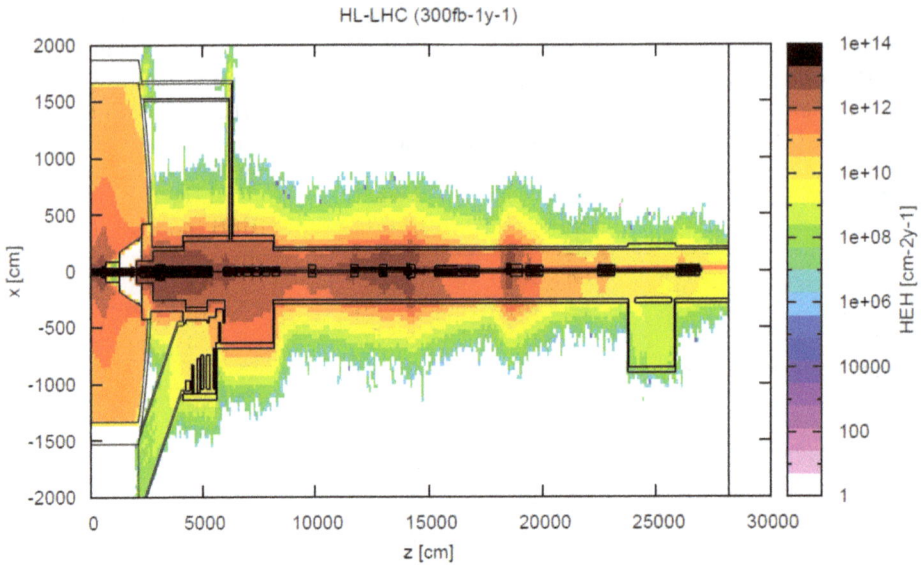

Fig. 7. Annual HEH fluence expected at LHC-P1 (with the present machine layout at nominal parameters) scaled to the HL-LHC annual luminosity of 300 fb^{-1}. The radiation levels in the UJ refer to the layout with only limited shielding, already improved along the R2E mitigation measures prior and during LS1.

to a detailed analysis process prior an exceptional installation can be granted under the following conditions:

- the equipment is not linked to any safety system,
- the failure of the equipment will not lead to a beam dump,
- the failure of the equipment does not require quick access (thus lead to down-time),
- there is no any other operational impact (loss of important data, etc.).

In all other cases requiring installation in critical areas, a respective radiation tolerant electronics development must be considered from the very early stage on-ward. Related expertise exists at CERN within the equipment groups, the R2E project and a dedicated working group [16].

In a first approximation and limiting the total number of exposed systems, the above mentioned annual radiation design level of 10^7 cm^{-2}y^{-1} can also be chosen as acceptable aiming to achieve an overall performance of less than one radiation induced failure per one or two weeks of HL-LHC operation.

For operation critical equipment, the HL-LHC project foresees respective radiation tolerant developments already at an early stage of the design phase, taking into account that:

- for the LHC-tunnel: in addition to SEEs also cumulative damage has to be considered for both existing and future equipment,
- for partly shielded areas (UJs, RRs, ULs): cumulative damage should be carefully analyzed but can most likely be mitigated by preventive maintenance (detailed monitoring mandatory), but radiation tolerant design is mandatory in order to limit SEE induced failures,
- the knowledge of radiation induced failures and radiation tolerant development within the equipment groups and in the overall A&T sector has to be maintained and further strengthened,
- the access and availability of radiation test facilities (CERN internal and external) has to be ensured providing efficient support to equipment groups,
- building on the experience obtained during the LHC R2E project and in view of the HL-LHC time-scale, it is important that the expertise of and support to radiation tolerant developments (currently available through the Radiation Working Group [16]) is maintained and ensured from the early project stage onwards.

References

[1] L. S. Esposito *et al.*, Power load from collision debris on the LHC Point 8 insertion magnets implied by the LHCb luminosity increase, IPAC2013, TUPFI022, p. 1382.

[2] R2E website, www.cern.ch/r2e.

[3] M. Brugger *et al.*, R2E experience and outlook for 2012, in *Proceedings of LHC Performance Workshop*, Chamonix 2012.

[4] G. Battistoni *et al.*, The FLUKA code: Description and benchmarking, *AIP Conf. Proc.* **896**, 31–49 (2007).

[5] A. Ferrari *et al.*, FLUKA: A multi-particle transport code (Program version 2005), CERN-2005-010.

[6] V. Vlachoudis, FLAIR: A powerful but user friendly graphical interface for FLUKA, in *Proc. Int. Conf. on Mathematics, Computational Methods & Reactor Physics*, Saratoga Springs, New York, 2009.

[7] A. Mereghetti *et al.*, The FLUKA linebuilder and element database: Tools for building complex models of accelerator beam lines, IPAC2012, WEPPD071, p. 2687.

[8] N. V. Mokhov *et al.*, Protecting LHC IP1/IP5 components against radiation resulting from colliding beam interactions, CERN-LHC-Project-Report-633.

[9] N. V. Mokhov, I. L. Rakhno, Mitigating radiation loads in Nb_3Sn quadrupoles for the CERN Large Hadron Collider upgrades, *Phys. Rev. STAB* **9**, 101001 (2006).

[10] K. Roed *et al.*, FLUKA simulations for SEE studies of critical LHC underground areas, *Nuclear Science, IEEE Transactions on* **58**(3), 932 (2011).

[11] M. Brugger *et al.*, FLUKA capabilities and CERN applications for the study of radiation damage to electronics at high-energy hadron accelerators, *Progress in Nuclear Science and Technology*, PNST10184-R1 (2010).

[12] K. Roed, M. Brugger, G. Spiezia, An overview of the radiation environment at the LHC in light of R2E irradiation test activities, CERN publication, CERN-ATS-Note-2011-077 TECH, (2011).

[13] G. Spiezia *et al.*, The LHC accelerator Radiation Monitoring System – RadMON, in *Proceedings of Science*, 2011.

[14] D. Kramer *et al.*, LHC RadMon SRAM detectors used at different voltages to determine the thermal neutron to high energy hadron fluence ratio, *IEEE Trans. Nucl. Sci.* **58**(3), 1117 (2011).

[15] G. Spiezia *et al.*, R2E experience and outlook, LHC Beam Operation Workshop, Evian, 2012.

[16] CERN Radiation Working Group (RadWG), www.cern.ch/radwg.

Chapter 11

Radiation Protection Considerations

C. Adorisio[1], S. Roesler[1], C. Urscheler[2] and H. Vincke[1]

[1]CERN, TE Department, Genève 23, CH-1211, Switzerland
[2]Bundesamt fuer Gesundheit, Direktionsbereich Verbraucherschutz,
Zchwarzenburgstrasse 165, 3003 Bern, Switzerland

This chapter summarizes the legal Radiation Protection (RP) framework to be considered in the design of HiLumi LHC. It details design limits and constraints, dose objectives and explains how the As Low As Reasonably Achievable (ALARA) approach is formalized at CERN. Furthermore, features of the FLUKA Monte Carlo code are summarized that are of relevance for RP studies. Results of FLUKA simulations for residual dose rates during Long Shutdown 1 (LS1) are compared to measurements demonstrating good agreement and providing proof for the accuracy of FLUKA predictions for future shutdowns. Finally, an outlook for the residual dose rate evolution until LS3 is given.

1. Radiological Quantities

The design phase of a new project includes an evaluation of radiological risks as well as their limitation and minimization by appropriate protection and optimization measures. The radiological quantities to assess comprise

- for periods of beam operation:
  (1) dose equivalent to personnel by stray radiation in accessible areas,
  (2) activation of effluents and air and their release into the environment as well as the resulting annual dose to the reference groups of the population,
  (3) dose equivalent to personnel and environment in case of abnormal operation or accidents,
- for beam-off periods:
  (4) radioactivity induced by beam losses in beam-line components and related residual dose equivalent rates,
  (5) dose equivalent to personnel during interventions on activated beam-line components or experiments,

- for decommissioning:
  (6) radionuclide inventory.

2. Regulatory Framework, Design Limits and Dose Objectives

The upgrade of the LHC studied within the HiLumi LHC Project must be optimized according to CERN's Radiation Protection rules and regulations of which the most relevant for the design and upgrade of accelerators are summarized in the following.

2.1. *Justification, optimization, limitation*

The CERN Radiation Protection legislation is detailed in the Safety Code F [1]. It stipulates that all activities involving ionizing radiation have to be *justified, optimized and limited* and defines respective limits:

- Any practice leading to an effective dose exceeding $100 \, \mu$Sv per year for individuals working on the CERN site or $10 \, \mu$Sv for members of the general public must be justified.
- It is obligatory to optimize radiation protection according to the As Low As Reasonably Achievable (ALARA) principle. Optimization can be considered as respected if the annual dose of a practice is below $100 \, \mu$Sv for persons exposed because of their professional activity and $10 \, \mu$Sv for members of the general public.
- The effective dose in any consecutive 12-month period is limited to 20 mSv for so-called Category A Radiation Workers, to 6 mSv for Category B Radiation Workers and to 1 mSv for not occupationally exposed personnel. The effective annual dose to any person outside of the CERN site boundaries must not exceed $300 \, \mu$Sv.

2.2. *Design constraints*

Design constraints for new or upgraded facilities ensure that the exposure of persons working on the CERN sites as well as the public will remain below the dose limits under normal as well as abnormal conditions of operation and that the optimization principle is implemented. In particular, the following design constraints apply:

- The design of components and equipment must be optimized such that installation, maintenance, repair and dismantling work does not lead to an effective dose, e.g., as calculated with Monte Carlo simulations, exceeding 2 mSv per person and per intervention [2]. The design is to be revised if the dose estimate exceeds this value for cooling times compatible with operational scenarios.

- The annual effective dose to any member of a reference group outside of the CERN boundaries must not exceed 10 μSv. The estimate must include all exposure pathways and all contributing facilities.
- The selection of construction material must consider activation properties to minimize the dose to personnel and the production of radioactive waste. In order to guide the user a web-based code (ActiWiz) is available for CERN accelerators [3].

2.3. *ALARA and dose objectives*

Detailed CERN-specific ALARA rules apply to any work implying risks due to ionizing radiation [4, 5]. Procedures define the optimization process to follow based on a risk-dependent classification scheme: The estimated individual and collective dose equivalent estimated for an intervention determines the so-called "ALARA category" (see Fig. 1); the dose equivalent rate at the worksite as well as contamination risks might be used as additional criteria in the definition of the category (see Fig. 2).

Individual dose equivalent	Level I	100 μSv	Level II	1 mSv	Level III
Collective dose equivalent		500 μSv		5 mSv	

Fig. 1. Criteria and threshold values that determine the ALARA category and, thus, the optimization process [5].

Ambient dose equivalent rate	Level I	50 μSv/hr	Level II	2 mSv/hr	Level III
Airborne activity		5 CA		200 CA	
Surface contamination		10 CS		100 CS	

Fig. 2. Criteria that provide further guidance to the definition of the optimization process [5]. The quantities CA (Concentration dans l'Air) and CS (Contamination Surfacique) are guidance levels for airborne and surface contamination defined in the Swiss radiation protection legislation [6] that are adopted by the CERN regulations. For example, the exposure to air with an activity concentration of 1 CA during 2000 hours results in a committed effective dose of 20 mSv. A similar definition exists for 1 CS.

Specific optimization procedures are associated with each ALARA category. For example, Level-II and -III interventions require a detailed work-and-dose planning, a documented optimization process and a formal approval. In addition, Level-III interventions have to be reviewed by an ALARA committee.

In order to evaluate their later impact on the accelerator operation it is useful to consider these ALARA rules already during the design phase for interventions that may lead to considerable individual or collective doses.

Furthermore, CERN defines personal dose objectives for consecutive 12-month periods (presently 3 mSv).[a]

3. The FLUKA Monte Carlo Code for Radiation Protection Studies

The use of general-purpose particle interaction and transport Monte Carlo codes is often the most accurate and efficient choice for assessing radiation protection quantities at accelerators. Due to the vast spread of such codes to all areas of particle physics and the associated extensive benchmarking with experimental data, the modeling has reached an unprecedented accuracy. Furthermore, most codes allow the user to simulate all aspects of a high energy particle cascade in one and the same run: from the first interaction of a TeV particle over the transport and re-interactions (hadronic and electromagnetic) of the produced secondaries, to detailed nuclear fragmentation, the calculation of radioactive decays and even of the electromagnetic shower caused by the radiation from such decays.

FLUKA [7, 8] is a general-purpose particle interaction and transport code with roots in radiation protection studies at high energy accelerators. It therefore comprises all features needed in this area of application:

- Detailed hadronic and nuclear interaction models cover the entire energy range of particle interactions at the LHC, from energies of thermal neutrons to interactions of 7 TeV protons. Moreover, the interface with DPMJET3 [9] also allows the simulation of minimum-bias proton–proton and heavy ion collisions at the experimental interaction points which enormously facilitates calculations of stray radiation fields around LHC experiments.
- Numerous variance reduction techniques are available, among others, weight windows, region importance biasing as well as leading particle, interaction and decay length biasing.
- FLUKA includes unique capabilities for studies of induced radioactivity, especially with regard to nuclide production, their decay and the transport of

---

[a]The values are not to be confused with the dose constraints applied during the design phase as defined in Section 2.2.

residual radiation. Particle cascades by prompt and residual radiation are simulated in parallel based on microscopic models for nuclide production and a solution of the Bateman equations [10] for activity build-up and radioactive decay. The decay radiation and its associated electromagnetic cascade are internally flagged as such in order to distinguish them from the prompt cascade. This allows the user to apply different transport thresholds and biasing options to residual and prompt radiation and to score both independently.

- Particle fluence can be multiplied with energy-dependent conversion coefficients to effective dose or ambient dose equivalent [11] at scoring time. Prompt and residual dose equivalent can thus be computed in three-dimensional meshes, the latter for arbitrary user-defined irradiation and cooling profiles.
- Integral part of the FLUKA code development is benchmarking of new features against experimental data. It includes both the comparison of predictions of individual models to measurement results (e.g., nuclide production cross sections) as well as benchmarks for actual complex situations as, for example, arising during accelerator operation.

4. Benchmark of Radiological Assessments with Measurements

Since the design phase of the LHC the FLUKA code has been extensively used in the assessment of radiological quantities up to ultimate parameters of operation. With the completion of the first operational period in early 2013 and the opening of the experimental detectors for maintenance comprehensive benchmarks of FLUKA predictions are now possible. For a most accurate comparison, FLUKA simulations have been performed for the operational parameters of the past three years, including beam energy, intensity as well as instantaneous and integrated interaction rates in the high-luminosity experiments ATLAS and CMS and annual shutdown periods.

As an example, Fig. 3 shows ambient dose equivalent rates around the ATLAS detector one week after completion of the proton physics run in 2012. As the simulations made use of the rotational symmetry of the detector around the beam axis, only its upper half is displayed. Dose rates are well below natural background level outside of the detector while they increase towards the beam axis ($r = 0$ in Fig. 3) reaching several hundreds of μSv/h at the most radioactive locations.

After several weeks of cooling the so-called "forward shielding" around the beam-pipes was opened in order to allow for maintenance and repair interventions. Detailed radiation surveys accompanied each step of the shielding removal and allowed first benchmarks of FLUKA simulations. Figure 4 shows the locations (labeled "1–5") for which measurements were compared to FLUKA predictions. The numerical values are presented in Table 1 and demonstrate remarkable agreement taking into account measurement uncertainties as well as geometrical approximations in the calculations.

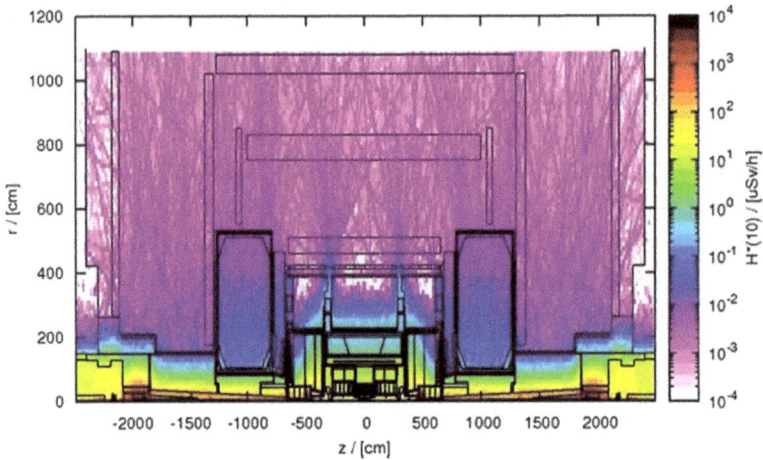

Fig. 3. Ambient dose equivalent rates (H*(10)) in μSv/h around the ATLAS detector one week after completion of the proton physics run in 2012.

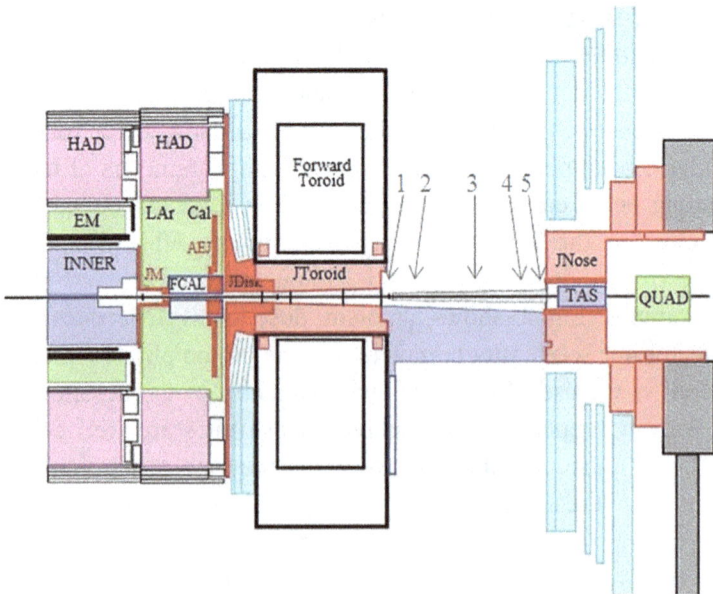

Fig. 4. Schematic representation of the ATLAS detector. Measurement locations are labelled by "1–5" (see also Table 1).

Table 1. Comparison of measured and calculated ambient dose equivalent rates along the ATLAS beam pipe at the locations as indicated in Fig. 4. The uncertainties indicated for the FLUKA results include statistical errors only.

Location	Measurement [μSv/h]	FLUKA [μSv/h]
1	19	13 ($\pm$0.3)
2	10	13 ($\pm$0.3)
3	7.2	10 ($\pm$0.2)
4	47	46 ($\pm$0.5)
5	42	72 ($\pm$0.5)

Similarly, detailed calculations were performed for the CMS detector and adjacent beam-line elements. Figure 5 shows the calculated ambient dose equivalent around the TAS absorber and Q1 magnet for one week of cooling after the LHC operation with proton beam in 2012. They can be compared to the radiation survey results presented for the long-straight section at LHC Point 5 (LSS5) in Fig. 6. The two measurement locations close to the TAS and inner triplet magnets showing values of 85 μSv/h and 20 μSv/h, respectively, are marked in Fig. 5 with crosses. From the color coding of the dose equivalent rates it can be seen that the simulations predict dose rates that agree well with those measured.

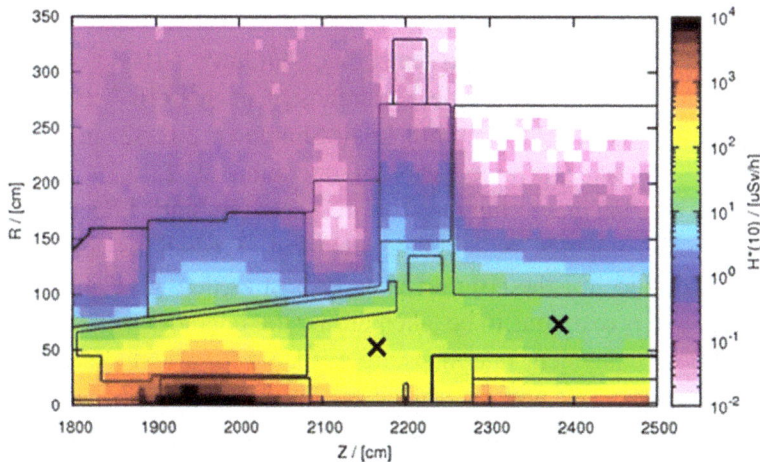

Fig. 5. Ambient dose equivalent rates (H*(10)) in μSv/h around the TAS absorber and Q1 magnet adjacent to the CMS experiment one week after completion of the proton physics run in 2012.

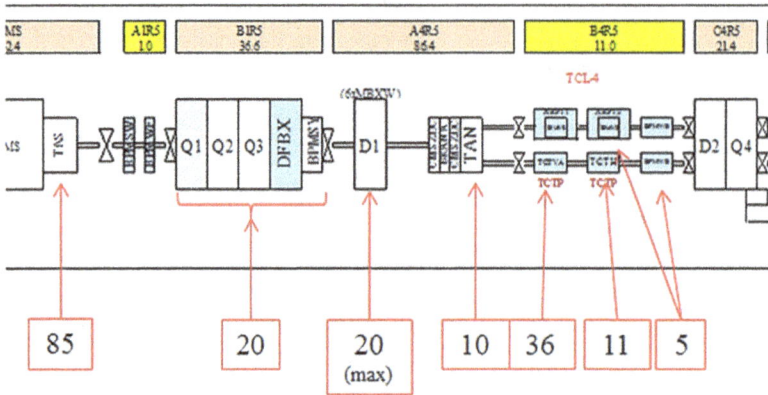

Fig. 6. Ambient dose equivalent rates in μSv/h measured in LSS5 at 40 cm distance to the respective components one week after completion of the proton physics run in 2012.

5. Estimation of Residual Dose Rates Around ATLAS Until LS3

The above mentioned comparisons between FLUKA results and radiation survey measurements support the reliability of predictions for future operational periods. Simulations have been performed for both ATLAS and CMS until Long Shutdown 3 (LS3), presently planned for the year 2022. As an example, the present Chapter presents the evolution of residual dose rates around the ATLAS experiment.

It has been assumed that the present Long Shutdown 1 (LS1) is followed by three years of operation with a peak luminosity of 1.0×10^{34} cm^{-2}s^{-1} and an integrated luminosity of 134 fb^{-1} until Long Shutdown 2 (LS2) in 2018. LS2 is then followed by another three years of operation at 2.0×10^{34} cm^{-2}s^{-1} and 249 fb^{-1} peak and integrated luminosities, respectively, until LS3 in 2022 when the HiLumi LHC upgrade will be implemented. Predictions for LS3 are of particular importance for the HiLumi LHC Project as they give an indication of waiting times and precautions to be taken during the upgrade.

Ambient dose equivalent rate maps are available for a wide range of cooling times between one week and several years; Figs. 7 and 8 show them for four months of cooling after the last high luminosity proton operation in the years 2017 and 2021, respectively. Calculating ratios between dose equivalent rates to be expected during the different LS for identical locations yields an increase of dose rates by a factor of about 4 until LS2 and by about a factor of 8.5 from LS1 until LS3. It follows that dose rates close to the beam-pipe or interaction point may reach several mSv/h until the installation of the HiLumi LHC upgrade which has to be taken into consideration in the design and planning.

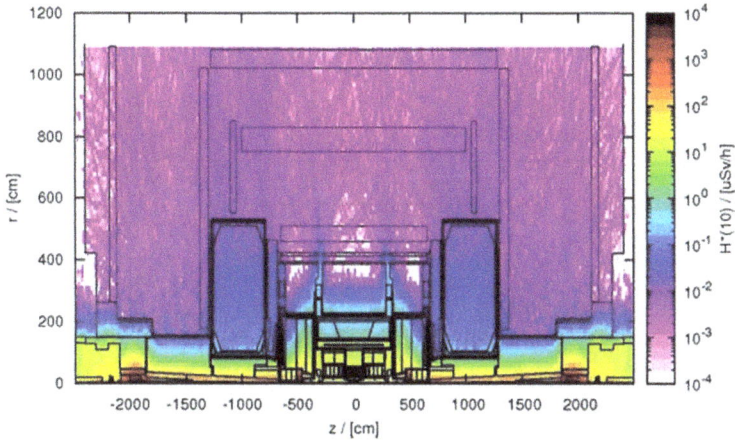

Fig. 7. Ambient dose equivalent rates (H*(10)) in μSv/h around the ATLAS detector four months after completion of the proton physics run in 2017 (Long Shutdown 2).

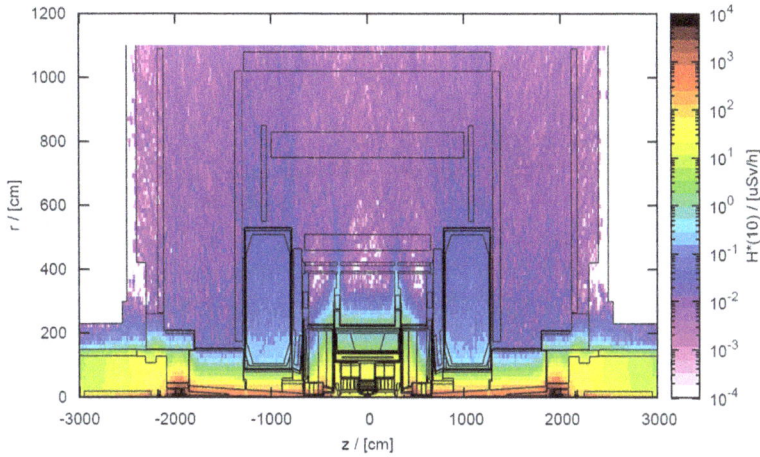

Fig. 8. Ambient dose equivalent rates (H*(10)) in μSv/h around the ATLAS detector four months after completion of the proton physics run in 2021 (Long Shutdown 3).

The FLUKA simulations for different cooling times indicate a decrease of residual dose rates due to radioactive decay by a factor of two from one month to four months of cooling and by another factor of two from four months to one year of cooling (see Table 2). Dose rate maps, as shown in this Chapter, along with intervention scenarios can be used to compute individual and collective doses that can then be compared to above mentioned design constraints. It gives an indication of the required cooling time before the intervention can start and may trigger design modifications for fast or remote handling if the work would otherwise be impossible during the planned shutdown period.

Table 2. Scaling factors for residual dose rates around the ATLAS detector in LS3. The factors have been obtained by dividing the dose rate at the respective cooling time by the dose rate at one month cooling.

Cooling time	Scaling factor
1 week	1.6
1 month	1.0
4 months	0.47
6 months	0.35
1 year	0.2

Furthermore, the methodology of residual dose rate and job dose predictions with FLUKA will be extended to operational periods beyond LS3 as soon as a first upgrade design for the detectors and adjacent beam-line components is available.

References

[1] Safety Code F, Radiation Protection, CERN (2006).
[2] M. Brugger *et al.*, The estimation of individual and collective doses for interventions at the LHC beam cleaning insertions, in *Proceedings of the LHC Project Workshop*, Chamonix XIV, 17–21 January 2005 (2005).
[3] H. Vincke and C. Theis, ActiWiz – optimizing your nuclide inventory at proton accelerators with a computer code, in *Proceedings of the ICRS-12 & RPSD 2012 Conference*, Nara, Japan, 2–7 September 2012, Progress in Nuclear Science and Technology, Volume 4 (Atomic Energy Society of Japan, 2014), pp. 228–232.
[4] General Safety Instruction, *ALARA criteria and requirements applicable to interventions*, RGE section 9/S5-GSI1, EDMS No. 810176 (2006).
[5] ALARA Review Working Group, *Final Report*, EDMS No. 1244380 (2012).
[6] Swiss Radiation Protection Ordinance, revision 1 January 2009.
[7] A. Fassò, A. Ferrari, J. Ranft, and P. R. Sala, FLUKA: a multi-particle transport code, CERN-2005-10 (2005), INFN/TC_05/11, SLAC-R-773.
[8] G. Battistoni, S. Muraro, P. R. Sala, F. Cerutti, A. Ferrari, S. Roesler, A. Fassò, J. Ranft, The FLUKA code: Description and benchmarking, in *Proceedings of the Hadronic Shower Simulation Workshop 2006*, Fermilab 6–8 September 2006, M. Albrow, R. Raja eds., *AIP Conference Proceeding* **896**, 31–49 (2007).
[9] S. Roesler, R. Engel, J. Ranft, The Monte Carlo event generator DPMJET-III, in *Proceedings of the Monte Carlo 2000 Conference*, Lisbon, 23–26 October 2000, A. Kling, F. Barao, M. Nakagawa, L. Tavora, P. Vaz eds. (Springer-Verlag Berlin, 2001) pp. 1033–1038.

[10] H. Bateman, Solution of a system of differential equations occurring in the theory of radioactive transformations, *Proc. Cambridge Philos. Soc.* **15**, 423–427 (1910).

[11] M. Pelliccioni, Overview of fluence-to-effective dose and fluence-to-ambient dose equivalent conversion coefficients for high energy radiation calculated using the FLUKA code, *Radiation Protection Dosimetry* **88**, 279–297 (2000).

Chapter 12

Machine Protection with a 700 MJ Beam

T. Baer[1], R. Schmidt[1], J. Wenninger[2], D. Wollmann[1] and M. Zerlauth[1]

[1]CERN, TE Department, Genève 23, CH-1211, Switzerland
[2]CERN, BE Department, Genève 23, CH-1211, Switzerland

After the high luminosity upgrade of the LHC, the stored energy per proton beam will increase by a factor of two as compared to the nominal LHC. Therefore, many damage studies need to be revisited to ensure a safe machine operation with the new beam parameters. Furthermore, new accelerator equipment like crab cavities might cause new failure modes, which are not sufficiently covered by the current machine protection system of the LHC. These failure modes have to be carefully studied and mitigated by new protection systems.

Finally the ambitious goals for integrated luminosity delivered to the experiments during the era of HL-LHC require an increase of the machine availability without jeopardizing equipment protection.

1. Introduction

The combination of high intensity and high energy that characterizes the nominal beam in the LHC leads to a stored energy of 362 MJ in each of the two beams. This energy is more than two orders of magnitude larger than in any previous accelerator — and about to increase by another factor of two following the luminosity upgrade for the LHC as shown in the comparisons of Fig. 1. With intensities expected to increase up to $2.3 \cdot 10^{11}$ p/bunch with 25 ns bunch spacing, respectively to $3.7 \cdot 10^{11}$ p/bunch with 50 ns bunch spacing [1], an uncontrolled beam loss at the LHC could cause even more severe damage to accelerator equipment than for today's nominal beam parameters. Recent simulations that couple energy-deposition and hydro-dynamic simulation codes show that already the nominal LHC beam can penetrate through the full length of a copper block 20 m long, in case the entire beam is deflected accidentally. Such an accident could happen if the beam extraction kickers deflect the beam by a wrong angle. Hence, it becomes necessary to revisit many of the damage studies in light of the new beam parameters [2]. In addition, new failure scenarios will have to be considered

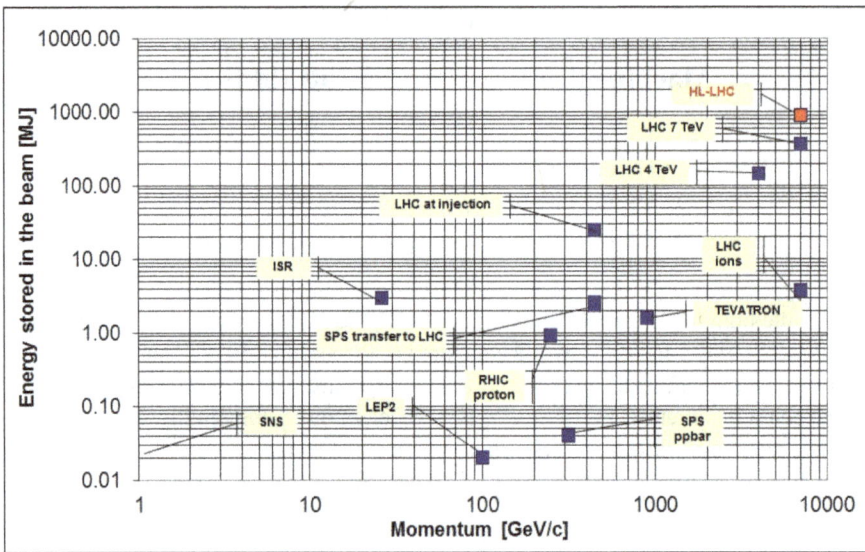

Fig. 1. Stored beam energy as a function of beam momentum of HL-LHC in comparison with other particle accelerators.

following the proposed optics changes and the installation of new accelerator components such as crab cavities and hollow electron beam lenses. Special care is required to find a trade-off between equipment protection and machine availability in view of the reduced operational margins (e.g. decreasing quench limits and beam loss thresholds versus increased beam intensity and tighter collimator settings, UFOs at higher energies and reduced bunch spacing, ...).

2. Present Performance of LHC Machine Protection and Future Challenges with HL-LHC Beams

Safe operation of the LHC currently relies on a complex system of equipment protection. The machine protection system (MPS) is designed for preventing the uncontrolled release of energy stored in the magnet system and beam-induced damage with very high reliability. An essential part of the MPS system, the active protection system, is the early detection of failures within the equipment, as well as monitoring of the beam parameters with fast and reliable beam instrumentation. This is required throughout the entire cycle, from injection to collisions. Once a failure is detected by any of the protection systems, the information is transmitted to the beam interlock system that triggers the extraction of the particle beams by the LHC beam dumping system. It is essential that the beams are always properly extracted from the accelerator via 700 m long transfer lines into large graphite dump blocks, as these are the only elements of the LHC that can withstand the impact of the full beams.

The current machine protection architecture is based on the assumption of three types of failure scenarios [3], namely:

- Ultra-fast failures: failures within three turns, e.g., during beam transfer from the SPS to the LHC, beam extraction into the LHC beam dump channel or the effect of missing beam–beam deflection during beam extraction (1 LHC turn = 89 μs).
- Fast failures: timescale of several LHC turns (< few milliseconds) as a result of certain equipment failures with fast effect on particle trajectories.
- Slow failures: multi-turn failures on timescales $\geq$ few milliseconds, e.g. powering failures, magnet quenches, RF failures, …

2.1. *Ultra-fast failures*

Failures occurring on the timescale of a single turn cannot be mitigated by active protection systems, but require a protection of the vacuum chamber and the accelerator equipment in the vicinity (magnets, cryogenics, instrumentation, …) by passive protection elements such as collimators and absorbers. In view of the increase in beam energy both for the injected as well as the circulating beams, several consolidation programs are already under way to upgrade the critical elements for injection protection (TDI), dump protection (TCDQ) as well as the LHC collimation system. Several promising novel materials, such as Copper-Diamond, are currently being tested to replace the existing jaws of tertiary collimators, with the aim of rendering them more robust for beam impact in case of asynchronous dumps. The jaws of other collimators could also profit from such materials. Several of the new materials have the additional advantage of reducing the impedance contribution of the collimator jaws, hence having a beneficial effect on beam stability. The simultaneous integration of button pickups into the new collimator jaws will allow for a more accurate, quicker and dependable positioning of the collimator jaws around the beam axis. This will allow maintaining the protection of the aperture while reaching smaller values of β^* in the high luminosity insertions. Operating with reduced β^* requires tighter settings of all LHC collimators with respect to the current operation.

An example of a single turn beam loss mechanism is the absence of the beam–beam deflection due to the non-simultaneous removal of the two LHC beams. Trajectory perturbations of the remaining LHC beam by as much as 230 μm = $0.60\sigma_{nom}$ within a single turn have been measured and are in good agreement with simulations, as illustrated in Fig. 2 [4]. When extrapolating the simulations to HL-

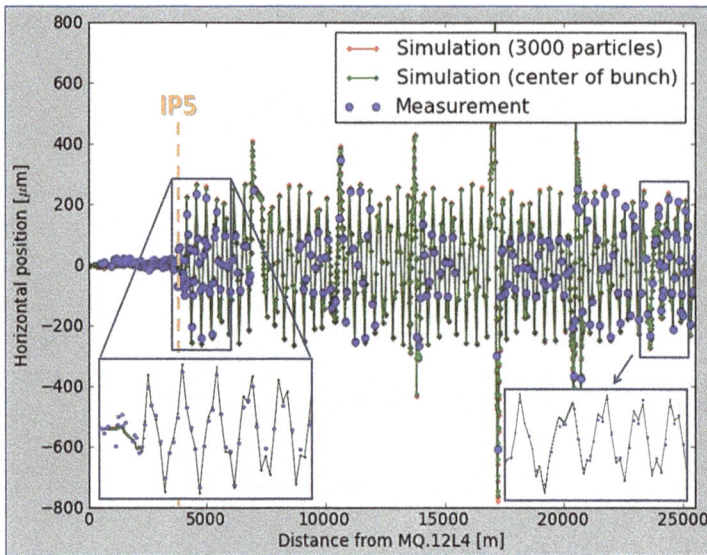

Fig. 2.   Horizontal trajectory perturbation of Beam1 as measured by the beam position monitors in the LHC ring (blue) and as predicted by simulation (red, green) in the turn directly after the Beam2 dump kickers were fired. The measurement is for bunches with full long-range encounters. Measurement on 13.12.2012 08:26:54. Beam energy: 4 TeV, bunch intensity: $0.9 \cdot 10^{11}$ protons, 84 bunches per beam, 25 ns bunch-spacing, crossing angle in IP5: $\approx 68 \,\mu$rad [4].

LHC beam parameters, the perturbation amplitudes due to this effect are expected to increase up to $0.9\sigma_{\mathrm{nom}}$–$1.1\sigma_{\mathrm{nom}}$. This displacement will lead to beam losses around the machine, namely at the primary collimators of IR7. With the present transverse beam distribution, such single turn trajectory perturbations with amplitudes of about $1\sigma_{\mathrm{nom}}$ can lead to beam losses far beyond the specifications of the collimation system and hence imply a significant damage potential.

Since this effect occurs regularly during normal LHC operation, a fast and reliable diagnostics (and interlocking) of the transverse tail population is essential for safe operation in the HL-LHC era. Such a diagnostics could, e.g., be based on the synchrotron light monitor (BSRT) and related studies are strongly encouraged. Furthermore a depletion of the transverse tail population to reduce the number of protons in the beam halo to an acceptable level may be required. A hollow electron-lens [5] would provide this functionality. Dedicated studies are presently on-going and are strongly supported [6].

2.2. Fast failures

Equipment failures or beam instabilities appearing on the timescale of multiple turns allow for dedicated protection systems to mitigate their effects on the circulating beams. The LHC Beam Loss Monitoring system (BLM) features the fastest failure detection time of 40 μs as illustrated in the comparison in Fig. 3. The BLM system is complemented with fast interlocks on the beam position in IR6, Fast Magnet Current Change Monitors and a beam lifetime monitor (currently under development by the beam instrumentation group at CERN). All of these systems feature similar failure detection times in the 100 μs–1 ms range, providing diverse redundancy to the BLM system.

Adding the additional time required to transmit the detected failure through the LHC beam interlock system, the time required to synchronize the firing of the beam dump kickers with the abort gap as well as the time needed to completely extract the beam from the LHC leads to an equivalent worst case MPS response time of three LHC turns as depicted in Fig. 4.

This reaction time is sufficient in the absence of failures occurring on timescales below some 10 LHC turns. The basis for the design of the current MPS system to date has been a failure of the normal conducting separation dipole D1 in

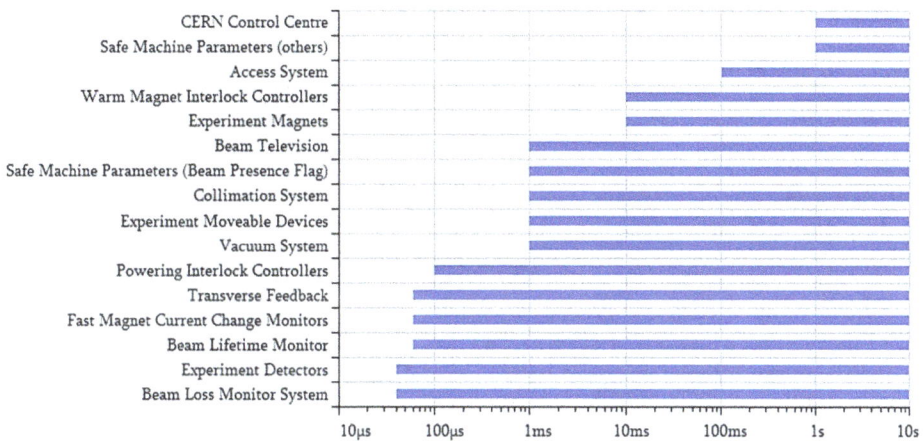

Fig. 3. Failure detection times at the LHC. Shortest failure detection time currently assured by the BLM system, with a fastest integration time of 40 μs = half a LHC turn.

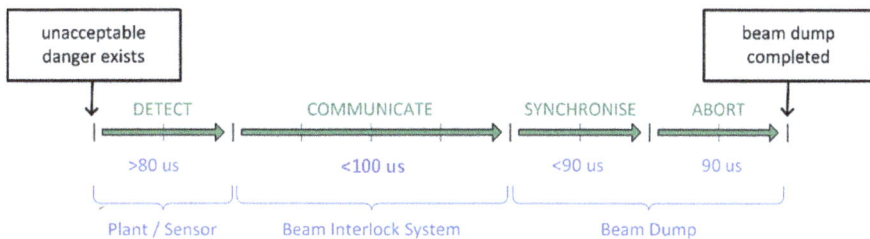

Fig. 4. Current MPS response time from failure detection to completion of beam dump.

P1 and P5 [7], considered the fastest possible failure with circulating beam. These normal conducting magnets induce, due to their location in areas with high beta functions and low time constants, fast changes of the particle trajectory in case of magnet powering failures, which in turn lead to rapidly increasing beam losses on the primary collimators in IR7. At nominal energy and intensity these losses can reach the damage level of collimators within several tens of turns only, hence a dedicated protection system — the so-called Fast Magnet Current Change Monitors (FMCM) — has been very successfully deployed on critical magnets in the LHC and its transfer lines in 2006 [8].

With the HL-LHC upgrade, the optics in the insertion regions will significantly change and the β-function at the D1 separation dipole magnets in IR1 and IR5 will increase up to $\sim 17,000$ m for certain ATS optics. At the same time a replacement of the D1 separation dipole magnets by superconducting magnets is currently considered the baseline for HL-LHC, which would significantly increase the time constants of these circuits, practically mitigating the potential of fast failures originating from these magnets.

In case the D1 separation dipole magnets remain normal-conducting, the increased β-functions imply an increased sensitivity of the beam to corresponding current changes. The expected orbit deviation in the arc would increase within a few tens of turns to $\Delta x_{\max} \sim 230$ μm $\sim 0.43\sigma_{\text{nom}}$ for a current change of 100 mA, i.e. about 25% more than in 2012 stable beams conditions. This increased sensitivity is still well within the operational reach of the present FMCM system.

For HL-LHC operation, the use of crab cavities will introduce failures that can affect the particle beams on timescales well below the fastest failures considered so far [9]. Studies of different failure scenarios are still underway. These studies require considering details of the design finally to be adopted for the crab cavity and the corresponding low-level RF system. Both have a significant impact on the effect on the circulating beams following, e.g., cavity quenches or trips of the RF power generator. In addition detailed measurements of the quench and failure behaviour of the chosen design have to be conducted. First experience with similar devices at KEK however shows that certain failures can happen within very few turns only as depicted in Fig. 5.

While the protection against failures with time constants > 15 ms is not expected to be of fundamental concern, voltage and/or phase changes of the crab cavities will happen with a time constant τ which is proportional to the Q_{ext}. For a 400 MHz cavity with a $Q_{\text{ext}} = 1\text{E}6$ this will result in a time constant as low as 800 μs. The situation becomes even more critical for cavity quenches, where the energy stored in the cavity can be dissipated in the cavity walls on ultra-fast timescales. Quenches observed in cavities at KEKB show a complete decay of the

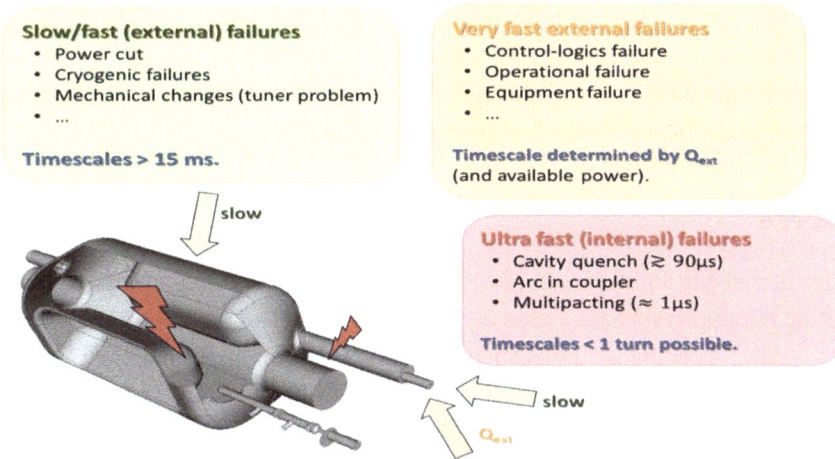

Fig. 5. Schematic overview of crab cavity failure categories [9].

cavity voltage in 100 μs, accompanied by an oscillation of the phase by 50 degrees in only 50 μs. Such crab cavity failures can imply large global betatron oscillations, which could lead to critical beam losses for amplitudes above about $1\sigma_{nom}$. Highly overpopulated transverse tails compared with Gaussian beams were measured in the LHC. Based on these observations the energy stored in the tails beyond 4σ are expected to correspond to ~ 30 MJ for HL-LHC parameters. These levels are significantly beyond the specification of the collimation system of up to 1 MJ for very fast accidental beam losses.

Mitigation techniques hence have to include a fast, dependable and redundant detection and interlocking of a crab cavity failure on these timescales as well as taking appropriate measures when designing the cavity and associated RF control to increase as much as possible the failure time constants, namely:

- Avoid correlated failures of multiple cavities (on one side of an IP) through mechanical and cryogenic separation of the individual modules and appropriate design of the low-level RF [10].
- Investigate the use of fast failure detection mechanisms such as RF field monitor probes, diamond beam loss detectors, power transmission through input coupler and head-tail monitors.
- Ensure the depletion of the transverse beam tails to reduce the energy stored in the beam halo which would potentially be deflected onto the collimation system beyond the design value of 1 MJ. For the current baseline this would correspond to an area of $1.7\sigma_{nom}$ (before reaching the closest primary collimator) as the possible transverse beam trajectory perturbation following an ultra-fast failure of a single crab cavity.

- Decrease the reaction time of the MPS system for such ultra-fast failures by, e.g., increasing the number of abort gaps, accepting to trigger asynchronous beam dumps with potential local damage, add direct beam dump links to IR6 and consider the installation of disposable absorbers.

2.3. UFOs

Besides increasing failure rates of equipment systems when approaching nominal operating parameters, beam losses due to macro particles interacting with the LHC beams could become a performance limitation for nominal and HL-LHC operation. These so-called (un)identified falling objects (UFOs) are most likely μm sized dust particles which lead to very fast beam losses on timescales of a few tens of turns (~ 1 ms) when interacting with the beams. Following their first identification in July 2010, UFOs have led during the first three years of operational running to 58 premature beam dumps following beam losses above the dump thresholds. As illustrated in Fig. 6 the occurrence of UFOs is, apart for some outliers around the MKIs, distributed all around the LHC ring.

While experience shows a conditioning effect of the UFOs along a machine run and a saturating effect on beam intensity above several hundred bunches, beam losses due to UFOs will most likely increase with beam energy. Extrapolating the current observations to higher beam energies leads to the prediction of expected beam dumps depicted in Fig. 7. Moreover, during initial operation with 25 ns

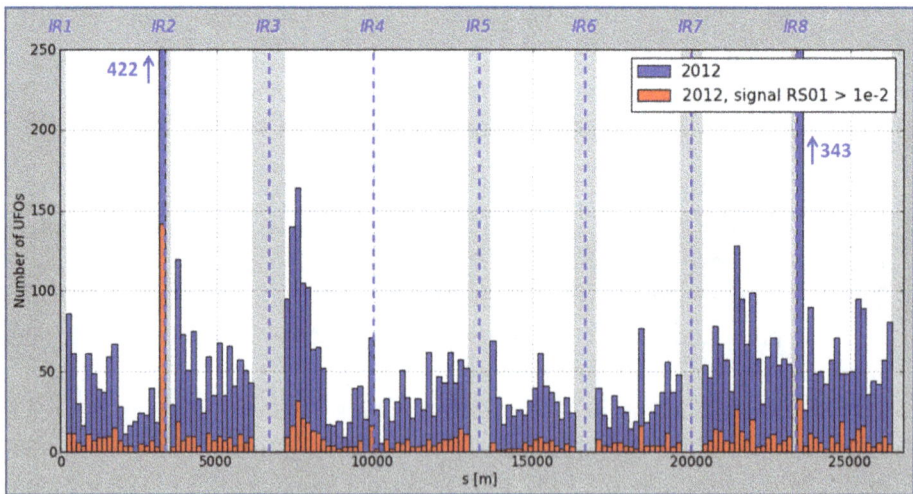

Fig. 6. Spatial distribution of UFO occurrence around the LHC ring (A total of 7171 UFOs at 4 TeV in 2012 as shown with the blue bars, the red bars show the UFOs for which the BLM signal in the first running sum is exceeding 1% of the dump threshold).

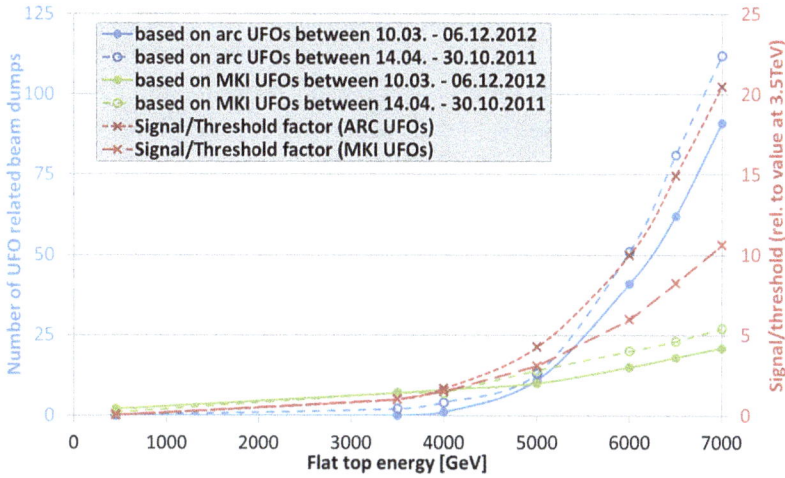

Fig. 7.   Expected number of beam dumps by arc and MKI UFOs and the expected scaling of the BLM signal versus threshold as a function of the beam energy. All extrapolations are done relative to the BLM dump thresholds as used for 3.5 TeV operation. Note that at higher energies the BLM thresholds would have to be further decreased to maintain the same level of protection for the superconducting elements.

bunch spacing their occurrence drastically increased up to a level unacceptable for machine operation. For these reasons it is of primary importance to further study the origin and consequences of UFOs to fully understand the limitations they might impose on future operation.

While for the MKI magnet a dedicated consolidation program is already underway during LS1, the main mitigation strategy for the arc UFOs remains the increase of the BLM thresholds towards the magnet quench limit and to profit from the conditioning effect. An additional relocation of BLM monitors from the quadrupole to the dipole magnets will allow for better protection against UFO events whilst maintaining the required BLM thresholds well within a sensible operational range.

A complementary approach is to add a few bunches with large emittance to the filling scheme for UFO detection well before the macro particle reaches the centre of the beam. This could allow a detection of very fast UFO events at higher energies in time to dump the beam before the beam losses exceed the magnet quench margin.

2.4. *Slow failures*

Failures on timescales beyond the millisecond range are not expected to significantly impact the machine protection considerations for the HL-LHC era. In several domains, machine protection considerations will however become an increasing challenge for machine availability. The enhanced luminosity of the HL-LHC will increase the radiation levels in certain underground areas like the RR and the dispersion suppressors to levels no longer compatible with the operation of radiation tolerant electronics based on components of the shelf (COTS) installed at present in those areas for the quench protection system (QPS). Based on the progress in electronics, it is probably feasible to re-locate a major part of such protection electronics to low radiation zones or eventually to surface buildings and use long instrumentation cables to link to the protected elements. In addition, equipment such as the new superconducting elements and superconducting links in the high luminosity insertions will require new dedicated protection systems, which are expected to be based on similar principles as already in use for the current LHC magnets and busbars.

Machine availability will be furthermore impacted by the defined quench thresholds of magnets, which will be significantly lower for higher beam energies.

The corresponding reduction of BLM thresholds will represent an increasing challenge in light of higher beam intensities and tighter collimator settings expected for HL-LHC operation. The potential absence of the transverse beam tails in case a hollow electron beam lens will be installed would in addition have a detrimental effect on the efficiency and latency of the BLM system to detect beam losses in the machine. This effect could be mitigated by leaving a low density of particles in the beam tails for detection and machine protection purposes or by leaving the halo along short lengths of the beam.

With the availability of equipment systems and the control of the particle beams already today playing a major role in the physics output of the LHC, the tighter operational margins in the coming years will certainly require to operate the machine with more flexible interlock conditions and less conservative protection thresholds. This has been confirmed by recent simulations which show that, assuming today's operational cycle and availability of the LHC systems, the ambitious goals of integrated luminosity for the HL-LHC cannot be met without a significant increase in machine availability [11].

References

[1] O. S. Brüning, HL-LHC parameter space and scenarios, in *Proceedings of Chamonix 2012 Workshop on LHC Performance*, CERN-2012-006, p. 315.
[2] N. A. Tahir, J. B. Sancho, A. Shutov, R. Schmidt and A. R. Piriz, Impact of high energy high intensity proton beams on targets: Case studies for Super Proton Synchrotron and Large Hadron Collider, *Phys. Rev. ST ACcel. Beams* **15**, 051003 (2012).

[3] R. Filippini, B. Goddard, M. Gyr, V. Kain, R. Schmidt, J. Uythoven and J. Wenninger, Possible causes and consequences of serious failures of the LHC machine protection systems in *Proceedings of 9th European Particle Accelerator Conference*, Lucerne, Switzerland, 5–9 July 2004, pp. 620 (2004).

[4] T. Baer, *Very Fast Losses of the Circulating LHC Beam, their Mitigation and Machine Protection*, PhD Thesis, CERN, University of Hamburg, 2013, CERN-THESIS-2013-233.

[5] G. Stancari *et al.*, Collimation with hollow electron beams, *Phys. Rev. Lett.* **107**, 084802 (2011).

[6] S. Redaelli *et al.*, in *Proceedings of Internal Review of Tevatron Hollow E-Lens Usage at CERN*, Nov. 2012.

[7] V. Kain and R. Schmidt, *Machine Protection and Beam Quality during the LHC Injection Process*, Ph.D. dissertation, Technical University Vienna/CERN, Oct. 2005, CERN-THESIS-2005-047.

[8] M. Werner, *Einrichtung zur Bestimmung der Starke des Magnetfeldes eines Elektromagneten*, Patent DE102005045537B3 28.12.2006, Sep. 2005.

[9] T. Baer *et al.*, Very fast crab cavity failures and their mitigation, in *Proceedings of IPAC'12*, May 2012, pp. 121–123, MOPPC003.

[10] P. Baudrenghien, LLRF for crab cavities, presentation at the 2nd HiLumi LHC/LARP Meeting, INFN Frascati, Nov. 2012.

[11] A. Apollonio *et al.*, HL-LHC: Integrated luminosity and availability, in *Proceedings of IPAC'13*, pp. 1352–1354, TUPFI012.

Chapter 13

Cleaning Insertions and Collimation Challenges

S. Redaelli[1], R. B. Appleby[2], A. Bertarelli[3], R. Bruce[1], J. M. Jowett[1],
A. Lechner[3] and R. Losito[3]

[1]*CERN, BE Department, Genève 23, CH-1211, Switzerland*
[2]*University of Manchester and the Cockcroft Institute, UK*
[3]*CERN, EN Department, Genève 23, CH-1211, Switzerland*

High-performance collimation systems are essential for operating efficiently modern hadron machine with large beam intensities. In particular, at the LHC the collimation system ensures a clean disposal of beam halos in the super-conducting environment. The challenges of the HL-LHC study pose various demanding requests for beam collimation. In this paper we review the present collimation system and its performance during the LHC Run 1 in 2010–2013. Various collimation solutions under study to address the HL-LHC requirements are then reviewed, identifying the main upgrade baseline and pointing out advanced collimation concept for further enhancement of the performance.

1. Present LHC Collimation

In this introductory section the present LHC collimation system is reviewed. Its main features are presented and the highlight performance achievements during the LHC Run 1 are recalled. The possible limitations and challenges for the collimation in the HL-LHC era are then discussed.

1.1. *Introduction to LHC multi-stage collimation*

The LHC collimation system is designed to safely dispose of beam losses in order to reduce the risk of quenches of superconducting magnets and damage of accelerator components. The cleaning functionality is required in case of unavoidable transverse betatron — as well as off-momentum losses. For this purpose, two dedicated warm LHC insertions address betatron (IR7) and momentum (IR3) cleaning separately [1]. A very efficient halo cleaning, as required to operate with unprecedented stored beam energies in a supercon-ducting collider, is achieved through a multi-stage cleaning. This is illustrated

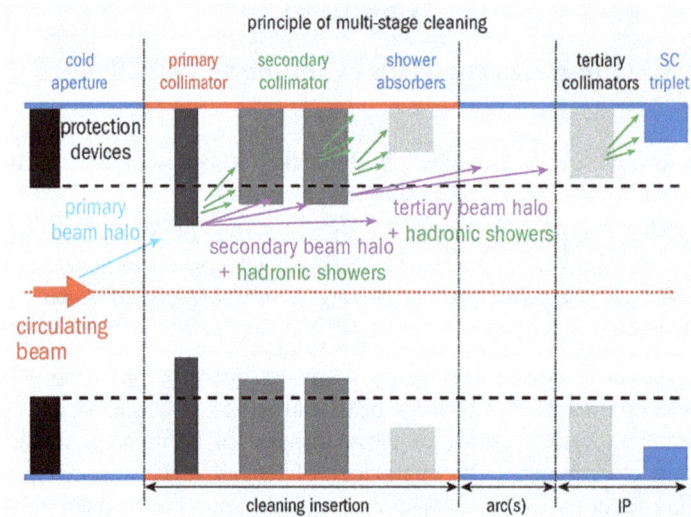

Fig. 1. Illustrative scheme of the multi-stage collimation cleaning at the LHC. Primary and secondary collimators (darkest grey) are the devices closest to the circulating beam and are made of robust carbon-fiber composites. Shower absorbers and tertiary collimators (lighter grey) sit at larger apertures and are made of a Tungsten alloy to improve absorption. Collimators of different families are ordered in a pre-defined collimation hierarchy that must be respected in order to ensure the required system functionalities. The collimator hierarchy is ensured by defining collimator settings in units of local beam size at the collimator location.

schematically in Fig. 1. The present system deployed for the LHC operation between 2010 and 2013 is designed to provide a cleaning efficiency above 99.99% [2], i.e. to ensure that less than 10^{-4} of the beam losses is lost in superconducting magnets. The collimation system includes 43 movable ring collimators per beam. The complete list including injection protection collimators in the transfer lines (built within the LHC collimation project) is given in Table 1. For completeness, the injection protection TDI blocks and the one-side beam dump collimator TCDQ are also listed (see Chapter 19). The full system comprises 108 collimators, 100 of which are movable.

Beam halo collimation is achieved by placing very precisely blocks of materials close to the circulating beams, while respecting a pre-defined collimator hierarchy that ensures optimum cleaning in a multi-stage collimation process. The list of collimator families with the main parameters such as material, orientation and total number of devices is given in Table 1. Since the collimator "jaws" sit close to the beam (e.g., the minimum collimator gap in 2012 was 2.1 mm, i.e. jaws were 1.05 mm apart from the circulating beam), the collimation system also has a critical role in the passive machine protection in case of beam failures that cannot be counteracted by active systems (see

Chapter 12). Primary and secondary collimators in IR7 are the closest to the beam and are made of robust carbon-fiber composites (CFC) that withstand the most critical failures. However they contribute significantly to the machine impedance because of the low conductivity of the CFC and this determines the smallest gaps that can be used. Other absorbers and tertiary collimators sit at larger gaps in beam size unit. They can be less robust than primary and secondary collimators because they are less exposed to beam losses. Thus, metal-based jaws with a higher stopping power can be used. The operation of LHC Run 1 proved that the collimation hierarchy constrains the LHC performance in terms of minimum achievable β^*, determined by the maximum aperture that can be protected [3]. For example, in 2012 the minimum machine aperture in the triplet magnets was about 11σ for a β^* of 60 cm. The betatron cuts from the collimators (see Fig. 1) ranged from 4.3σ (primary cut) to 9σ (tertiary cut). This 2σ is required to ensure adequate magnet protection in presence of transient orbit and optics drifts in the IRs. The momentum cut from the IR3 collimators was 0.2% for the reference particle with zero betatron amplitude.

In addition to the beam halo cleaning, the collimation system has also other important roles that can be summarized as follows:

- **Passive machine protection**: the collimators are the closest elements to the circulating beam and represent the first line of defence in case of various normal and abnormal loss cases, see also Chapter 12. Due to the damage potential of the LHC beams, this functionality has become one of the most critical aspects for the LHC operation and commissioning. In particular, it must be ensured that the triplet magnets in the experiments are protected during the betatron squeeze [3].
- **Active cleaning of collision debris products**: this is achieved with dedicated (TCL) collimators located on the outgoing beams of each high-luminosity experiment that catch the debris produced by the collisions keeping losses below the quench limit of the superconducting magnets in the matching sections and dispersion suppressors around the interaction points.
- **Experiment background optimization**: this is one of the classical roles of collimation systems in previous colliders like ISR, SppS and Tevatron. For the LHC, the contribution to background from beam halo has always been expected to be small, thanks to the good IR7 collimation cleaning that induces only limited losses close to the experiments. The initial run confirmed this expectation [4].
- **Concentration of radiation losses**: for high power machines, it is becoming increasingly important to be able to localize beam losses in confined and optimized "hot" areas rather than having a distributed activation of equipment

along the machine. This is an essential functionality to allow easy access for maintenance in the largest fraction of machine.

- **Local protection of equipment and improvement of lifetime**: Dedicated movable or fixed collimators are used to shield equipment. For example, eight passive absorbers are used in the collimation insertions in order to reduce total doses to warm dipoles and quadrupoles that otherwise would have a short lifetime in the high-radiation environment foreseen during the nominal LHC operation.

- **Beam halo scraping and halo diagnostics**: Collimator scans in association to the very sensitive LHC beam loss monitoring system proved to be a powerful way to probe the population of beam tails [5, 6], otherwise too small compared to the beam core to be measured by conventional emittance measurements. Thanks to their robustness, the present primary collimators can also be efficiently used to scrape and shape the beams, like in [7].

In order to fulfil all these functionalities, the LHC collimation system features an unprecedented complexity compared to previous state-of-the-art in particle accelerators. Table 1 (right) lists, for example, the degrees of freedom for collimator movements and the number of interlock functions of the 2012 system [8]. As a comparison, the Tevatron collimation system had less than 30 degrees of freedom. For this reason, the possibility to operate reliably the collimation system has always been considered as a major concern for the LHC performance.

Table 1. Left: Collimators for the LHC Run 1 in 2010–2013. For each type, acronyms, rotation plane (horizontal, vertical or skew), material and number of devices are given. Right: Various degree of freedom for collimation movements as deployed for the LHC operational cycle in 2012–13. About 400 motors are moved in discrete steps or according to functions of time in order to ensure optimum collimator settings in all phases of the operational cycle.

Functional type	Name	Plane	Num.	Material
Primary IR3	TCP	H	2	CFC
Secondary IR3	TCSG	H	8	CFC
Absorbers IR3	TCLA	H,V	8	W
Primary IR7	TCP	H,V,S	6	CFC
Secondary IR7	TCSG	H,V,S	22	CFC
Absorbers IR7	TCLA	H,V	10	W
Tertiary IR1/2/5/8	TCT	H,V	16	W/Cu
Physics absor. IR1/5	TCL	H	4	Cu
Dump protection IR6	TCSG	H	2	CFC
	TCDQ	H	2	C
Inj. prot. (lines)	TCDI	H,V	13	CFC
Inj. prot. IR2/8	TDI	V	2	C
	TCLI	V	4	CFC
	TCDD	V	1	CFC

Parameters	Number
Movable collimators in the ring	85
Transfer line collimators	13
Stepping motors	392
Resolvers	392
Position/gap measurements	584
Interlocked position sensors	584
Interlocked temperature sensors	584
Motor settings: functions / discrete	448/1180
Threshold settings versus time	9768
Threshold settings versus energy	196
Threshold settings versus β^*	384
Temperature thresholds	490

1.2. *Brief recapitulation of collimation performance in LHC Run 1*

The cleaning performance of the LHC collimation system is measured by the so-called *local cleaning inefficiency*, η_c, defined as the number of proton lost $N_{lost}(s \rightarrow s + \Delta s)$ per unit length at a longitudinal s position in the ring, normalized by the total losses in the collimators, N_{abs}:

$$\eta_c = \frac{N_{lost}(s \rightarrow s + \Delta s)}{N_{abs}} \frac{1}{\Delta s}.$$

In simulations, losses are sampled using Δs bins of 10 cm. In practice, losses in the machine are measured at the discrete locations of beam loss monitors (BLMs). An example of cleaning measured at the LHC during the 2012 run at 4 TeV is given in Fig. 2 [9]. In this case, losses measured at the ~ 3600 BLMs around the ring are normalized by the peak loss at the horizontal primary collimator in IR7 (horizontal loss map). The local IR7 losses, showing the limiting cleaning locations in the dispersion suppressor, DS, (right side of IR7 for B1), are given in Fig. 3. In this example, a cleaning efficiency above 99.993% is achieved. Note that, with the exception of a few isolated peaks in the DS, the rest of the cold machine experiences losses that are more than one order of magnitude smaller.

In Fig. 4, the achieved cleaning inefficiency as a function of time is given for the different loss map campaigns carried out in 2012 [9]. These are validation tests performed regularly during the run to monitor the system performance by generating on purpose high losses on the primary collimators in controlled conditions. The highest (worst) inefficiency measured at cold locations is given

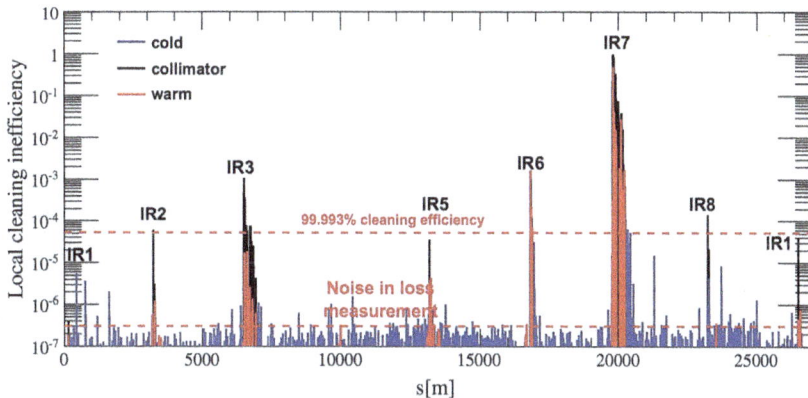

Fig. 2. Collimation cleaning measured at 4 TeV with $\beta^* = 60$ cm in IR1/5 in case of horizontal beam losses. Courtesy of B. Salvachua for the collimation team [9].

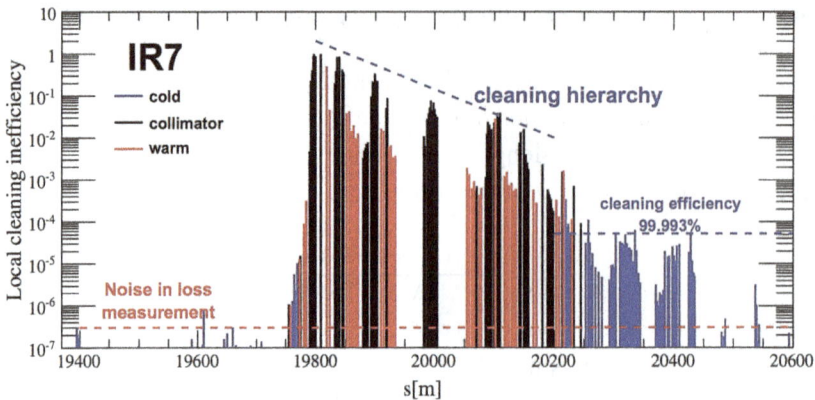

Fig. 3. Local IR7 losses from the graph in Fig. 2. Courtesy of B. Salvachua for the collimation team [9].

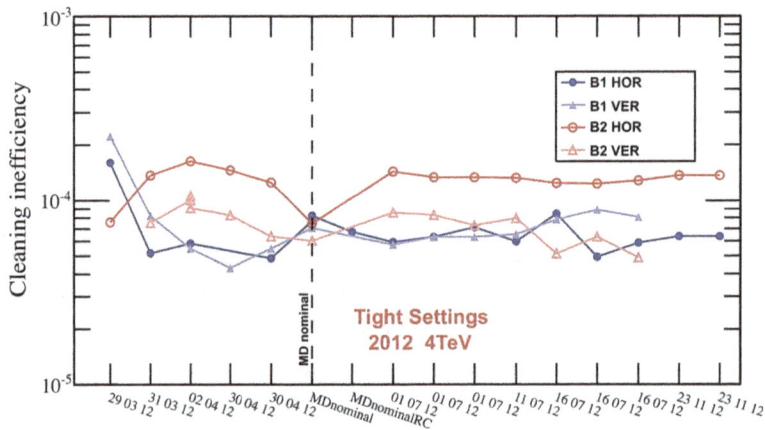

Fig. 4. Collimation cleaning inefficiency at the worst location in the DSs at either side of IR7 for both beams and planes as measured throughout the 2012 operation with protons (4 TeV, $\beta^* = 60$ cm). Courtesy of B. Salvachua [9].

for each plane and beam. This defines the system performance reach in terms of its capability to protect cold magnets from quenches. Highest losses are always recorded at the DSs around IR7, consistently with the simulation predictions. The cleaning inefficiency is very stable throughout the year and remains typically below a few 10^{-4}.

It is important to note that this performance was achieved with one single beam-based alignment per year for the collimators in IR3 and IR7. This is a major achievement for a large and distributed system like the one deployed at the LHC. This result was achieved thanks to the excellent stability of the machine (orbit, optics, etc.) and of the collimator settings. On the other hand, work to improve the alignment speed remains important: note that some 15 alignment campaigns were required in 2012 in order to setup the collimators in the experimental regions to match the requirements of new machine configurations requested by the experiments. This aspect represented an important constraint to the LHC operation in the Run 1 that is being addressed by a system upgrade during LS1 by adding collimators with integrated beam position monitors for orbit control and fast alignment, see below.

In 2012, the collimators were operated with full gaps as small as 2.1 mm (case of the vertical primary collimators in IR7), as required to push the β^* performance reach down to 60 cm [3]. Primary collimators were set in millimetres to their 7 TeV reference settings of 5.7σ. This is another important commissioning milestone illustrating that the collimator mechanics and control design choices (see next section) are adequate for the LHC small beam challenge.

On the other hand, the operation with small gaps increases the impedance of the machine (see Chapter 15), dominated by primary and secondary collimators in IR3/7. The 2012 operation was significantly affected by beam losses throughout the operational cycle [10]. The interplay between impedance and beam–beam effects is one of the possible sources of instabilities, which were instead not observed in 2011 with same normalized beam–beam separation and larger collimator settings. The operation at smaller gaps also causes naturally larger losses because the primary betatron collimators cut closer into the beam core. Studies are on-going to understand the performance limitations after LS1 from the collimation losses but it is already clear that an important ingredient for the performance in the HL-LHC era will be the reduction of the collimator impedance [11], in particular in view of the operation with larger single-bunch intensities. The backup solution to open further the collimators is always available but this has a non-negligible cost in terms of achievable β^*.

1.3. *Preliminary LHC intensity reach from collimation*

For a given collimation cleaning, the performance estimate in terms of total intensity reach, I_{max}, before quenching the magnets can be calculated for a given quench limit of superconducting magnets, R_q, and for a minimum allowed beam lifetime throughout the operational cycle, τ_b^{min}:

$$I_{max} \leq \frac{R_q \tau_b^{min}}{\eta_c}.$$

Here, the quench limit R_q is expressed in protons lost per metre per second. In reality, the performance reach estimates rely on, apart from dedicated measurements with beam, complex integrated simulations that combine multi-turn tracking of halo particles and energy deposition studies to compute the loss distribution in the magnet coils. This is then used as input for dedicated quench analysis tools. The most recent simulations were presented and discussed in detail at the collimation project review in May 2013 [12]. It is important to note that the performance reach estimates are based on experimentally achieved losses during LHC quench tests (see for example [13]).

Putting together the best knowledge of the various inputs, and assuming conservatively a minimum allowed lifetime of 0.2 h as suggested by an external review panel [14], one can estimate a total intensity reach for proton operation between a factor of 1.5 and 3 more than the present HL-LHC baseline (i.e., 3 to 6 times more than the nominal LHC intensity). It is important to realize that these estimates are based on the operational experience at lower LHC energy and at reduced total beam intensity (50 ns bunch spacing instead than the nominal 25 ns). Estimates are therefore intrinsically affected by uncertainties, in particular:

- extrapolation of quench limits to higher energies (margins in superconducting magnets may not follow the expected scaling laws);
- simulated cleaning inefficiency versus energy;
- assumption that the minimum lifetime does not degrade at higher energies, with reduced bunch spacing and increased collimator impedance;
- scaling of simulation results to higher energies.

It is therefore important to prepare alternative solutions in case of unexpected limitations and in order to ensure appropriate safety margins for the HL-LHC operation. The uncertainties listed above will be addressed by monitoring the performance in the post-LS1 operation.

1.4. Challenges of HL-LHC parameters

For higher luminosity operation of the LHC, the challenges for the collimation system are pushed forward in various respects. For the same collimation cleaning and primary beam loss conditions, the factor ~ 2 increase in total stored beam energy foreseen by the HL-LHC parameters requires a corresponding improvement of cleaning performance to achieve the same losses in cold magnets. Total losses might also exceed the robustness limit of collimators. The system is designed to withstand without damage lifetime drops down to 0.2 h during 10 s, corresponding to peak losses up to 500 kW. The larger stored energy also imposes more severe challenges for the collimator robustness against standard loss scenarios. Brighter beams impose potentially higher demands on the collimator material and design. The higher peak luminosity challenges entails the definition of new concepts for physics debris cleaning and an overall redesign of the IR collimation layouts. For example, in the present layout the inner triplet represents the IR aperture bottleneck and is protected by two dedicated tertiary collimators per plane per beam. Future optics scenarios might add critical aperture restrictions at magnets further away from the IP, requiring additional cleaning and protection.

2. Present and Future Collimator Design Concepts

In this section, the present design of the LHC collimator is recapitulated and the on-going studies for possible improvements are reviewed. This covers design features for improved operation of the system, a reduced impedance design and new material studies for future HL-LHC challenges.

2.1. *Collimator design for precision and robustness*

Two photographs of the present LHC collimator are given in Fig. 5, where a horizontal and a 45° tilted collimator are shown. An example of the tunnel installation layout for a IR7 collimator is given in Fig. 6. The LHC collimators are built as high precision devices in order to ensure the correct hierarchy of devices along the 27 km ring with beam sizes as small as 200 microns. Details of the collimator design can be found in [20]. Key features of the design are (1) a jaw flatness of about 40 microns along the 1 m-long active jaw surface, (2) a surface roughness below 2 microns, (3) a 5 micron positioning resolution (mechanical, controls), (4) an overall setting reproducibility below 20 microns

Fig. 5. Photograph of a horizontal (left) and a skew (right) LHC collimator. The latter has the vacuum tank open to show the two movable CFC jaws.

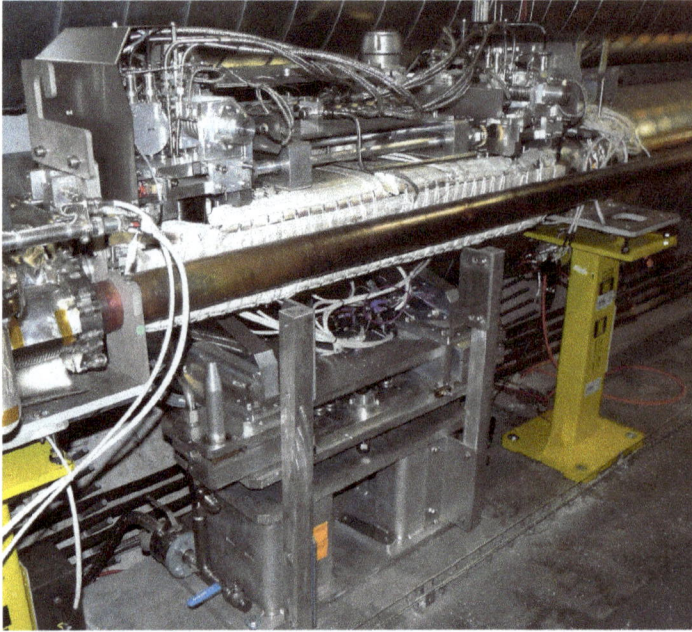

Fig. 6. Photograph of the active absorber TCLA.B6R7.B1 as installed in the betatron cleaning insertion.

[21], (5) a minimal gap of 0.5 mm, (6) evacuated heat loads of up to 7 kW in steady-state regime and of up to 30 kW in transient conditions. Primary and secondary collimator are made of robust carbon-fiber reinforced carbon composite (CFC) that is designed to withstand without significant permanent damage beam impacts for the worst failure cases such as impacts of a full injection batch of $288 \times 1.15 \times 10^{11}$ protons at 450 GeV and of up to $8 \times 1.15 \times 10^{11}$ protons at 7 TeV [22]. Other collimators made of tungsten heavy alloy or copper, obviously, do not have the same robustness and are only operated at larger distances from the circulating beams.

2.2. *Collimator with embedded beam position monitors*

The collimator design has been recently improved by adding two beam position monitors (BPMs) on either extremity of each jaw [23]. An example of a CFC jaw prototype with this new design is shown in Fig. 7. This concept will allow a fast collimator alignment as well as a constant monitoring of the beam orbit at the collimator as opposed to the BLM-based alignment that presently can only be performed during dedicated low-intensity commissioning fills. The BPM buttons will improve significantly the collimation performance in terms of operational

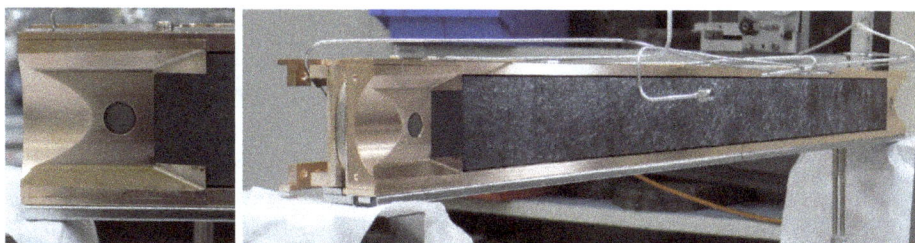

Fig. 7. New carbon/carbon collimator jaw with integrated BPMs at each extremity to be installed as secondary collimator in the dump insertion IR6. A detail of the BPM is given on the left side. A variant of this design, made with a Glidcop support and Tungsten inserts on the active jaw part, will be used for the tertiary collimators in all IRs.

flexibility and β^* reach [3]. The BPM-embedded design is considered as the baseline for future collimation upgrades. The concept has been tested extensively at the CERN SPS with a collimator prototype with BPMs [24, 25, 26]. Based on these results, 18 new collimators with integrated BPMs will already be installed during LS1, replacing the present tertiary collimators in all experiments (critical for β^* reach) and the secondary collimators in the dump region. Note that the BPM design is equally applicable to all collimators regardless of the jaw material.

2.3. *Rotatory collimator design*

The rotatory collimator design developed at SLAC proposes a "consumable collimator" concept based on two round jaws with 20 flat facets that can be rotated to offer to the beam a fresh collimator material in case a facet is damaged [27]. This concept provides a low-impedance design that is based on standard non-exotic materials. It is conceived for a high-power operation, with a performing 12 kW active cooling system to withstand the extreme power loads experienced by the secondary collimators in IR7. A photograph of this device before closing the vacuum tank is given in Fig. 8, where the rotatory glidcop (a copper allow) jaws are visible. The first full-scale prototype of this advanced collimator concept has recently been delivered to CERN [28] and is being tested in preparation of beam tests. The ultimate goal is to validate the rotation mechanism after high-intensity shock impacts at the HiRadMat facility, aimed at demonstrating that the concept of consumable collimator surface can indeed work for the works LHC beam load scenarios. The precision accuracy of this prototype and the impedance are also being tested together with its vacuum performance.

Fig. 8.   Photograph of the SLAC rotatory collimator prototype jaws before assembly in the vacuum tank. Courtesy of T. Markiewicz (SLAC).

2.4. *Status of R&D on novel advanced collimator materials*

One key element to ensure that next-generation collimators meet their challenging requirements lies in the development and use of novel advanced materials for the collimator jaws as no existing metal-based or carbon-based material possesses the combination of physical, thermal, electrical and mechanical properties which are required by the extreme working conditions. A rich R&D program has been launched to find optimum materials to improve robustness and impedance of the collimators. Several families of novel materials have been studied and developed, also in the frame of the FP7 EU programs EuCARD and EuCARD2 and in partnership with an Italian SME (Brevetti Bizz, San Bonifacio, Verona, Italy).

The driving requirements for new material's development are: (1) low resistive-wall impedance in order to avoid beam instabilities, (2) high cleaning efficiency, (3) high geometrical stability to maintain the extreme precision of the collimator jaw during operation despite temperature changes and (4) high structural robustness in case of accidental events like single-turn losses (see Chapter 12). It is interesting to note that several of these requirements are shared with other advanced thermal management applications, so that the object of this R&D program may have interesting spin-offs on industries for Aerospace, Medical, Nuclear, Electronics, etc.

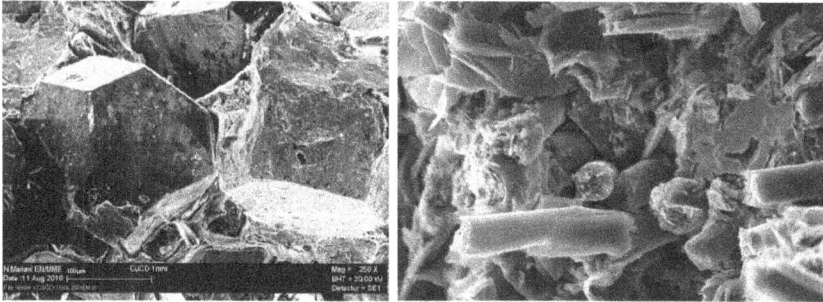

Fig. 9. Left: SEM view of CuCD: 175 μm diamonds surrounded by the Cu phase. The white spots on diamond surfaces are boron carbides (right). Right: Molybdenum-Graphite composite reinforced with carbon fibers.

A new family of materials, with promising features, has been identified: metal-carbon composites. These materials combine the outstanding thermal and physical properties of two carbon allotropes, diamond and graphite, with the electrical and mechanical properties of metals. The best candidates are Copper-Diamond (Cu-CD) and Molybdenum-Graphite (Mo-Gr), shown in Fig. 9. In particular, Mo-Gr may provide interesting properties regarding operating temperature, thermal shock resistance and, thanks to its availability in a wide range of mass density, also energy absorption capability. Additionally, this material may be effectively coated with pure molybdenum, dramatically decreasing the RF impedance contribution of future collimators. The addition of carbon fibers increases the mechanical strength of Mo-Gr.

A complex and comprehensive experiment was carried out at CERN HiRadMat facility [29, 30] to assess the consequences of highly energetic particle pulses impacting on sample collimator materials. Tests were also performed on a fully functional LHC collimator with Inermet180 jaws [31]. This is the heavy tungsten alloy that the present tertiary collimators are made of. The experiment aimed at the characterization, mostly in real time, of six different materials impacted by 440 GeV intense proton pulses. Chosen materials were a combination of relatively conventional metals for collimation applications, such Inermet, dispersion-strengthened copper (Glidcop) and molybdenum, and of novel composites under development including Cu-CD and carbon-fiber reinforced Mo-Gr. The design of the test set-up required innovative solutions in terms of lighting, support stabilization, radiation resistance and noise control.

Preliminary post-irradiation observations indicate that both Cu-CD and fiber reinforced Mo-Gr survived the high intensity impacts. Copper-Diamond (Cu-CD) has been developed by RHP-Technology, Seibersdorf, Austria, and studied for particle accelerator applications in the frame of the EuCARD collaboration.

Cu-CD is made of 60% synthetic diamonds, 39% copper powder and 1% boron powder, mixed and sintered by rapid hot pressing: diamonds enhance the material thermal conductivity while decreasing density. Boron is added to create a bridging between Cu matrix and diamond re-enforcement by forming boron carbides (B_4C) at interfaces. However, the strength of the resulting composite is limited because the boron carbide links are brittle and they are present only on a limited fraction of the diamond surface. An additional drawback is posed by the difficulty to machine these materials.

In the case of Mo-Gr, the preferential recrystallization of graphite planes during rapid hot pressing at temperatures in excess of 2500 °C leads to a compact structure, assuring outstanding thermal properties in the principal direction (even more than 700 W/m-K of thermal conductivity) and fair mechanical properties. In addition, coating Mo-GR jaws with a layer of pure Mo is being developed as a way to reduce the collimator impedance. The coating would reduce the surface resistivity by about a factor of 20 compared to Mo-GR (while maintaining sufficient robustness) and by more than a factor 100 compared to CFC. The benefit on the impedance budget of the collimation system would be significant: in the relevant frequency range, impedance would be reduced to 10% of the one of the CFC jaws (Fig. 10).

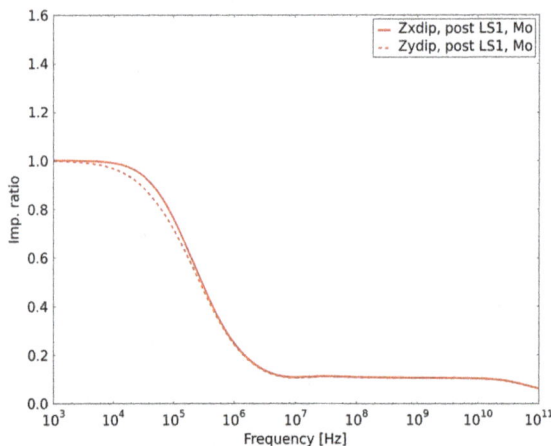

Fig. 10. Collimation impedance versus frequency: impedance ratio between Mo coating on Mo-Gr (50 μm layer) and present CFC jaw. A secondary collimator is considered. Courtesy of N. Mounet.

3. Improved Cleaning of Dispersion Suppressor Losses

In this section, the present baseline solution for improving the collimation cleaning, based on adding local collimation at the high-loss locations in the dispersion suppressors, is described for the different relevant LHC insertions. This is a technically challenging solution due to the tricky integration into cold areas but otherwise very robust from the beam physics view point. It relies on intercepting losses before they hit the magnets. This solution applies both for experimental and collimation insertions.

3.1. *Introduction to local DS collimation*

The limiting locations for collimation losses, both in the cleaning insertions and in the experimental regions, are the cold dispersion suppressors immediately downstream of the straight sections. This is the first high dispersion location seen by the outgoing particles that change their rigidity in the insertion, from inter-actions with the collimator materials (cleaning insertions) or from the collision with the other beam (experimental insertions). The dedicated momentum clean-ing insertion in IR3 cannot catch local single-turn effects: if the change of a particle's rigidity is larger than the acceptance of the arc, these particles are lost before reaching IR3. The DSs around IR7 might limit the performance both for proton and for heavy ion operation. The proton limitations were discussed in the introductory section of this chapter. In addition, ion losses in the DSs around the experiments might limit the achievable peak luminosity if the DSs are not adequately protected.

A possible solution to this problem is to add local collimators in the dispersion suppressors, which is only feasible with a major change of the cold layout at the locations where the dispersion start rising. Indeed, the present system's multi-stage cleaning is not efficient at catching these dispersive losses. Clearly, the need for local collimation depends on the absolute level of losses achieved in operation and the quench limit of superconducting magnets. In this design phase when the quench limits and the operational performance are not yet known accurately enough at energies close to 7 TeV, it is important to take appropriate margins to minimize the risk of being limited in the future (post-LS1 operation and even more in the HL-LHC era).

Our present best guess on the needs for DS collimation in the different IRs, for proton and heavy-ion beam operation, is summarized in Table 2 [32]. For the betatron cleaning, the present performance reach estimates indicate that the intensity goal should be within reach albeit with reduced safety margins. The situation changes for HL-LHC due to some specific features of the loss maps with the Achromatic Telescopic Squeeze (ATS) optics [33]. These interim conclusions are subject to a re-evaluation of the beam performance and of the quench limits during the post-LS1 operation in 2015.

Table 2. Summary of need for DS collimation in the different insertion
points, for the operation until LS3 and beyond (HL-LHC era).

		Until HL-LHC (before LS3) [L=2.5x10³⁴cm⁻²s⁻¹, I_{tot}=3.2x10¹⁴p]		HL-LHC era (after LS3) (L=5x10³⁴cm⁻²s⁻¹, I_{tot}=6.2x10¹⁴p)	
		Protons	Ions	Protons	Ions
IR7	Betatron cleaning	**Needed?** *(unlikely)*	**Needed?** *(unlikely)*	Needed TBC for ATS	Needed?
IR3	Momentum cleaning	Not needed	Not needed	Not needed	Not needed
IR1/5	ATLAS/CMS	Not needed	**Needed**	Needed? *(unlikely)*	Needed
IR2	ALICE	Not needed	**Needed**	Not needed	Needed
IR8	LHCb	Not needed	Not operating	Not needed	Not needed *LHCb operating?*

The driving factor that calls for an implementation of DS collimation already in LS2 is the ion case. This applies to IR1, 2 and 5 even though the priority shall be given to ALICE during ion physics time. This is described in detail below in this section. On the other hand, thanks to an upgraded layout of the physics debris collimation in IR1 and IR5 that takes place in LS1 [16, 17], no limitations from luminosity losses are expected in the high-luminosity points for proton operation with the present layout that will remain until LS3. This must be re-evaluated for the final HL-LHC layout [19].

3.2. DS collimation solutions for proton and ion cases

In the past, a solution was conceived that relied on moving the position of a number (24 per IR) superconducting dipoles and quadrupoles, together with associated cold powering elements like DFBs and shuffling modules, in order to free enough space for installing collimators in cells 8 and 10 [34]. This major layout change was made possible by using the space of the connection cryostat in cell 11, just upstream of the Q11. In its first concept, this solution was considered for IR3 for the so-called combined momentum and betatron cleaning. The solution based on displacing magnets was the only viable — though clearly very challenging — option for improving the cleaning during LS1, also taking into account constraints from radiation to electronics that favored an installation in IR3 rather than in IR7. The evaluation of the collimation operational performance in 2011 and 2012 indicates, as discussed above, that no immediate limitations from betatron cleaning should be expected in the post-LS1 era. Earliest actions for DS collimation are therefore postponed until LS2 [35, 36]. This has the advantage of allowing a much more elegant solution based on two shorter, higher field dipoles that could be used to replace one present 15 m long dipole by

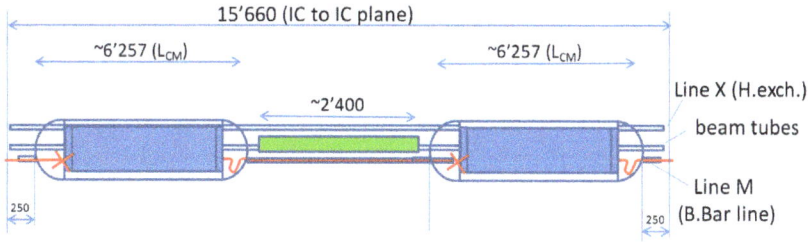

Fig. 11. Schematic view of the assembly of two shorter 11 T dipoles with a collimator in between, which can replace one standard main dipole. Courtesy of V. Parma.

making space for a warm collimator, as schematically shown in Fig. 11. This is a modular solution that can be applied to any dipole without additional changes to the adjacent superconducting magnets or other cold elements [37].

Presently, three cases with DS collimation have been studied in detail [38]: ion losses from collision products in IR2 (1) and proton losses from collimation cleaning downstream of IR7 in case of standard (2) and HL-LHC baseline (3) optics scenarios. Ion losses around IR1 and IR5 are not studied in detail in the assumption that similar conclusions drawn for IR2 apply if the peak luminosity is the same. The proposed layout in the IR7 DS is shown in Fig. 12. Because of the profile of the dispersion function, two DS collimators are required in this case to efficiently clean the losses. Tracking simulations of cleaning performance have demonstrated that this proposed layout is effective both for the present optics and for the HL-LHC case. The effect of DS collimators for the latter case is shown in Fig. 13, where simulated loss maps are given for the cases without (top) and with (bottom) new collimators [39]. It is seen that the new layout significantly improves the cleaning by reducing losses immediately downstream of IR7 as well as loss peaks around the ring that occur for the new telescopic squeeze

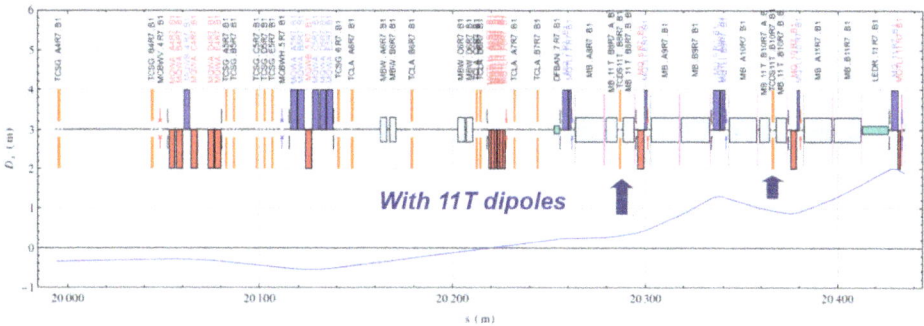

Fig. 12. Proposed locations in the DS near the betatron cleaning insertion where dipoles might be replaces by the new assembly in Fig. 11. The periodic dispersion function versus the longitudinal coordinate *s* is also given.

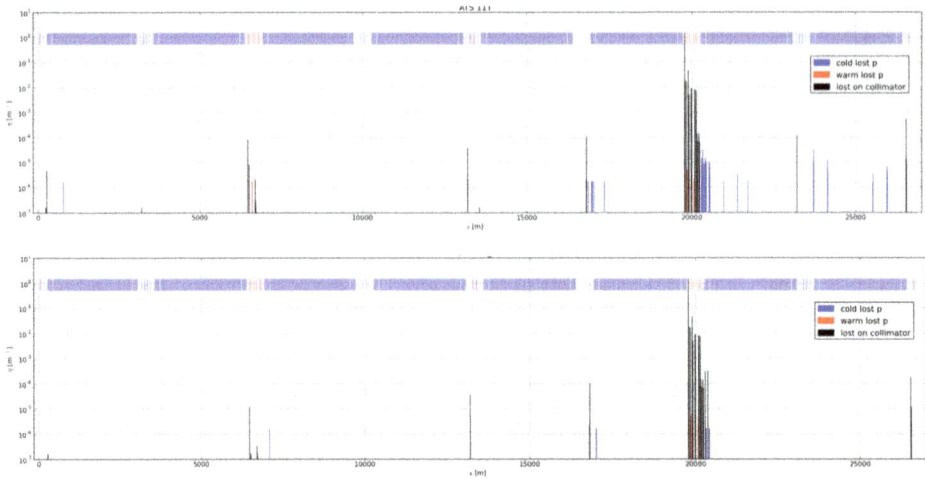

Fig. 13. Simulated loss maps around the 27 km long LHC ring without (top) and with (bottom) local DS collimators around IR7. The baseline HL-LHC ATS optics with $\beta^* = 15$ cm in IP1/5 is used. From [39].

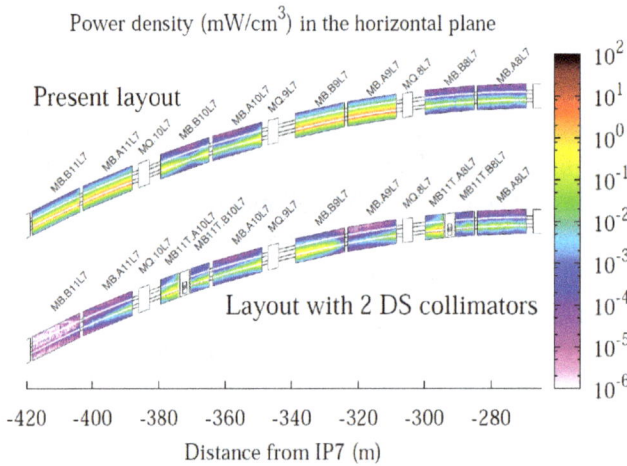

Fig. 14. Simulated power density map in the horizontal plane of DS dipoles for nominal 7 TeV operation and a beam lifetime of 0.2 h (4.5e11 protons lost per second). Comparison of the present layout and a layout with two TCLDs. Results correspond to relaxed collimator settings. Beam direction is from the right to the left. From [40].

optics. The improved performance must be compared against the expected quench limits at 7 TeV for a certain assumed beam lifetime. This is done by detailed energy deposition studies that are used to quantify the energy deposited in the coil of the superconducting magnets [40]. For example, the case of 0.2 h lifetime for the nominal LHC beam is illustrated in Fig. 14. It is seen that the presence of local DS collimators as in Fig. 12 reduces the peak energy deposition by about a factor of 10 compared to the present layout with standard dipoles. If the LHC total intensity reach were limited by collimation losses with the present layout without DS collimation, this solution would allow increasing the intensity reach by the same factor.

Although the loss maps generated by collimation of heavy-ion beams in IR3 and IR7 are different because of the wider range of nuclear interactions that can occur in the primary collimator material [51], it has been shown [52] that these collimators will also achieve a substantial reduction of the losses in the IR7 dispersion suppressors.

The case of ion losses from collision products in IR2 [52, 53, 54, 12] is treated differently. The magnets in the DS might quench in this case due to the production of "beams" with different rigidities from ultraperipherical electromagnetic interactions of the counter-rotating beams at the collision point. The dominating processes are bound-free pair production (BFPP) where electron-positron pairs are created and an electron is caught in a bound state by one (BFPP1) or both (BFPP2) nuclei, thus changing their charge, and 1- or 2-neutron electromagnetic dissociation (EMD1 and EMD2) where one of the colliding ions

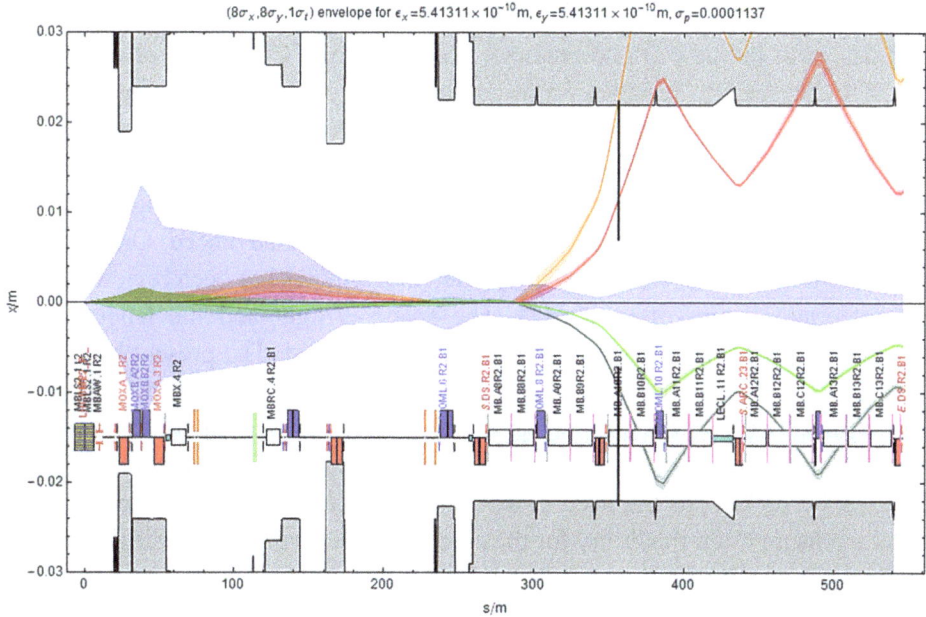

Fig. 15. 1σ envelope of the main Pb^{82+} beam (violet) together with the dispersive trajectories of ions undergone BFPP1 (red), BFPP2 (orange), EMD1 (light green) and EMD2 (dark green) coming out of the ALICE experiment in nominal optics. The DS collimator appears as a black line. Varying its opening allows different secondary beams to be intercepted.

emits one or two neutrons, respectively, thus changing mass. Further photo-induced processes also take place, but the four ones mentioned here have the higher cross sections. An example of ion beams produced in collisions of $^{208}Pb^{82+}$ nuclei in IR2 is given in Fig. 15. The BFPP1, producing $^{208}Pb^{81+}$, is the dominant process for this ion and the corresponding beam can carry about 150 W for the foreseen upgrade ALICE luminosity peak of 6×10^{27} cm^{-2}s^{-1}. Detailed energy deposition simulation for the cases without and with DS collimators replacing the dipole MB.A10R2.B1 indicate that 0.5 m of copper are sufficient to reduce the peak losses in the cold magnet by a factor of 25. This would ensure a safe operation with the upgraded ALICE luminosity. Without local collimation, losses would be up to a factor of 2 above the quench limits, depending on the models used for the quench analysis [12].

During heavy-ion operation, a similar situation prevails around the ATLAS and CMS experiments in IR1 and IR5 and further installations of DS collimators may eventually have to be considered. Thanks to optical differences in these insertions, there is some flexibility in positioning of the collimator assemblies [12].

3.3. *Status of prototyping and design*

The design of the new DS collimators, designated as TCLDs, and of the bypass cryostat necessary to install a warm collimator are well advanced thanks to the preparatory work done for the possible implementation in LS1, based on moving magnets [36]. A bypass cryostat prototype was built (see Fig. 16) to perform the necessary tests at cold to validate this concept. These tests are ongoing at the time of writing. The TCLD collimator design was also very advanced. On the other hand, the new baseline that relies on shorter 11 T dipoles has been reviewed from the integration point of view [12]. The space is tight and the length of all components and transitions must be carefully optimized. The present baseline is that the TCLD will have an active jaw length of 80 cm that proved to be sufficient to improve the cleaning in all relevant cases. Tungsten heavy alloy is assumed for the material because the TCLD will hardly be exposed to large beam load, so we do not see the need at this stage to consider advanced materials. In order to ensure more flexibility for the case of ion beams, where positive changes of the beam rigidity also occur, a 2-jaw design is considered for the moment.

From the RF view point, designs with transverse RF finger (as in the present system) as well as a with ferrite blocks to absorb high order modes (as in the collimators with BPMs) are being comparatively assessed. The latter design is shown in Fig. 17, where a detail of the collimator jaw corner is given.

Fig. 16. Photograph of the prototype bypass cryostat (QTC) designed to install a warm collimator in the cold DSs.

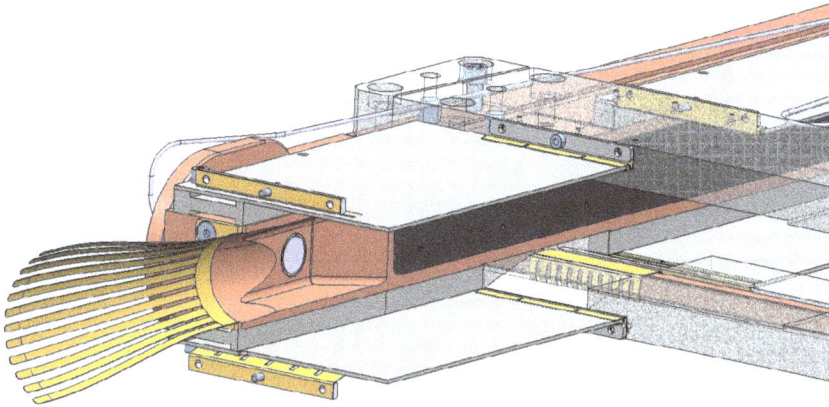

Fig. 17. Detail of one corner of the TCLD collimator to be installed in the DS between two new 11 T dipoles. The present design foresees an 80 cm long jaw made of tungsten (the first of 4, 20 cm tungsten tiles is shown) and will have two jaws. Designs with transverse RF fingers or ferrite tiles are being comparatively assessed to reduce the detrimental effects of trapped RF modes.

4. Advanced Collimation Concepts for HL-LHC

Other advanced collimation concepts that still require R&D and therefore cannot be considered yet as a baseline are discussed in this section.

4.1. *Halo diffusion control techniques*

The 2012 operational experience indicates that the LHC collimation would profit from halo control mechanisms. These were used in other machines like HERA and Tevatron. The idea is that, by controlling the diffusion speed of halo particles, one can act on the time profile of the losses, for example by reducing rates of losses that otherwise would take place in short time, or simply by controlling the static population of halo particles in a certain aperture range. These aspects were recently discussed at a collimation review on the possible usage of the hollow e-lens collimation concept at the LHC [41], where it was concluded that hollow e-lenses could be used at the LHC for this purpose. In this case, a hollow electron beam, running parallel to the proton or ion beam, is used to generate an annular beam in the transverse (x, y) plane. This hollow beam induces a field affecting halo particles above a certain transverse amplitude and can change their transverse speed. The conceptual working principle is illustrated in the left part of Fig. 18. A solid experimental basis achieved at the Tevatron indicates that this solution is promising for the LHC ([42] and reference therein).

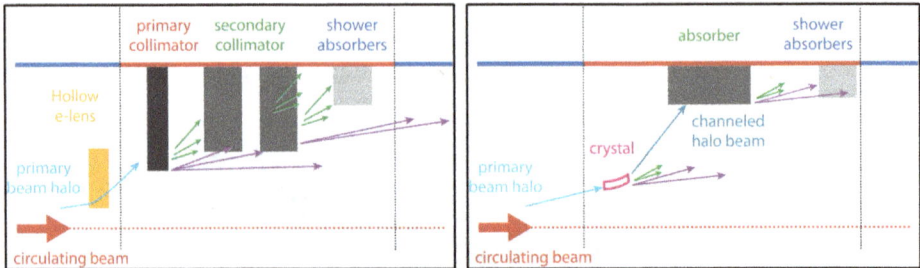

Fig. 18. Illustrative view of the collimation system with integrated hollow e-lens or equivalent halo diffusion mechanism (left) and of an ideal crystal-based collimation (right). A reduced collimator layout than the one in Fig. 1 is adopted to show the betatron cleaning functionality only (one side only). Halo control technics are used to globally change the diffusion speed of halo particles and rely on the full collimation system remaining in place. Crystals entail a change of concept where the whole beam losses are concentrated, ideally, in one single beam absorber per plane.

At the review [41], it also became clear that, if loss spikes were limiting the LHC performance after LS1, the hollow e-lens solution would not be viable because it could only be implemented in a next long shutdown at the earliest (driven by time for the integration into the cryogenics system). It is therefore crucial to work on viable alternatives that, in case of need, might be implemented on an appropriate time scale. Two alternatives are presently being considered:

- Tune modulation through noise in the current of lattice quadrupoles;
- Narrow-band excitation of halo particles with the transverse damper system.

Though very different from the hardware point of view, both these techniques rely on exciting tail particles through resonances induced in the tune space by appropriate excitations. This works in the assumption of a presence of correlation between halo particles with large amplitudes and corresponding tune shift in tune space (de-tuning with amplitude). Clearly, both methods require a solid experimental verification in a very low noise machine like the LHC, in particular to demonstrate that this type of excitations do not perturb the beam core emittance. Unlike for hollow e-lenses that act directly in the transverse plane by affecting particles at a amplitudes above the inner radius of the hollow beam, resonance excitations methods required a good knowledge of the beam core tune even in dynamic phases of the operational cycle, so the possibility to use these techniques at the LHC remains to be demonstrated. For this purpose, simulation efforts are on going with the aim of defining the required hardware interventions during LS1 that might enable beam tests of these two halo control methods early on in 2015. Ideally, these measurements would profit from appropriate halo diagnostic tools, see Chapter 14. But we are confident that conclusive measurements could be achieved with the techniques describe for example in [5].

4.2. *Crystal collimation*

Highly pure bent crystal can be used to steer high-energy particles that get trapped between the potential of parallel lattice planes. Equivalent bending fields up to hundreds of Tesla can be achieved in crystals with a length of only 3–4 mm, which allows in principle to steer halo particles to a well-defined point. As opposed to a standard collimation system based on amorphous materials, requiring several secondary collimators and absorbers to catch the products developed through the interaction with matter (see Fig. 2, right), one single absorber per collimation plane is in theory sufficient in a crystal-based collimation system [43]. This is shown in the scheme in Fig. 18 (right). Indeed, nuclear interactions with well-aligned crystals are much reduced compared to a primary collimator, provided that high channeling efficiencies of halo particles can be achieved (particles impinging on the crystal to be channelled within a few turns). This is expected to reduce significantly the dispersive beam losses in the DS of the betatron cleaning insertion compared to the present system that is limited by the leakage of particles from the primary collimators. Simulations indicate a possible gain between 5 and 10 [44] even for a layout without an optimized absorber design. The crystal collimation option is particularly interesting for collimating heavy-ion beams thanks to the reduced probability of ion dissociation and fragmentation compared to the present primary collimators. SPS test results are promising [45].

Another potential of crystal collimation is a strong reduction of the machine impedance due to the facts that (1) only a small number of collimator absorbers is required and that (2) the absorbers can be set at much larger gaps thanks to the large bending angle from the crystal (40–50 μrad instead than a few μrad from the multiple-Coulomb scattering in the primary collimator). On the other hand, an appropriate absorber design must be conceived in order to handle the peak loss rates in case of beam instabilities: the absorber must withstand continuous losses up to 1 MW during 10 s while ensuring the correct collimation functionality. This is a change of paradigm compared to the present system where such losses are distributed among several collimators. Other potential issues concern the machine protection aspects of this system (what happens if the crystal is not properly aligned and channels an important fraction of the total stored energy to the wrong place?) and the operability of the system that requires mechanical angular stability in the sub-μrad range to be ensured through the operational cycle of the LHC (injection, ramp, squeeze and collision).

Promising results have been achieved in dedicated crystal collimation tests at SPS performed from 2009 within the UA9 experiment [45, 46, 47]. On the other hand, some outstanding issues about the feasibility of the crystal collimation concept for the LHC can only be addressed by dedicated beam tests at high energy in the LHC. For this purpose, a study at the LHC has been proposed that might already take place in the LHC Run 2 after LS1 [48]. The main purpose of this test with LHC beams is to demonstrate the feasibility of the crystal-collimation concept in the LHC environment, in particular to demonstrate that such a system can provide a better cleaning of the present high-performance system throughout the operational cycle. Until a solid demonstration is achieved, this scheme cannot be considered for future HL-LHC baseline scenarios.

4.3. *Improved optics scenarios for the collimation insertions*

Alternative optics concepts in IR7 can be conceived in order to improve some present collimation limitations without major hardware changes. For example, non-linear optics schemes derived from the linear collider experience [49] were considered for IR7. The idea is that one can create a "non-linear bump" that deforms the trajectories of halo particles and effectively increases their transverse amplitudes in a way that allows opening the gaps of primary and secondary collimators. These studies are well advanced from the optics point of view but for the moment it was not easily possible to find a layout solution providing the same cleaning as the present system [50]. These studies, and other aimed at increasing the beta functions at the collimators, are on-going.

References

[1] O. Brüning (ed.) *et al.*, LHC Design Report Vol. 1, CERN-2004-003-V-1.

[2] R. Assman *et al.*, The final collimation system for the LHC, in *Proc. of EPAC2006*, Edinburgh (UK), 2006. Also as CERN-LHC-PROJECT-REPORT-919 (2006).

[3] R. Bruce *et al.*, LHC beta* reach in 2012, LHC Operation Workshop, EVIAN2011, Evian (FR), 2011. http://indico.cern.ch/event/155520.

[4] R. Bruce *et al.*, Sources of machine-induced background in the ATLAS and CMS detectors at the CERN Large Hadron Collider, *Nucl. Instrum. Meth. A* **729**, 825–840 (2013).

[5] G. Valentino *et al.*, Beam diffusion measurements using collimator scans at the LHC, *Phys. Rev. Spec. Top. Accel. Beams* **16**, 021003 (2013).

[6] K. H. Mess and M. Seidel, Collimators as diagnostic tools in the proton machine of HERA, *Nucl. Instrum. Meth. A* **351**, 279–285 (1994).

[7] H. Burkhardt, S. Redaelli, B. Salvachua, G. Valentino, Collimation down to 2 sigmas in special physics runs in the LHC, in *Proc. of IPAC2013*, Shanghai (China). Also as CERN-ACC-2013-0144 (2013).

[8] S. Redaelli *et al.*, LHC collimator controls for a safe LHC operation, in *Proc. of ICALEPSC2011*, Grenoble (FR), 2011. http://accelconf.web.cern.ch/AccelConf/icalepcs2011/papers/wepmu020.pdf.

[9] B. Salvachua *et al.*, Cleaning performance of the LHC collimation system up to 4 TeV, in *Proc. of IPAC2013*, Shanghai (China). https://cds.cern.ch/record/1574583?ln=en.

[10] B. Salvachua *et al.*, Lifetime analysis at high intensity colliders applied to the LHC, in *Proc. of IPAC2013*, Shanghai (China). http://cds.cern.ch/record/1574586/files/CERN-ACC-2013-0072.pdf.

[11] N. Mounet *et al.*, Beam stability with separated beams at 6.5 TeV, in *Proc. of LHC Operations Workshop, EVIAN2012*, Evian (Fr). https://indico.cern.ch/getFile.py/access?contribId=18&sessionId=10&resId=0&materialId=paper&confId=211614.

[12] 2013 Collimation Project Review, http://indico.cern.ch/event/251588.

[13] S. Redaelli *et al.*, Quench tests at the large hadron collider with collimation losses at 3.5 Z TeV, in *Proc. of HB2012*, Beijing (CH). http://accelconf.web.cern.ch/AccelConf/HB2012/papers/mop245.pdf.

[14] Recommendation from the external review panel of the 2013 collimation review. Available here.

[15] A. Marsili *et al.*, Simulations and measurements of physics debris losses at the 4 TeV LHC, in *Proc. of IPAC2013*, Shanghai (China).

[16] CERN EDMS document 1283867 (2013), also available at http://lhc-collimation-project.web.cern.ch/lhc-collimation-project/LS1/default.php.

[17] CERN EDMS document 1283826 (2013), also available at http://lhc-collimation-project.web.cern.ch/lhc-collimation-project/LS1/default.php.

[18] L. Esposito, presentation at the 2nd HiLumi Annual meeting, Frascati (I), 2012. http://indico.cern.ch/event/183635.

[19] L. Esposito, presentation at the 3rd HiLumi Annual meeting, Daresbury (UK), 2012. http://indico.cern.ch/event/257368.

[20] A. Bertarelli *et al.*, The mechanical design for the LHC collimators, in *Proc. of EPAC2004*, Lucern (CH), EPAC-2004-MOPLT008.

[21] S. Redaelli *et al.*, Final implementation and performance of the LHC collimator control system, in *Proc. of PAC09*, Vancouver (CA).

[22] A. Bertarelli *et al.*, Mechanical design for robustness of the LHC collimators, in *Proc. of PAC2005*, Knoxville (USA). http://accelconf.web.cern.ch/AccelConf/p05/PAPERS/TPAP004.PDF.

[23] F. Carra *et al.*, LHC collimators with embedded beam position monitors: a new advanced mechanical design, in *Proc. of IPAC2011*, San Sebastian (E), IPAC-2011-TUPS035.

[24] D. Wollmann *et al.*, First beam results for a collimator with in-jaw beam position monitors, in *Proc. of IPAC2011*, http://accelconf.web.cern.ch/AccelConf/IPAC2011/papers/thpz027.pdf.

[25] D. Wollmann *et al.*, Experimental verification for a collimator with in-jaw beam position monitors, in *Proc. of HB2012*, http://accelconf.web.cern.ch/AccelConf/HB2012/papers/mop242.pdf.

[26] G. Valentino *et al.*, Successive approximation algorithm for BPM-based LHC collimator alignment, submitted to PRST-AB (2013).

[27] J. Smith *et al.*, Design of a rotatable copper collimator for the LHC phase II collimation upgrade, in *Proc. of EPAC2008*, Genova (I). http://accelconf.web.cern.ch/AccelConf/e08/papers/mopc096.pdf.

[28] T. Markiewiz, Status of SLAC RC, presentation at the 3rd HiLumi annual meeting, Daresbury (UK), 2013. http://indico.cern.ch/event/257368.

[29] A. Bertarelli *et al.*, First results of an experiment on advanced collimator materials at CERN HiRadMat facility, in *Proc. of IPAC2013*, Shanghai (China). https://cds.cern.ch/record/1635957/files/CERN-ACC-2013-0268.pdf.

[30] A. Bertarelli *et al.*, An experiment to test advanced materials impacted by intense proton pulses at CERN HiRadMat facility, *Nucl. Instr. Meth. B* (2013) http://dx.doi.org/10.1016/j.nimb.2013.05.007.

[31] M. Cauchi *et al.*, High energy beam impact tests on a LHC tertiary collimator at CERN HiRadMat facility, submitted to PRST-AB (2013).

[32] S. Redaelli, The LHC collimation baseline for HL-LHC, presentation at the 3rd HiLumi annual meeting, Daresbury (UK), 2013. http://indico.cern.ch/event/257368.

[33] A. Marsili, Simulations of collimation cleaning performance for HL-LHC optics, in *Proc. of IPAC2013*, Shanghai (China).

[34] R. Assmann *et al.*, Accelerator physics concept for upgraded LHC collimator performance, in *Proc. of PAC2009*, Vancouver (CA). https://cds.cern.ch/record/1307562/files/EuCARD-CON-2009-047.pdf.

[35] S. Redaelli *et al.*, Do we really need a collimation upgrade? LHC Performance Workshop, Chamonix2012, https://indico.cern.ch/event/164089.

[36] 2011 Collimation Project review, https://indico.cern.ch/event/139719.

[37] Review of 11 T dipoles and cold collimation, https://indico.cern.ch/event/155408.

[38] HiLumi-WP5 (collimation) section at the 3rd Joint HiLumi LHC-LARP Annual Meeting, Daresbury (UK), 2013. http://indico.cern.ch/event/257368.

[39] A. Marsili, Simulated cleaning for HL-LHC layouts with errors, presentation at the 3rd Joint HiLumi LHC-LARP Annual Meeting, Daresbury (UK), 2013. http://indico.cern.ch/event/257368.

[40] A. Lechner, Energy deposition with cryo-collimators in IR2 (ions) and IR7, presentation at the 3rd Joint HiLumi LHC-LARP Annual Meeting, Daresbury (UK), 2013. http://indico.cern.ch/event/257368.

[41] Review of hollow e-lens for the LHC collimation, https://indico.cern.ch/event/213752.

[42] G. Stancari, Progress towards the conceptual design of a hollow electron lens for the LHC, presentation at the 3rd Joint HiLumi LHC-LARP Annual Meeting, Daresbury (UK), 2013. http://indico.cern.ch/event/257368.

[43] W. Scandale, Crystal-based collimation in modern hadron colliders, *Int. J. Mod. Phys. A* **25**(S1), 70–85 (2010).

[44] D. Mirarchi *et al.*, Layouts for crystal collimation tests at the LHC, in *Proc. of IPAC2013*, Shanghai (China) https://cds.cern.ch/record/1573725?ln=en.

[45] W. Scandale *et al.*, Comparative results on collimation of the SPS beam of protons and Pb ions with bent crystals, *Phys. Lett. B* **703**, 547–551 (2011).

[46] W. Scandale *et al.*, Optimization of the crystal-assisted collimation of the SPS beam, *Phys. Lett. B* **726**, 182–186 (2013).

[47] W. Scandale *et al.*, Strong reduction of the off-momentum halo in crystal assisted collimation of the SPS beam, *Phys. Lett. B* **714**, 231–236 (2012).

[48] CERN EDMS document 1329235, LHC-TEC-EC-0001 (2013).

[49] A. Faus-Golfe *et al.*, Non-linear collimation in linear and circular colliders, in *Proc. of EPAC2006*, Edinburgh (UK). http://accelconf.web.cern.ch/AccelConf/e06/PAPERS/WEXFI03.PDF.

[50] L. Lari *et al.*, Studies for an alternative LHC non-linear collimation system, in *Proc. of IPAC2012*. http://accelconf.web.cern.ch/AccelConf/IPAC2012/papers/moppd077.pdf.

[51] H.-H. Braun *et al.*, Collimation of heavy ion beams in LHC, in *Proc. of EPAC 2004*, Lucerne, http://accelconf.web.cern.ch/AccelConf/e04/PAPERS/MOPLT010.PDF (2004).

[52] LHC Collimation Review 2009, http://indico.cern.ch/conferenceDisplay.py?confId=55195.

[53] J.M. Jowett *et al.*, Heavy ion beams in the LHC, in *Proc. of PAC 2003*, Portland (2003), http://accelconf.web.cern.ch/AccelConf/p03/PAPERS/TPPB029.PDF.

[54] R. Bruce *et al.*, Beam losses from ultraperipheral nuclear collisions between $^{208}Pb^{82+}$ ions in the Large Hadron Collider and their alleviation, *Phys. Rev. ST Accel. Beams* **12**, 071002 (2009).

Chapter 14

Long-Range Beam–Beam Compensation Using Wires

F. Zimmermann and H. Schmickler

CERN, BE Department, Genève 23, CH-1211, Switzerland

At the LHC, the effect of unavoidable long-range beam–beam collisions reduces the dynamic aperture, calling for a minimum crossing angle. A wire compensator partially cancels the effect of the long-range collisions, and may allow operation with reduced crossing angle or decreased beta function at the interaction point, thereby increasing the (virtual) peak luminosity. In this chapter, we describe the proposed compensation scheme, previous validation experiments with a single beam and multiple wires at the SPS, simulations for the LHC high-luminosity upgrade, a demonstrator project with real long-range encounters foreseen in the LHC proper, and the possible use of a low-energy electron beam as a future ultimate "wire".

1. Motivation

Following earlier studies investigating the effect of long-range collisions for the SSC [1] and LHC [2, 3], in 1999 weak-strong beam–beam simulations for the LHC — using the modeling recipe of Ref. [4] — revealed the existence of a diffusive aperture at a transverse amplitude of 6–7σ, which is induced by the nominal long-range beam–beam encounters [5]. An example simulation result is illustrated in Fig. 1.

This chapter describes in detail the proposed compensation scheme, previous validating studies in the CERN SPS, simulations for the LHC high luminosity upgrade and, briefly, a demonstrator project in the LHC foreseen in the years 2016–2018.

The demonstrator project will use DC powered wires integrated into collimator jaws. Assuming the related experiments with two beams in the LHC will be successful, first studies have been launched on using low energy electron beams as ultimate "wires", which could be integrated into the LHC machine close to the high-luminosity insertion points. Using an electron beam would also allow modulating the "wire" current at 40 MHz, enabling individual bunch-by-bunch

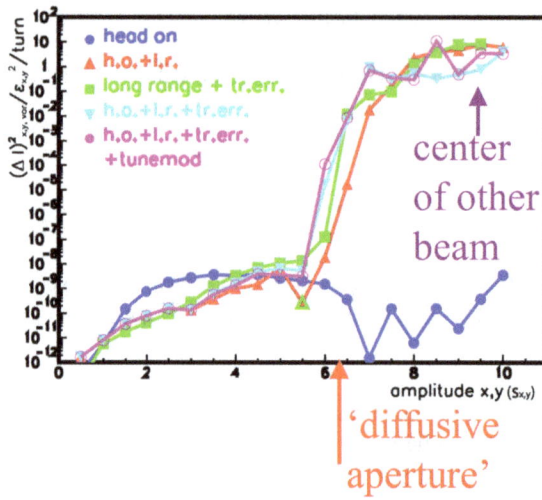

Fig. 1. Transverse action diffusion rate $\Delta I^2_{\mathrm{rms}}/\varepsilon^2_{x,y}/$turn as a function of transverse amplitude in units of σ under various conditions, obtained from a weak-strong beam–beam simulation [5].

compensation schemes. This latter proposal and the long-term perspective of an electron-beam wire are only briefly mentioned at the end of this chapter.

2. Compensation Scheme

The simulated strong effect of the LHC long-range collisions inspired the search for mitigation, and in 2000 J.-P. Koutchouk proposed a long-range beam–beam compensation for the LHC based on current-carrying wires [6]. At a sufficiently large transverse distance, the wires generate the same transverse force of shape $1/r$, as the field of the opposing beam at the parasitic long-range encounters [6]. In order to correct all non-linear effects the correction must be local. For this reason, there needs to be at least one wire compensator (in the CERN internal naming convention called 'BBLR') on one side of each primary interaction point (IP) for either beam, i.e. one compensator per beam per IP. The compensator should be installed in a region where the two beams are already physically separated, but otherwise as close as possible to the common region where the long-range encounters occur. An originally proposed layout features the compensators 41 m upstream of the separation dipole D1, on both sides of IP1 and IP5, where the horizontal and vertical beta functions are equal, as is shown in Fig. 2. Figure 3 illustrates how one wire cancels the effect of all 16 long-range encounters occurring on one side of the IP. The betatron phase difference between the BBLR and the average LR collision is 2.6° (ideally it should be zero).

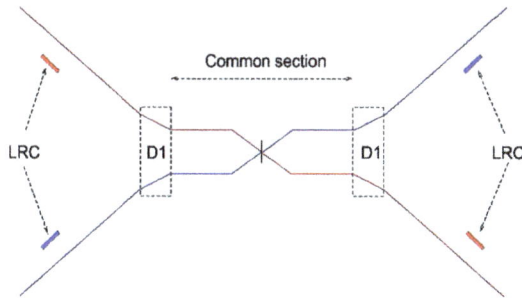

Fig. 2.   Schematic location of proposed LHC wire compensators [6, 7]. In this picture the compensators are called LRC, and there are two times more than the strict minimum.

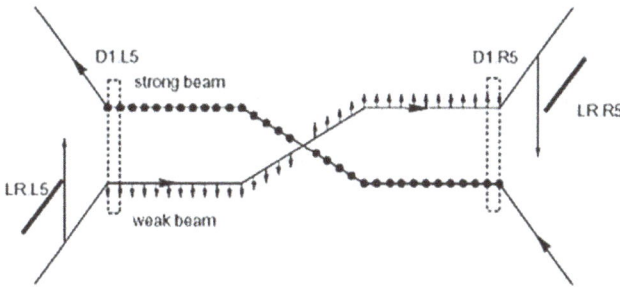

Fig. 3.   Illustration of the compensation principle [6, 7]. In this schematic the compensators are designated as LRL15 and LRR5 (referring to left and right of Point 5). Two compensators for the "weak beam" are shown, while only a single one, at twice the strength, would be minimally required. The equivalent compensators for the strong beam are not displayed.

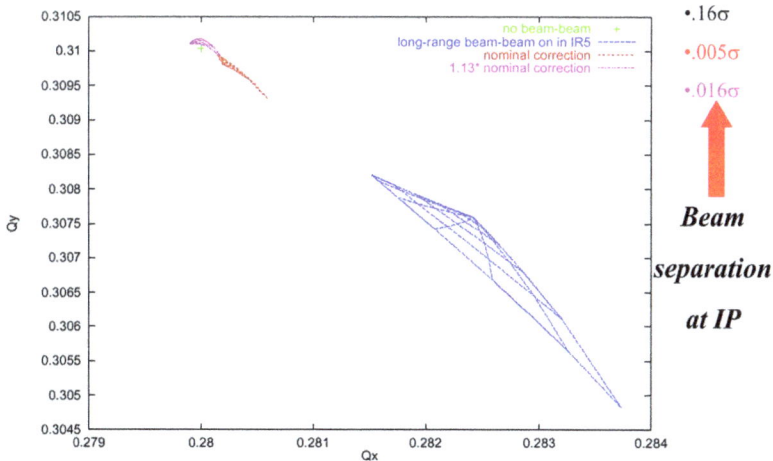

Fig. 4.   Simulated LHC tune footprint due to long-range collisions with and without wire compensator [6]. The abscissa shows the vertical tune from 0.3045 to 0.3105, the ordinate the horizontal tune from 0.279 to 0.284.

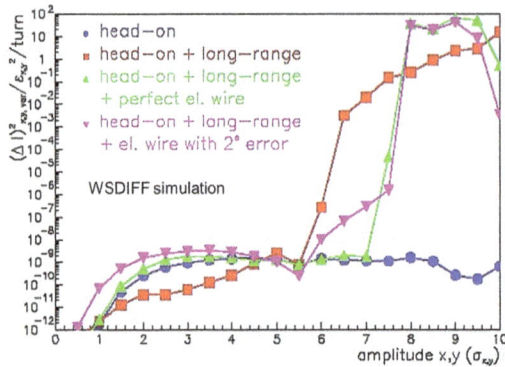

Fig. 5. Simulated LHC diffusive aperture with ideal (green) and realistic wire compensator (pink) compared with the case of no compensation (red) and head-on collisions only (blue) [8].

In simulations the wire compensator was shown to effectively shrink the tune spread caused by the long-range collisions to essentially zero [6] (Fig. 4) and to gain about 1.5σ in diffusive aperture [8] (Fig. 5).

Strong-strong beam–beam simulations including wire compensators were reported in Ref. [9], and further analytical studies of the onset of chaos due to the long-range collisions in Ref. [10].

3. CERN SPS Wire Compensators

In order to explore the 'simulated' effect of long-range encounters and to benchmark the simulations with the SPS beam, in 2002 a first prototype compensator was fabricated and installed. This BBLR consisted of two 80-cm long units (each with a wire length equal to 60 cm), installed one behind the other, and each containing a single water-cooled wire, vertically displaced from the beam centre. Two years later, a second BBLR was constructed, equipped with three wires of different transverse orientation. The second BBLR also consisted of two units of the same length, like the first one, but mounted on a movable support so that their vertical position could be varied over a range of 5 mm through remote control. A primary purpose of this second BBLR, installed at a betatron phase advance of about 3° from the first one (hence similar to the phase advance between the proposed location of the LHC wire compensator and the centre of the long-range collisions), was to demonstrate the effectiveness of a realistic compensation scheme, which could be simulated by powering the vertical wire(s) of the two BBLRs with opposite polarity. In addition to the vertical wire, a horizontal wire and a wire at 45° were added to allow for experimental studies and comparisons of various crossing schemes (horizontal-vertical, vertical-vertical, and 45°).

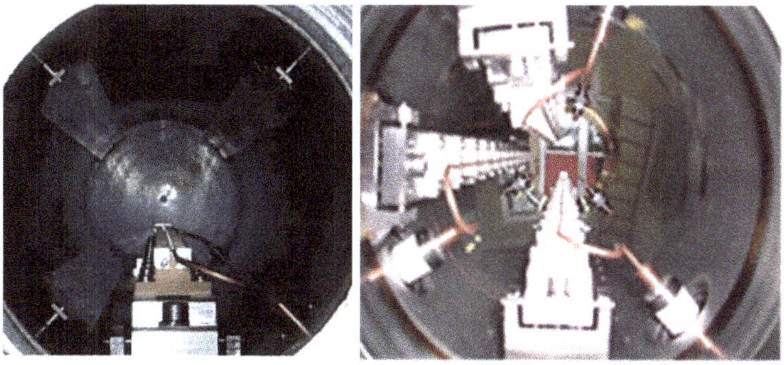

Fig. 6. The first (left) and the second prototype wire compensator (right) installed in the CERN SPS in 2002 and 2004, respectively.

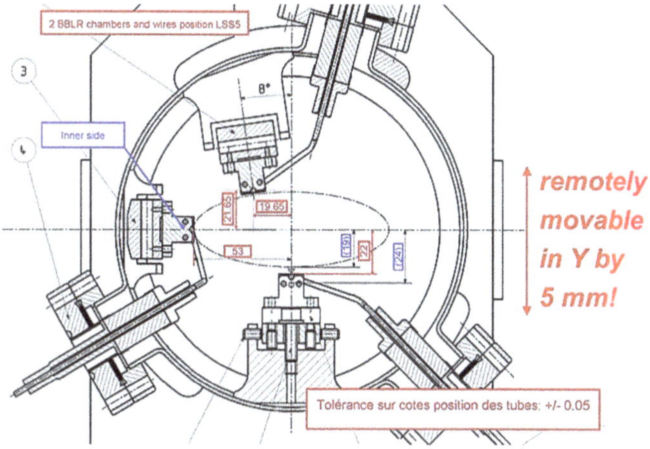

Fig. 7. Technical drawing of the second SPS wire compensator (2004).

Photographs of both devices are shown in Fig. 6, as well as a technical drawing in Fig. 7. The wire of the first BBLR is mounted at a fixed nominal vertical distance of 19 mm from the centre of the chamber (so that it is in the shadow of the SPS arc aperture). More details and documentation on the SPS wire compensator proto-types (and the experiments conducted in the SPS using these devices) can be found on a dedicated web site [11].

The needed wire current I_w is related to the number of long range collisions #*LR*, the length of the l_w and the bunch population N_b via $I_w = N_b ec \, \#LR/l_w$, where e denotes the elementary charge and c the speed of light. The two 60-cm long wires of one unit can be excited with up to 267 A of current, which, according to the above equation, produces an effect equivalent to 60 LHC LR collisions (e.g. roughly the combined effect of all nominal long-range encounters around IPs 1 and 5).

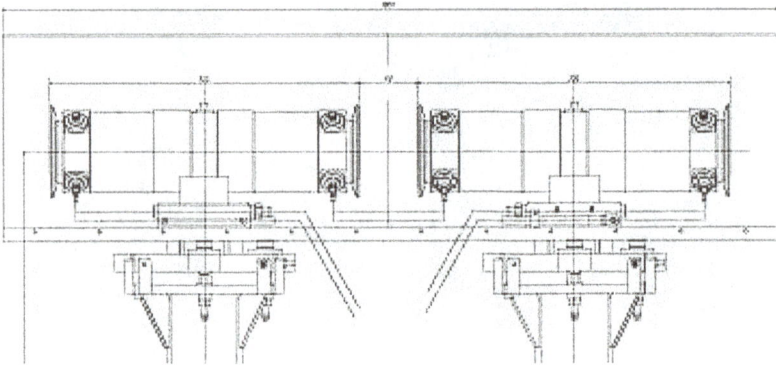

Fig. 8.   Side view of SPS BBLR no. 1.

Fig. 9.   SPS BBLRs nos. 1 and 2 (two pairs of Plexiglas boxes) installed in SPS Straight Section 5.

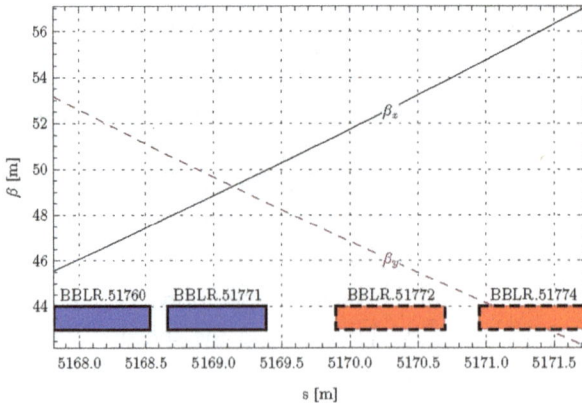

Fig. 10.   Horizontal and vertical beta functions across the two SPS BBLRs (each consisting of two units) [21].

Figure 8 presents a side view of the first BBLR device. Each BBLR, consisting of two units, has a total length of $(2 \times 0.8 + 0.25)$ m $= 1.85$ m. A photograph shows BBLRs 1 and 2 installed in the SPS tunnel (Fig. 9). Figure 10 illustrates the horizontal and vertical beta functions along the two $\times$ two BBLR units. The average value of the beta functions is about 50 m.

Additional compensator wire units are available at CERN. A complete BBLR consisting of two units with water-cooling, similar to BBLR no. 2, is ready (repaired after an earlier leak). Two air-cooled BBLRs from the Relativistic Heavy Ion Collider (RHIC) have been shipped from Brookhaven National Laboratory and are in store at CERN [12]. Thus, including the two BBLRs presently installed in the SPS, a total of five sets are (or have been) available.

4. Scaling Laws

The perturbation by the wire compensator at a distance d from the beam center is

$$\Delta y' = \frac{2 r_p l_w I_w}{\gamma e c (y - d)},$$

with γ denoting the relativistic Lorentz factor, from which the relative perturbation at the dynamic aperture becomes

$$\frac{\Delta y'}{\sigma_{y'}} = \left(\frac{2 r_p l_w}{ec} \right) \left(\frac{I_w}{(\gamma \varepsilon) \tilde{n}_{da}} \right),$$

where $\tilde{n}_{da}$ denotes the dynamic aperture in units of the rms beam size, l_w the wire length and I_w the wire current. This equation shows that, for constant normalized emittance, the effect in units of sigma is independent of energy and beta function. In *scaled experiments* the wire current is varied in direct proportion to the factor by which the emittance differs from the desired emittance.

5. History of SPS BBLR Studies

The SPS BBLRs were used to perform the following beam studies:

- perturbation by single wire as LHC LR simulator (2002 to 2003) [13, 14];
- two wire compensation, scaled experiments, distance scan (2004) [15, 16];
- tests of crossing schemes (2004) [15, 16, 17];
- one and two wires at different energies: 26, 37, and 55 GeV/c; scans of Q', distance, current (2007) [18, 19, 20];

- two-wire compensation with varying Q, I_w, Q' scans at 55 GeV/c (2008) [21, 22];
- two-wire compensation and excitation in coasts at 120 GeV/c (2009) [22]; and
- two-wire compensation and excitation in coasts at 55 GeV/c (2010) [23].

Figure 11 illustrates typical SPS cycles used towards the end of the last decade for BBLR studies at three different beam energies. During dedicated machine studies, at the target energy the SPS cycle could be stopped and the beam be made to 'coast' for, e.g. ten minutes for measurements of the beam lifetime in steady-state conditions and parameter scans.

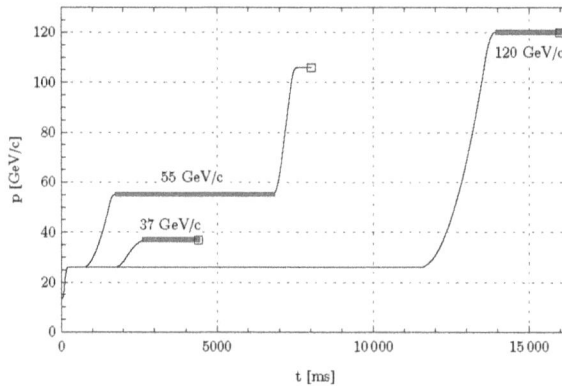

Fig. 11.   SPS cycles during experiments in 2008 and 2009 [21].

6. Technical Issues

A number of practical or technical issues had to be addressed, especially in the early days of the SPS BBLR studies. These included:

- installation of dedicated ion chambers and photomultiplier tubes (PMTs) near the BBLR;
- the addition of an inductive coil to suppress wire-current ripple;
- computation and experimental verification of wire heating;
- emittance blow-up by means of the transverse damper or by injection mismatch together with resonance crossing (equalizing the vertical and horizontal emittances) so as to achieve the nominal LHC parameters or to increase sensitivity;
- use of fast wire scanners and scrapers;
- installation of a dedicated dipole near the BBLR to correct the induced orbit change locally;
- continuous tune corrections;

- preparation and use of multiple superimposed orbit bumps to vary the beam-wire distance;
- (later) choice of higher beam energy: 37, 55 or 120 GeV/c (for good lifetime without wire excitation); and
- (later) experiments in coast (to avoid transient data).

Figure 12 illustrates the combination of orbit-corrector bumps used to vary the beam-wire distance at higher beam energy. The resulting minimum normalized distance in units of rms beam size depends on the beam energy and on the normalized emittance as shown in Fig. 13. In case the emittance was too small the beam could be blown up with transverse feedback and resonance crossing.

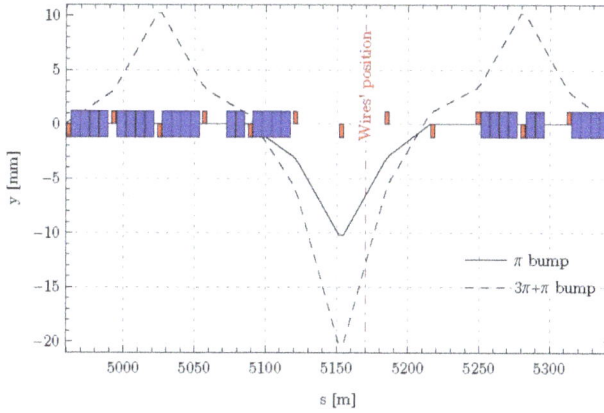

Fig. 12.   Superimposed 3+5 corrector bumps at the SPS wire compensator [21].

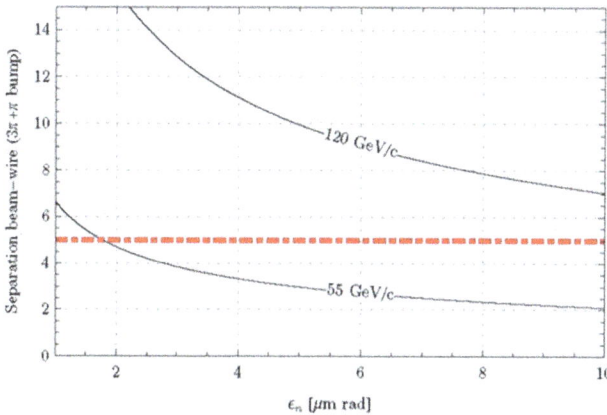

Fig. 13.   Minimum normalized distance in units of rms beam size as a function of normalized emittance for two beam energies [21]. The red line indicates a separation of 5σ, which corresponds to the minimum separation considered for levelling at the HL-LHC [21].

The natural SPS beam lifetime was about 30 h at 55 GeV/*c*, but only 5–10 min at 26 GeV/*c* (where the physical aperture was only about 4σ).

7. Single BBLR 'Excitation' Studies

Changes in orbit and tunes allow for a precise determination of the beam-wire distance. Example data from 2002 are shown in Figs. 14–16.

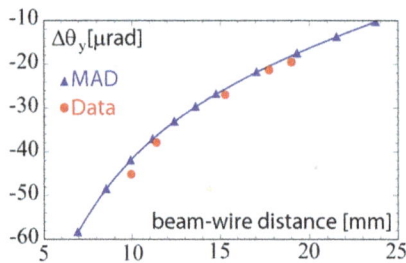

Fig. 14. Deflection angle at the wire compensator as a function of beam-wire distance, comparing data and measurements [14].

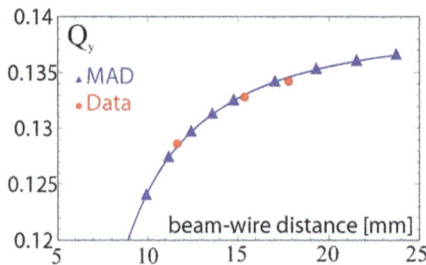

Fig. 15. Vertical tune change as a function of beam-wire distance, comparing data and measurements [14].

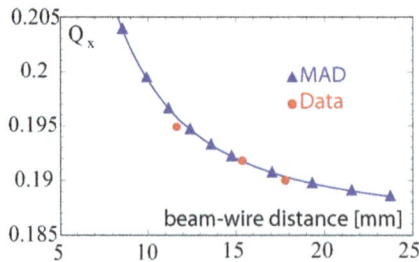

Fig. 16. Horizontal tune change as a function of beam-wire distance, comparing data and measurements [14].

The change in the beam orbit at the wire compensator, Δd, follows from the self-consistent equation

$$\Delta d = \frac{\beta_y I_{\text{wire}} l_{\text{wire}} r_p}{\gamma e c (d + \Delta d) \tan\left(\pi\left(Q_y + \Delta Q_y\right)\right)},$$

with Δd appearing on both sides, while the tune changes are given by

$$\Delta Q_{x,y} = \mp \frac{r_p \beta_{x,y} I_{\text{wire}} l_{\text{wire}}}{2\pi\gamma e c} \frac{1}{(d + \Delta d)^2}.$$

In most of the later studies only the tune change was monitored.

The effect of the BBLR wire on the nonlinear optics has also been studied, by acquiring turn-by-turn beam-position monitor (BPM) data after kicking the beam. The nonlinearity of the wire field resulted in a reduced decoherence time, due to an increased tune shift with amplitude, and in (additional) spectral resonance lines. The measured tune shift was consistent with the theoretical predictions

$$\Delta Q_x \approx \frac{3}{4} \frac{I_{\text{wire}} l_{\text{wire}} r_p}{\gamma e c} \frac{\beta_x}{d^4} \hat{y}^2$$

and

$$\Delta Q_y \approx -\frac{3}{8} \frac{I_{\text{wire}} l_{\text{wire}} r_p}{\gamma e c} \frac{\beta_x}{d^4} \hat{y}^2,$$

where $\hat{y}$ denotes the peak vertical betatron oscillation amplitude of the beam after a kick.

The resonance lines introduced by the BBLR are illustrated in Fig. 17.

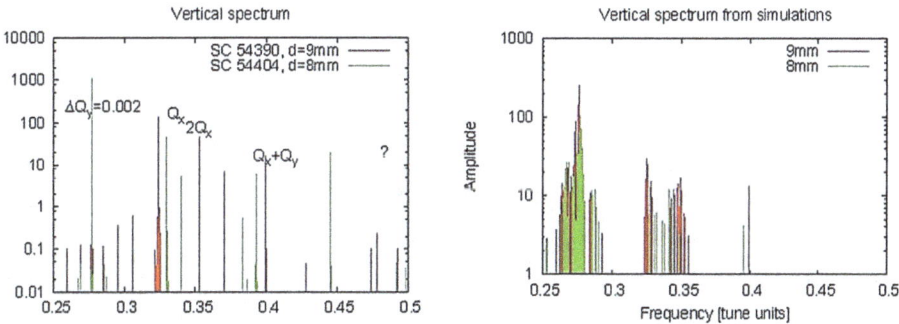

Fig. 17. Resonance spectra with wire excitation: experimental data with 240 A wire current at 9 (red) and 8 mm (green) beam-wire distance (left) and the corresponding simulation data (right) [18].

A strong effect of chromaticity was noticed when the compensator was excited. Figure 18 shows the beam intensity evolution during $Q_{x,y'}$ scans at 37 GeV/c.

Figure 19 compares the measured (left) and simulated beam loss (right) for two different values of the vertical chromaticity as a function of the integrated wire strength.

Various attempts were made to directly measure the 'diffusive' or dynamic aperture. To this end, three types of signals were used: (1) lifetime and background, (2) beam profiles and final emittance, and (3) local diffusion rate inferred by scraper-retraction experiments. Figures 20 and 21 present example measurements of lifetime and background at 55 GeV/c. A drop in the lifetime and increased losses are observed for separations less than 9σ; at 7–8σ separation the lifetime decreases to 1–5 h. These results indicated that the LHC nominal separation of 9.5σ for the encounters between the IP and the first quadrupole Q1 is well chosen,

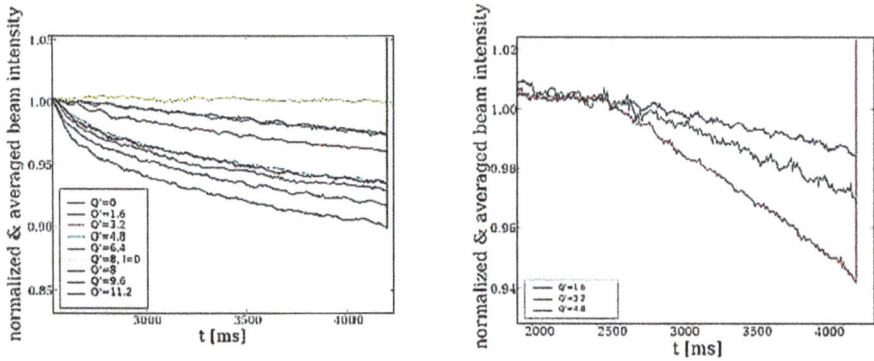

Fig. 18. Beam intensity as a function of time for various values of the horizontal ($Q_{x'}$, left) or vertical chromaticity ($Q_{y'}$, right) [20]. The wire excitation was 180 A-m, the beam momentum 37 GeV/c and the normalized beam-wire separation about 6.5σ (9 mm).

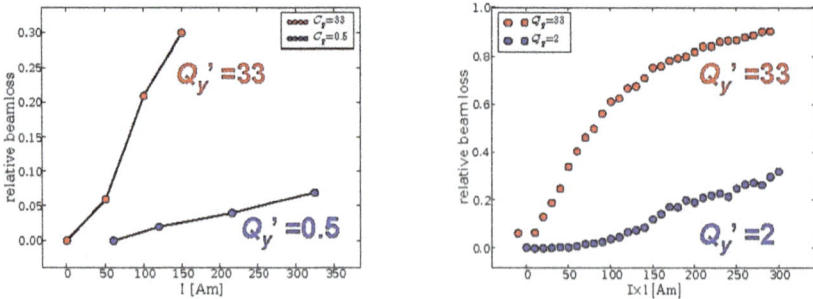

Fig. 19. Relative beam loss for two different values of the vertical chromaticity as a function of wire excitation in units of A-m, comparing experimental data (left) and simulations (right) [20]. The beam-wire separation was ~ 6.6σ.

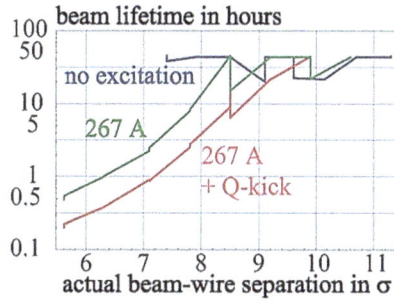

Fig. 20. Lifetime as a function of the wire-beam separation in units of rms beam size with (green) and without (blue) wire excitation at 267 A, which corresponds to the nominal total number of LHC long-range encounters at IPs1 and 5. The red data were also taken with the wire excited, while in addition firing the (weak) tune kicker to add a further perturbation.

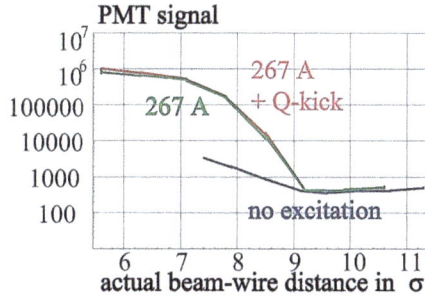

Fig. 21. Local relative beam loss rate measured by a photomultiplier as a function of the wire-beam separation in units of rms beam size with (green) and without (blue) a wire excitation of 267 A. As in Fig. 20 for the red data set the (weak) tune kicker was repeatedly fired while the wire was excited.

but 'close to the edge', assuming that the wire current of 267 A does indeed correspond to the long-range beam–beam effect in the LHC.

Beam profiles before and after wire excitation, measured with an SPS 'wire scanner' (fully unrelated to the wire compensator), reveal that the particles at large transverse amplitude are lost due to the wire excitation; see Fig. 22. These measurements confirmed that the wire compensator or the equivalent set of long-range encounters, acts as a highly effective diffusion enhancer and after some turns has an effect quite similar to the one of a physical scraper.

This type of measurement allows for an estimate of the diffusive/dynamic aperture. Specifically, an Abel transformation of the wire-scan data of the form [23, 24]

$$\rho(A) = -2A \int_A^R d\eta \frac{g'(\eta)}{\sqrt{\eta^2 - A^2}}$$

can be used to compute the change in the (normalized) amplitude distribution due to the wire excitation. For the data of Fig. 22 the results are presented in Fig. 23, indicating that in this particular example (with intentionally small separation) the dynamic aperture is at about 1σ).

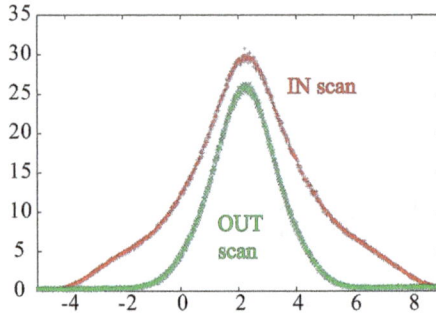

Fig. 22.   Beam profile from a wire scan before and after compensator-wire excitation measured at 26 GeV/c. The abscissa shows the wire-scan detector signal, the ordinate the (un-calibrated) wire-scanner position. The inferred initial and final emittances were 3.40 μm and 1.15 μm, respectively.

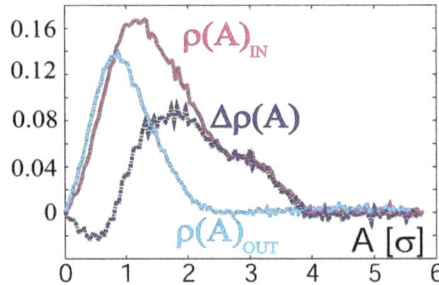

Fig. 23.   Abel transformation of the beam-profile data from Fig. 22, revealing the change in the (normalized) amplitude distribution. The abscissa shows the normalized density or density difference, the ordinate the amplitude in units of the initial rms beam size σ.

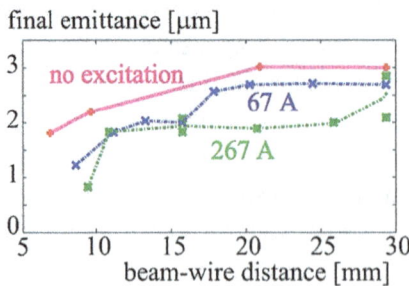

Fig. 24.   Final emittance without (red) and with wire excitation (blue 67 A, green 267 A) at a beam momentum of 26 GeV/c as a function of beam-compensator distance.

Figure 24 displays the final emittance inferred from the beam profiles as a function of beam-wire distance without wire excitation and for wire currents of 67 A and 267 A (the latter corresponding to 60 LHC long-range encounters). The reduction of the final emittance without wire excitation at smaller distances is due to mechanical scraping of the beam by the edge of the wire.

With the Abel-transformation technique it was not always possible to obtain a clean result for the diffusive aperture. Therefore, a different technique was also employed to infer the variation of the diffusive/dynamic aperture. Namely, without wire excitation, a known aperture restriction was introduced using a dedicated mechanical 'scraper,' and wire scans were then executed to determine the 'final emittance' corresponding to a given known aperture determined by the scraper position. This calibration measurement is presented in Fig. 25 — the curve of measured final emittance as a function of scraper position allows estimation of the effective aperture due to the wire excitation from the associated 'final emittance' value.

Following this plan and using the calibration line of Fig. 25, measurement results for different wire currents were converted into normalized diffusive apertures. The result, shown in Fig. 26, suggests a linear dependence of the

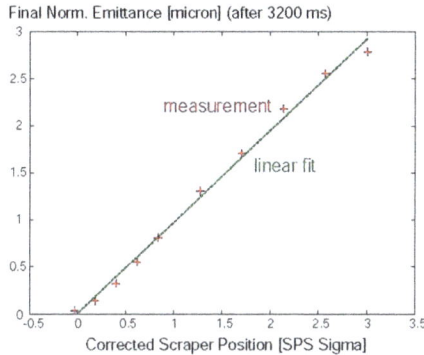

Fig. 25. Calibration of the final emittance values by a mechanical scraper.

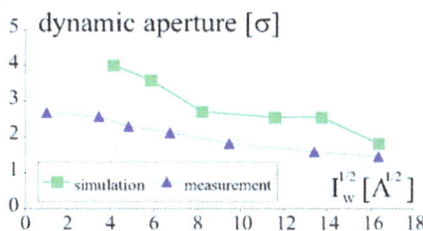

Fig. 26. Effect of wire current on SPS dynamic aperture (26 GeV/c), inferred from final emittance and the calibration of Fig. 26.

dynamic aperture on the square root of the wire current, which is consistent with a scaling law first pointed out by Irwin [4]. In the figure, the measured dynamic aperture is smaller than the simulated diffusive aperture, especially at lower current, hinting at additional effects not included in the simulations or at a systematic error in the calibration method.

Yet another approach to measuring the diffusive aperture is to directly detect the diffusion rates at various transverse amplitudes, by inserting a scraper to remove particles in a small area around the target amplitude article, then retracting this scraper by a small step, and observing how the loss signal reappears as particles diffuse outwards to the new position of the scraper. This type of measurement was previously used at HERA (and elsewhere) to determine the local diffusion coefficients [25]. Unfortunately, scraper retraction attempts for the SPS wire-compensator studies were not very successful.

One of the most interesting results from the SPS wire measurements is the measured dependence of the 'beam lifetime' τ_{beam}, as inferred from the beam loss during a cycle at 26 GeV/c, on the beam-wire distance d [15], illustrated in Fig. 27. The measured dependence extremely well follows a 5th order power law as seen from the fitting result embedded in the figure (another fit, with an exponential law, is shown as well). It has been suggested [26] that a nearby low-order resonance of order n should cause a dependence $\tau_{\text{beam}} \sim 1/d^{n+1}$ and that the power in the exponent should, therefore, depend on the betatron tunes. Indeed at the Tevatron (with an electron lens applied as 'wire') [27] and at RHIC [28], operating at other working points in the tune diagram, different power laws were observed (third power and linear dependence, respectively). Figure 28, presenting SPS data for three different sets of tunes, taken several years later at a higher energy, confirms that the losses due to the wire are strongly tune dependent.

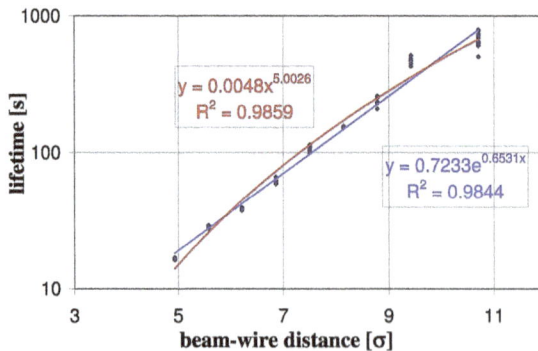

Fig. 27. Beam lifetime as a function of beam-wire distance at 26 GeV/c, for betatron tunes of $Q_x = 0.321$ and $Q_y = 0.291$ [14].

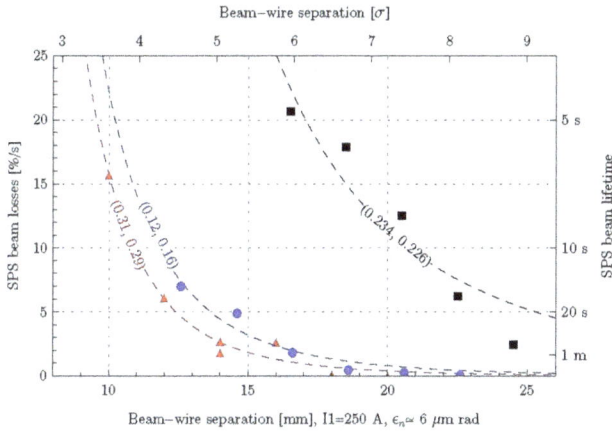

Fig. 28.   Beam losses as a function of beam-wire distance for three different pairs of tunes at 37 GeV/*c* with 1.1 s cycle [21].

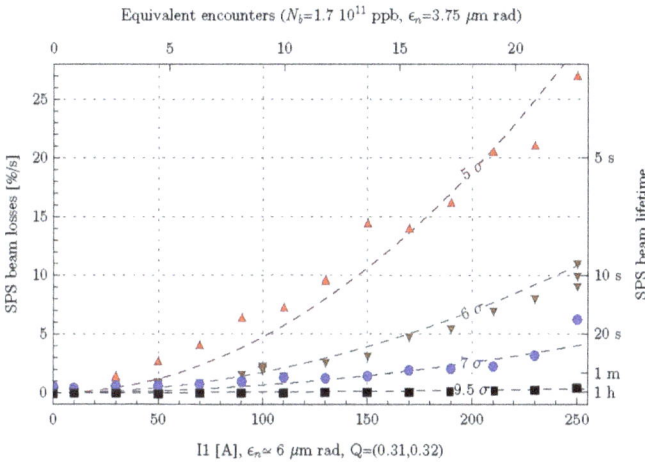

Fig. 29.   Beam losses as a function of wire current at 37 GeV/*c* with a 1.1 s cycle, for betatron tunes of $Q_x = 0.31$ and $Q_y = 0.32$ (nominal values for LHC collisions) [21].

Extrapolating the measurement of Fig. 27 to the nominal LHC beam–beam distance, $\sim 9.5\sigma$, predicts a 6 min lifetime. This result was one of the motivations for raising the SPS beam energy and for performing measurements with coasting (non-cycling machine) beams in later studies, where the beam lifetimes were significantly higher.

Figure 29 shows beam losses as a function of wire current I_w for different normalized beam-wire separations d_n (in units of σ). These later results were fitted as [21]:

$$\text{beam loss } (\%) = 0.07\, e^{-d_n} I_w^2 .$$

8. Studies for Wire Compensators in the LHC

For a successful compensation of long range beam–beam effects the location of the compensator must fulfil the following requirements:

- No phase advance to the origin of the effect (or a phase advance equal to a multiple of π); ideally the wire should be located in the drift space before the final-focus quadrupoles, i.e. the longitudinal space where the parasitic long range encounters occur, which is not possible (exactly due to the presence of the second beam).
- A region where the values of the horizontal and vertical beta functions are equal.
- In order to get the compensation correct for all multipoles the transverse location of a wire compensator must be on the **inside** of the compensated beam, i.e. between the two circulating beams. This poses a significant constraint, since in the ideal longitudinal position the transverse separation of the beams is only a few centimeters. [The strength the nth multipole scales as I_w/d^n, i.e. linearly with the wire current I_w, and with the nth power of the inverse separation d, including the sign of the separation. Hence, placing the wire compensator on the outside with opposite current would not be a solution — this would double the magnitude of every second multipole field excited by the long-range collisions, instead of cancelling it.]

In a thesis [29] the compensation scheme for the LHC was studied in detail using simulations with the BBTRACK code. Figure 30 shows the layout of an LHC high luminosity insertion (top) and the related beta functions for the nominal optics

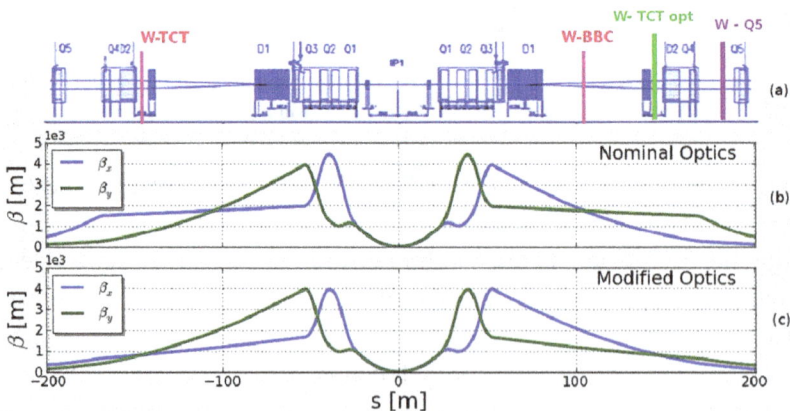

Fig. 30.   Schematic layout of a LHC high luminosity insertion (top) and the related beta functions for the nominal optics and another proposed modified optics [29].

and another proposed modified optics. The ideal position for a wire compensator is indicated as "W-BBC". In this location a final compensation scheme would be implemented. Since the compensation will be local, one compensator for each beam and in each high luminosity IP will be needed, hence a total of four compensators.

Furthermore, Fig. 30 indicates another location (W-TCT), the location of the TCT collimators. In the planned upgrade of the LHC collimation system at these locations new collimators will be installed in the time frame between LS1 and LS2 (2015–2018). In order to validate the concept of the BBLRs a demonstrator project is planned, for which wires will be integrated into collimator jaws. Therefore the positions of the collimators have also been included in the simulations.

9. Simulation Results

As indicators for the effectiveness of the compensation results two independent methods were used in [29]:

(a) the stability analysis of particle trajectories over many terms expressed by the value of the Lyapunov exponent [30]; and
(b) the tune footprint of particles in the lattice resonance diagram.

Typical examples of these studies are shown in Figs. 31 and 32.
As main results we can report the following [29]:
Considering a 1 m long wire at the TCT location, the best compensation of the BBLR effect can be obtained with a DC wire current of 177 A, per IP, at a distance of 9.5σ to the beam. By geometrical scaling a current of 237 A is needed at a

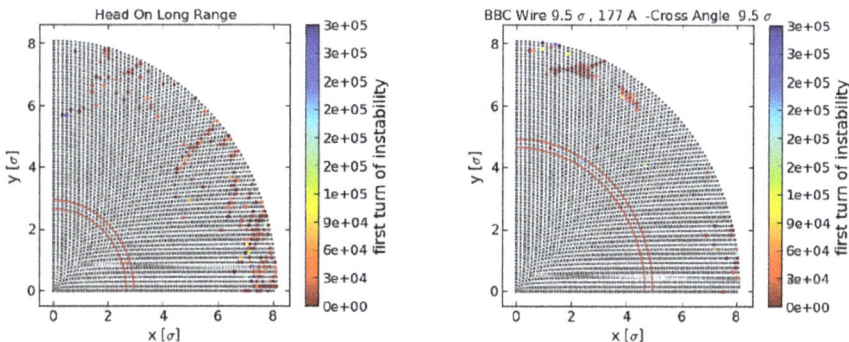

Fig. 31.   Two dimensional stability diagram for simulated collisions in the LHC (left) and for collisions with additional BBLR compensation (right). Unstable particles are identified by red dots in the figures [29]. The colour indicates the number of turns after which a trajectory is diagnosed as unstable through the irregular, exponential evolution of the phase-space distance between two twin particles. Points without colour are considered to be stable.

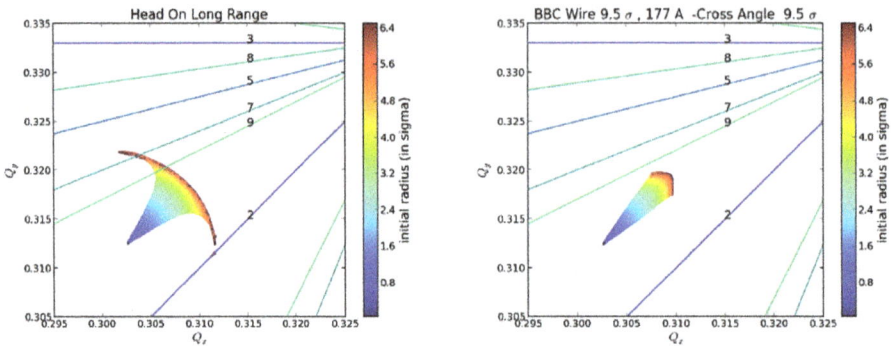

Fig. 32. Two dimensional tune diagram for simulated collisions in the LHC (left) and for collisions with additional BBLR compensation (right). The BBLR compensation changes the tune of the particles, which are affected by the long range beam force, back to the tune of the unperturbed beam [29].

distance of 11σ. These wire current values correspond to a symmetric layout with one compensator left of the IP and another — for the second IP — on the right side. This setup will be necessary in the TCT locations, since the ratio of the horizontal and vertical beta functions is not equal to one (see Fig. 30). [The LHC currents computed here are lower than the 267 A quoted above for the SPS experiments, performed with a 1.2 m long wire, since the latter represented the accumulated nominal LHC long-range beam–beam effect from two IPs.]

The LHC wire compensators at the TCT location should allow reducing the crossing angle from 12σ without wire to 9.5σ with wire, or from 9.5σ without wire to 8σ with wire, at comparable stability in phase space (and, consequently, at similar beam lifetimes) [29].

10. Demonstrator Setup

Due to machine-protection considerations, the integration of DC powered wires into collimator jaws seems to be the only possibility to enable some beam tests before embarking on a final implementation of the wires for the LHC high luminosity operation. This integration requires the solution of several important technical issues:

- No interference of the wires with the nominal operation of the collimators;
- Transfer of 1 kW resistive heat loss in the wire by heat conduction to the water cooled collimator jaw;
- Shielding of the wire inside the collimator through a thin metallic layer towards the beam for impedance reasons.

Figure 33 sketches the implementation principle, where the 1 mm thick wire is shown inside a small tungsten block, which is embedded in the collimator jaw [31]. Figures 34 and 35 present further integration details of the complex mechanical engineering aspects of the wire integration.

At the earliest the wire equipped collimators will be installed in the LHC in the winter shutdown 2015/2016.

After the installation machine experiments are planned, which aim to validate the coherence of predictions by simulation with machine experiments. Beam transfer functions, transverse particle distributions and bunch lifetimes will be used as observables. This setup will be the first time, at CERN, that one attempts to observe a compensation of the long-range beam–beam effects by observing an improvement of critical beam operation parameters (e.g. lifetime and losses). All past experiments in the SPS "only" demonstrated that a wire can deteriorate beam operation parameters and that two neighbouring wires can compensate each other.

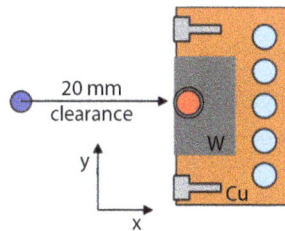

Fig. 33. Schematic layout of a wire integrated into a collimator jaw [31].

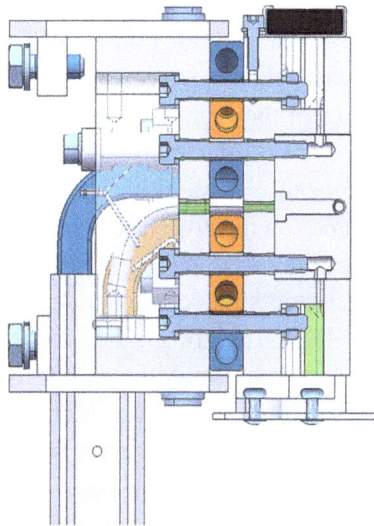

Fig. 34. Technical drawing of the wire (right side) in the collimator jaw [31].

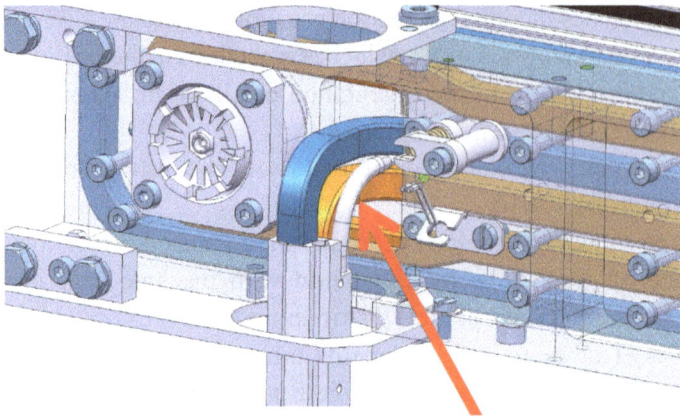

Fig. 35.   Technical detail of the integration: Outside the space in which the wire is cooled by heat conduction, the wire diameter had to be increased in order to avoid the burning of the wire at nominal current (arrow) [31].

Some previous attempts to compensate long-range beam–beam effects with a wire have been made at RHIC [32] (considering a single long-range collision) and at DAFNE [33] (with lepton beams, and successfully used in actual operation).

11. Conclusions and Outlook

From the SPS (and similar plus even two-beam experiments in RHIC) we have first indications of BBLRs affecting the particle beams in the predicted way.

Simulations show a potential gain of using these wires for gaining dynamic aperture, hence giving the possibility to reduce the crossing angles, and therefore producing higher luminosity with the same beams.

For future wire BBLRs in the high luminosity period of the LHC, 8 m long sections are reserved at a distance of about 100 m from each IP1 and IP5. In terms of beam dynamics these locations would be ideal for BBLRs, whereas the high flux of secondary neutrons will pose severe constraints for any technical realization. In order to be able to place (and align) a "wire" in these locations, in particular **in between** the two counter-rotating proton beams only an electron beam similar to well established electron coolers is envisaged at the present moment. Machine Protection considerations would be another argument for using electron beams as wires, since these would be non-destructible devices that can be placed at amplitudes smaller than that of the tertiary and secondary collimators.

An implementation sketch is shown in Fig. 36. Obviously the usage of an electron beam would allow a pulsed operation and hence even individual compensation of each proton bunch.

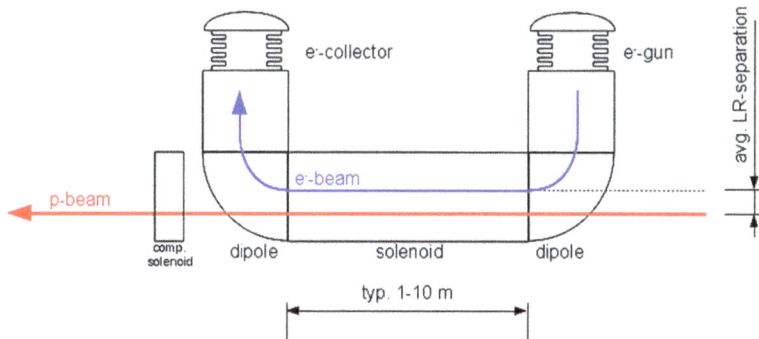

Fig. 36. Sketch of an electron cooler type layout for a non-destructible wire BBLR [31].

On the other hand such an implementation is beyond what has been achieved so far: Assuming an effective length of 6 m on both sides of the IP an electron beam current of about 30 A would be needed in order to produce the field equivalent to a 1 m long wire powered at 177 A. This is about an order of magnitude larger than what has been achieved for the Tevatron Electron Lens [27, 34] and with the electron lenses at RHIC [35]. A factor of 2 could be gained by placing a second electron beam wire on the other side of the IP, resulting in a total of eight installations.

Acknowledgments

The reported SPS wire experiments would not have been possible without the ideas, help and important contributions of Gerard Burtin, Rama Calaga, Jackie Camas, Gijs de Rijk, Octavio Dominguez, Ulrich Dorda, Jean-Pierre Koutchouk, Elias Métral, Yannis Papaphilippou, Federico Roncarolo, Tanaji Sen, Vladimir Shiltsev, Guido Sterbini, Rogelio Tomas, and Jörg Wenninger.

For the proposed BBLR in the LHC the contributions of Tatiana Rijoff, Ralph Steinhagen, Stephane Fartoukh, Tatiana Pieloni, Stefano Redaelli, Alessandro Bertarelli, Guillaume Maitrejean, Luca Gentini, Gianluigi Arduini and Yannis Papaphilippou are warmly acknowledged.

References

[1] D. Neuffer and S. Peggs, Beam-Beam Tune Shifts and Spreads in the SSC: Head-On, Long Range and PACMAN conditions, SSC-63 (1986).
[2] W. Herr, Tune Shifts and Spreads due to the Long-Range Beam-Beam Effects in the LHC, CERN/SL/90-06 (AP) (1990).
[3] W. Chou and D. M. Ritson, Dynamic aperture studies during collisions in the LHC, CERN LHC Project Report 123 (1998).

[4] J. Irwin, Diffusive Losses from SSC Particle Bunches due to Long Range Beam-beam Interactions, SSC-223 (1989).

[5] Y. Papaphilippou and F. Zimmermann, Weak-strong beam-beam simulations for the Large Hadron Collider, *Phys. Rev. ST Accel. Beams* **2**, 104001 (1999).

[6] J.-P. Koutchouk, Principle of a Correction of the Long-Range Beam-Beam Effect in LHC using Electromagnetic Lenses, LHC Project Note 223 (2000).

[7] J.-P. Koutchouk, Correction of the Long-Range Beam-Beam Effect in LHC using Electromagnetic Lenses, SL Report 2001-048 (2001).

[8] F. Zimmermann, Weak-Strong Simulation Studies for the LHC Long-Range Beam-Beam Compensation, Beam-Beam Workshop 2001 FNAL; LHC Project Report 502 (2001).

[9] J. Lin, J. Shi and W. Herr, Study of the Wire Compensation of Long-Range Beam-Beam Interactions in LHC with a Strong-Strong Beam-Beam Simulation, EPAC 2002, Paris (2002).

[10] Y. Papaphilippou and F. Zimmermann, Estimates of diffusion due to long-range beam-beam collisions, *Phys. Rev. ST Accel. Beams* **5**, 074001 (2002).

[11] CERN Beam-Beam Compensation web site at http://cern-ab-bblr.web.cern.ch/cern-ab-bblr.

[12] The RHIC wires were kindly provided by M. Minty and T. Curcio.

[13] J.-P. Koutchouk, J. Wenninger and F. Zimmermann, Compensating Parasitic Collisions using Electromagnetic Lenses, presented at ICFA Beam Dynamics Workshop on High-Luminosity e+e- Factories ('Factories'03') SLAC; in CERN-AB-2004-011-ABP (2004).

[14] J.-P. Koutchouk, J. Wenninger and F. Zimmermann, Experiments on LHC Long-Range Beam-Beam Compensation in the SPS, EPAC'04 Lucerne (2004).

[15] F. Zimmermann, J.-P. Koutchouk, F. Roncarolo, J. Wenninger, T. Sen, V. Shiltsev and Y. Papaphilippou, Experiments on LHC Long-Range Beam-Beam Compensation and Crossing Schemes at the CERN SPS in 2004, PAC'05 Knoxville (2005).

[16] F. Zimmermann, Beam-Beam Compensation Schemes, in *Proc. First CARE-HHH-APD Workshop (HHH-2004)*, CERN, Geneva, Switzerland, CERN-2005-006, p. 101 (2005).

[17] F. Zimmermann and U. Dorda, Progress of Beam-Beam Compensation Schemes, in *Proc. 2nd CARE-HHH-APD Workshop on Scenarios for the LHC Luminosity Upgrade*, Arcidosso, Italy, 2005, CERN-2006-008 (2005).

[18] U. Dorda, J-P. Koutchouk, R. Tomas, J. Wenninger, F. Zimmermann, R. Calaga and W. Fischer, Wire Excitation Experiments in the CERN SPS, in *Proc. EPAC08*, Genoa (2008).

[19] U. Dorda and F. Zimmermann, Wire Compensation: Performance, SPS MDs, Pulsed System, in *Proc. IR07*, p. 98, CERN-2008-006 (2007).

[20] U. Dorda, Compensation of long-range beam-beam interaction at the CERN LHC, PhD Thesis, Vienna TU., CERN-THESIS-2008-055 (2008).

[21] G. Sterbini, An Early Separation Scheme for the LHC Luminosity Upgrade, PhD Thesis EPFL, CERN-THESIS-2009-136 (2009).

[22] G. Sterbini, R. Calaga *et al.*, unpublished; see CERN BBLR web site in Ref. [11].

[23] P. W. Krempl, The Abel-type Integral Transformation with the Kernel (t2-x2)-1/2 and its Application to Density Distributions of Particle Beams, CERN Note MPS/Int. BR/74-1 (1974).

[24] C. Carli, A. Jansson, M. Lindroos and H. Schönauer, A Comparative Study of Profile and Scraping Methods for Emittance Measurements in the PS Booster, *Particle Accelerators* **63**, 255–277 (2000).

[25] M. Seidel, The Proton Collimation System of HERA, PhD Thesis, U. Hamburg, DESY-94-103 (1994).

[26] V. Shiltsev, private communication, 28 November 2004.

[27] F. Zimmermann, P. Lebrun, T. Sen, V. Shiltsev and X. L. Zhang, Using the Tevatron Electron Lens as a Wire and Other TEL Studies at FNAL, CERN AB-Note-2004-041 (2004).

[28] W. Fischer, R. Calaga, U. Dorda, J.-P. Koutchouk, F. Zimmermann, V. Ranjbar, T. Sen, J. Shi, J. Qiang and A. Kabel, Observation of Long-Range Beam-Beam Effect in RHIC and Plans for Compensation, EPAC'06, Edinburgh (2006).

[29] T. Rijoff, Testing long range beam-beam compensation for the LHC luminosity upgrade, master thesis, Milan U., CERN-THESIS-2012-377 (2012).

[30] F. Schmidt, F. Willeke and F. Zimmermann, Comparison of methods to determine long-term stability in proton storage rings, *Part. Accel.* **35**, 249 (1991).

[31] R. Steinhagen, LHC Beam-Beam Compensator – Status Update, 3rd Joint HiLumi LHC LARP Annual Meeting, Daresbury Laboratory (2013).

[32] W. Fischer *et al.*, Long-range and head-on beam-beam compensation studies in RHIC with lessons for the LHC, in *Proc. Final CARE-HHH Workshop on Scenarios for the LHC Upgrade and FAIR*, Chavannes-de-Bogis, Switzerland, 24–25 Nov. 2008, CERN Yellow Report CERN-2009-004, pp. 92–101 (2009).

[33] C. Milardi, D. Alesini, M. A. Preger, P. Raimondi, M. Zobov and D. Shatilov, DAFNE Lifetime Optimization with Octupoles and Compensating Wires, in *Proc. CARE-HHH-APD Workshop on Interaction Regions for the LHC Upgrade, DAFNE, and SuperB*, Frascati, Italy, 6–9 Nov. 2007, CERN Yellow Report CERN-2008-006, pp. 92–97 (2008).

[34] V. Shiltsev, Y. Alexahin, K. Bishofberger, V. Kamerdzhiev, V. Parkhomchuk, V. Reva, N. Solyak, D. Wildman, X.-L. Zhang and F. Zimmermann, Experimental Studies of Compensation of Beam-Beam Effects with Tevatron Electron Lenses, *New J. Phys.* **10**, 043042 (2008).

[35] W. Fischer *et al.*, First Experience with Electron Lenses for Beam-Beam Compensation in RHIC, in *Proc. IPAC14*, Dresden (2014).

Chapter 15

Impedance and Component Heating

E. Métral, F. Caspers, N. Mounet, T. Pieloni and B. Salvant

CERN, BE Department, Genève 23, CH-1211, Switzerland

The impedance is a complex function of frequency, which represents, for the plane under consideration (longitudinal, horizontal or vertical), the force integrated over the length of an element, from a "source" to a "test" wave, normalized by their charges. In general, the impedance in a given plane is a non-linear function of the test and source transverse coordinates, but it is most of the time sufficient to consider only the first few linear terms. Impedances can influence the motion of trailing particles, in the longitudinal and in one or both transverse directions, leading to energy loss, beam instabilities, or producing undesirable secondary effects such as excessive heating of sensitive components at or near the chamber wall, called beam-induced RF heating. The LHC performance limitations linked to impedances encountered during the 2010-2012 run are reviewed and the currently expected situation during the HL-LHC era is discussed.

1. Introduction

As the beam intensity increases, the beam can no longer be considered as a collection of non-interacting single particles: in addition to the "single-particle phenomena", "collective effects" become significant [1]. At low intensity a beam of charged particles moves around an accelerator under the Lorentz force produced by the "external" electromagnetic fields (from the guiding and focusing magnets, RF cavities, etc.). However, the charged particles also interact with their environment, inducing charges and currents in the surrounding structures, which create electromagnetic fields called wake fields. In the ultra-relativistic limit, causality dictates that there can be no electromagnetic field in front of the beam, which explains the term "wake". It is often useful to examine the frequency content of the wake field (a time domain quantity) by performing a Fourier transform on it. This leads to the concept of impedance (a frequency domain quantity), which represents, for the plane under consideration (longitudinal, horizontal or vertical), the force integrated over the length of an element, from a "source" to a "test" wave (as a function of their frequency), normalized by their charges. In general, the impedance in a given plane is a non-linear function of the test and source transverse coordinates, but it is most of the time sufficient to consider only the first few linear terms.

The wake fields (or impedances) can influence the motion of trailing particles, in the longitudinal and in one or both transverse directions, leading to energy loss, beam instabilities, or producing undesirable secondary effects such as excessive heating of sensitive components at or near the chamber wall (called beam-induced RF heating). Therefore, in practice the elements of the vacuum chamber should be designed to minimise the self-generated (secondary) electro-magnetic fields. For example, chambers with different cross-sections should be connected with tapered transitions; unnecessary cavities should be avoided; bellows should preferably be separated from the beam by shielding; plates should be grounded or terminated to avoid reflections; poorly conductive materials should be coated with a thin layer of very good conductor (such as copper) when possible, etc. In the case of beam instabilities, fortunately some stabilizing mechanisms exist, such as Landau damping, electronic feedback systems and linear coupling between the transverse planes if, for a transverse coherent instability, one plane is more critical than the other. Moreover, several beam or machine parameters can partly mitigate instabilities. All this translates into knobs which can be used in the control room to damp these instabilities: (i) transverse chromaticities, (ii) Landau octupoles current, (iii) gain(s) of the electronic feedback system(s) and (iv) bunch length (and/or longitudinal profile). In the case of the beam-induced RF heating, the bunch length (and sometimes longitudinal profile) is the main parameter (once the bunch intensity and number of bunches have been fixed): usually, the longer the bunch, the better.

Despite the excellent performance of the LHC in 2012, with a record peak luminosity at 4 TeV corresponding to 77% of the 7 TeV design luminosity of 10^{34} cm^{-2}s^{-1} (mainly thanks to the 50 ns bunch spacing beam with a brightness higher than nominal by more than a factor of two), the intensity ramp-up was perturbed by several types of transverse instabilities as well as beam-induced RF heating in many equipment. All these limitations need to be fully understood to allow future operation during the HL-LHC era [2]. This work is ongoing [3, 4, 5].

The performance limitations (linked to the impedances) encountered in 2010–2012 are reviewed in Section 2, while the currently expected situation during the HL-LHC era is discussed in Section 3.

2. 2010–2012 Experience

2.1. *Transverse impedance model and beam instability*

A transverse instability remained at the end of the 2012 run, at the end of the betatron squeeze (at 4 TeV), with about maximum Landau octupoles current (550 A) and maximum transverse damper (called ADT) gain (corresponding to a damping time of 50 turns at 4 TeV), after having increased the transverse chromaticities to about 15 units. This high value for the chromaticities was suggested after a new analytical approach, which includes the effect of the transverse damper [6, 7]. Some tests, changing the bunch length and/or the longitudinal profile, were performed but the instability did not disappear. Furthermore, increasing the beam energy from 4 TeV to 7 TeV will help for the (head-tail) instability rise-times which are proportional to the energy but it will be much more critical for the Landau octupoles current needed to introduce the required incoherent tune spread, which is proportional to the square of the beam energy. Taking the two effects together, increasing the beam energy from 4 TeV to 7 TeV means that more Landau octupoles current (by a factor $7/4 = 1.75$) is needed. Moreover, it should be also more difficult to produce the same damping time for the transverse damper at a higher energy. This un-cured instability is therefore a potential worry for future operation at higher energies (and higher beam intensities) and a major concern for the HL-LHC (even if for the HL-LHC era the collimators might be coated with Molybdenum, which should help for the transverse beam stability). Work is still on-going to try and better understand what happened but the current lessons from 2012 are the following:

(1) The impedance model and Landau damping mechanism with one beam only is reasonably well understood as measurements revealed a mismatch of a factor around 2 (on average) between predictions and measurements over the last few years (at 3.5 or 4 TeV) [5, 8, 9], a feature that is also observed in impedance models of several other machines. Furthermore, a new global instability model including the transverse damper is now available, which gives us a better understanding of the single-beam phenomena. This factor ~ 2 needs, however, to be better understood. This work is ongoing in re-simulating in particular the geometric contribution to the impedance of the collimators.

(2) The main problem concerns the two-beam operation, for which much more Landau octupole current than predicted is needed and the reason has not been identified yet [10]. Several observations have been made, some of which are clear and summarized below:

(i) Instabilities are observed only for β^* smaller than a few meters.

(ii) Increasing the Landau octupole current helps. As we should be limited at higher energies, it would be good to have more octupoles current in the future. It seems that a factor ~ 2 could be gained with the spool piece correctors MCO and MCOX, depending on the available dynamic aperture [11].

(iii) Once in collision, no instability is observed anymore due to the large beam–beam head-on tune spread [12]. This is why the current idea to solve the issue of the beam instability at the end of the squeeze for after the 2013–2014 long shutdown (LS1) is to collide the two beams before the end of the squeeze.

As a less clear observation, whereas some beam dumps have been observed when putting the beams into collision with the negative Landau octupoles polarity, no beam dumps have been observed anymore with the positive Landau octupoles polarity, as suggested in Ref. [13] (and higher chromaticities and ADT gain, which have been modified at the same time). Later, the collision beam process was also optimized to go faster through the critical points [12]. Increasing chromaticities to ~ 15–20 units seems to help (according to the new theory [7] a plateau has been reached and no further stability gain can be expected by increasing the chromaticity) but the change was done at the same time as other modifications and it is difficult to conclude. A detailed chromaticity scan should be performed.

(3) The plan for the future is to continue the data analyses and work more on interplays between the different mechanisms (incoherent and coherent): impedance, nonlinearities (machine and Landau octupoles), space charge (at low energy), transverse feedback, longitudinal bunch distribution, beam–beam when the beams start to see each other [14], electron cloud [15], etc.

2.2. *Beam-induced RF heating*

Beam-induced RF heating has been observed in several LHC components during the 2011 and 2012 runs when the bunch/beam intensity was increased and/or the bunch length reduced. This caused beam dumps and delays for beam operation (and thus less integrated luminosity) as well as considerable damages for some equipment. Furthermore, the rms bunch length used was ~ 10 cm in 2012 (it was ~ 9 cm in 2011) whereas the nominal value (which is also the value required by HL-LHC) is 7.5 cm. A shorter bunch means a wider power spectrum (see below). The beam-induced RF heating of some equipment is therefore also worrisome for HL-LHC and it is closely followed up [4, 16, 17].

Some successful impedance reductions have however been already achieved, as for instance on one of the modules forming the injection kicker [18], but further modifications will be required for the HL-LHC era [19]. Indeed, the design of the most critical module was improved during the second half of 2012 and the result was that it moved from the highest temperature measured to the lowest one.

The power loss, which is due to the real part of the longitudinal impedance, is always proportional to the square of the number of particles per bunch but depending on the shape of the impedance, it can be linear with the number of bunches (when the bunches are independent, i.e. for a sufficiently short-range wake-field — or broad-band impedance — which does not couple the consecutive bunches) or proportional to the square of the number of bunches (when the bunches are not independent, i.e. for a sufficiently long-range wake-field — or narrow-band impedance — which couples the consecutive bunches) [17].

Considering the latter case of a sharp resonance (i.e. when only one line of the bunch spectrum is significant) at a resonance frequency f_r (just to illustrate), the power loss is given by

$$P_{loss} = 2\,R\,I^2 \times F\,,\tag{1}$$

where R is the (maximum) value of the real part of the impedance (resistance) at the resonance frequency and I is the total beam current. The factor F describes the frequency dependence of the power loss, which depends on the longitudinal bunch length and profile. It converges to 1 at zero frequency and it is between 0 and 1 for any frequency. Assuming for instance the bunch profile of Fig. 1 (left), close to a Gaussian distribution but with finite tails, the power spectrum (in dB) P_{dB} is depicted in Fig. 1 (right), from which the factor F is deduced by

$$F = 10^{\frac{P_{dB}(f_r)}{10}}\,.\tag{2}$$

As the power loss is quadratic with the total intensity, if we compare to the situation of 2012 (with 1380 bunches of 1.6×10^{11} p/b spaced by 50 ns), the nominal case (2808 bunches of 1.15×10^{11} p/b spaced by 25 ns) will correspond to an increase by a factor ~ 2.1, the 25 ns HL-LHC beam (2808 bunches of 2.2×10^{11} p/b) will correspond to an increase by a factor ~ 7.8, and the 50 ns HL-LHC beam (1404 bunches of 3.5×10^{11} p/b) will correspond to an increase by a factor ~ 5. Assuming now for instance an impedance $R = 5$ kΩ (as an example of a single moderate cavity mode with long memory) at the resonance frequency $f_r = 1.4$ GHz (i.e. $F = 10^{-4}$ from Fig. 1 (right)) and a total beam current $I = 1$ A

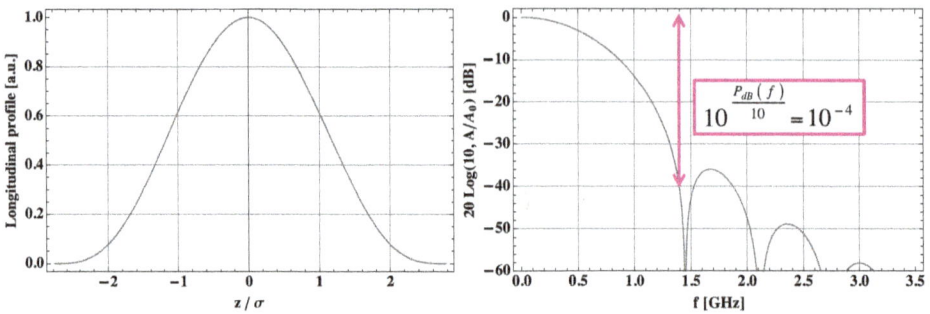

Fig. 1. (Left) Example of longitudinal bunch profile (with an rms bunch length $\sigma = 9$ cm, which was used in 2011) and (right) corresponding power spectrum (in dB) P_{dB}.

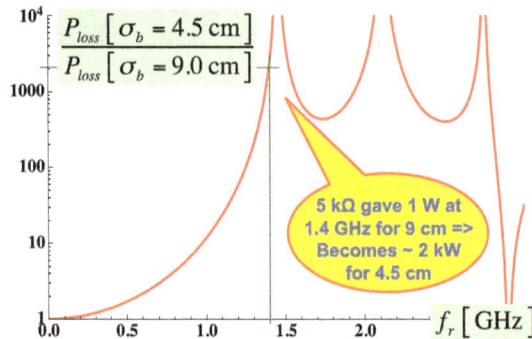

Fig. 2. Increased power loss factor (vs. frequency) between 4.5 cm and 9 cm rms bunch length.

(close to the HL-LHC value), the power loss would be therefore 1 W. To illustrate the (huge) effect of the bunch length, let us consider the case of an rms bunch length of 4.5 cm (i.e. two times smaller), assuming the same longitudinal bunch profile. The increased factor in power loss is represented in Fig. 2: reducing the bunch length by a factor of 2 increases the power loss by a factor ~ 2000!

In the opposite situation of a broad-band impedance, consider for instance the case of the resistive-wall impedance, and, as an example, the particular case of the beam screen (neglecting the holes, whose contribution has been estimated to be small in the past, and the longitudinal weld). Assuming a Gaussian longitudinal profile (other similar distributions would give more or less the same result), the power loss (per unit of length) is given by

$$P_{\text{loss/m}} = \frac{1}{C} \Gamma\left(\frac{3}{4}\right) \frac{M}{b} \left(\frac{N_b\, e}{2\pi}\right)^2 \sqrt{\frac{c\, \rho\, Z_0}{2}}\, \sigma_t^{-3/2}, \tag{3}$$

where $C = 26658.883$ m is the LHC circumference, Γ the Euler gamma function, M the number of bunches, b the beam screen half height (assumed here to be 18.4 mm), N_b the number of particles per bunch, e the elementary charge, c the speed of light, ρ the resistivity (assumed here to be 7.7×10^{-10} Ωm for copper at 20 K and 7 TeV), Z_0 the free-space impedance and σ_t the rms bunch length (expressed in unit of time). Assuming the nominal beam parameters ($M = 2808$, $N_b = 1.15 \times 10^{11}$ p/b and $\sigma_t = 0.25$ ns), Eq. (3) yields ~101 mW/m. For the 25 ns beam in the HL-LHC era ($M = 2808, N_b = 2.2 \times 10^{11}$ p/b and $\sigma_t = 0.25$ ns), this would give ~368 mW/m and for the 50 ns beam in the HL-LHC era ($M = 1404$, $N_b = 3.5 \times 10^{11}$ p/b and $\sigma_t = 0.25$ ns), this would give ~466 mW/m.

The usual solutions to avoid beam-induced RF heating are the following, depending on the situation:

(i) Increase the distance between the beam and the equipment.
(ii) Coat with a good conductor if the heating is predominantly due to resistive losses and not geometric losses.
(iii) Close large volumes (which could lead to resonances at low frequency) and add a smooth transition. This is why beam screens and RF fingers are installed.
(iv) Put some ferrite with high Curie temperature and good vacuum properties (close to the maximum of the magnetic field of the mode and not seen directly by the beam) or other damping materials. Adding a material with losses (the type of ferrite should be optimized depending on the mode frequency), the width of the resonance will increase (the impedance will become broader) and the (maximum) impedance will decrease by the same amount. The power loss will therefore be (much) smaller. However, the ferrite will then have to absorb the remaining power. Even if much smaller, the heating of the ferrite can still be a problem if the temperature reached is above the Curie point, or is above the maximum temperature allowed by the device. To cool the ferrite one should try and improve the thermal conduction from the ferrite as most of the time only radiation is used (given the general brittleness of the ferrite it is difficult to apply a big contact force).
(v) Improve the subsequent heat transfer:
  - Convection: there is none in vacuum.
  - Radiation: usually, the temperature is already quite high for the radiation to be efficient. One should therefore try and improve the emissivities of surrounding materials.
  - Conduction: good contacts and thermal conductivity are needed.

- Active cooling: the LHC strategy was to water cool all the near beam equipment.
(vi) Try and design an All Modes Damper (AMD) if possible, to remove the heat as much as possible to an external load outside vacuum, where it can be more easily cooled away. This can also work together with a damping ferrite.
(vii) Increase the bunch length, but then the luminosity will be decreased due to the geometric reduction factor in the absence of crab cavities (and possible losses from the RF bucket). The longitudinal distribution can also play a very important role for some devices, and it should be kept under tight control (in particular during the ramp when a controlled longitudinal emittance blow-up is applied and for operation at 7 TeV during Run II when radiation damping might reduce the bunch length during physics runs).
(viii) Install temperature monitoring on critical devices to avoid possible damages. It is worth mentioning that the mirror and support of the beam 2 synchrotron light monitor (used to measure the beam transverse emittances) suffered from damage in 2012 and that there is currently a heavy effort to find a robust design for after LS1. The precise knowledge of the material properties (dielectric losses in the microwave range) of the glass used for the mirror is required as well as a need to keep in mind the possibility of dielectric structure resonances in the microwave range.

Following some issues with RF fingers on some equipment (double-bellow modules, called VMTSA), a task force was set up during 2012 to review the design of all the components of the LHC equipped with RF fingers. The lessons learnt and the mitigation measures for the CERN LHC equipment with RF Fingers were reported in Ref. [20] (the important item of the Plug-In Modules, which were studied in great detail in the past, was also reviewed). For all the cases studied, no problem with impedance was revealed for conforming RF fingers. Therefore, no (big) problem is expected for HL-LHC bunch populations (i.e. up to 2.2×10^{11} p/b for the 25 ns beam and 3.5×10^{11} p/b for the 50 ns beam). But the top priority for the future should be to try and reach robust mechanical designs to keep the contacts of all the RF fingers (e.g. with the concept of funnelling as for the Plug-In Modules or using fixed extremities or any other robust designs) and to do a very careful installation. An impedance police is in place to estimate and follow the evolution of the impedance(s) when some equipment are removed, modified or added. The impedance(s) of all these devices should be simulated and measured on a bench to confirm the predictions. Finally, despite all these efforts, if impedance issues are discovered in the future, an impedance reduction campaign can/should be envisaged.

3. Expected Situation During the HL-LHC Era

Within the framework of the HL-LHC project, Work Package 2, Task 2.4 on collective effects [21], a first impedance estimation was done concerning the new high beta region of IR1 and IR5, i.e. in and around the triplets close to CMS and ATLAS experiments [22]. From this preliminary study the impedance in these regions could increase by around a factor of 2 or 3 with respect to the current layout, accounting for up to 10% of the total impedance budget (close to the main bunch spectrum frequency).

Moreover, the impedance of the three potential design options for the crab cavities was studied and, besides the higher — lower for one of the options — order modes that should be taken care of by *ad hoc* mode dampers at the design stage, the low frequency imaginary impedances (i.e. the constant values from 0 Hz to typically ~100 MHz, which give the upper limits for the effective imaginary impedances) were computed [23]. Both longitudinal and transverse contributions turned out to be significant for 12 crab cavities in total (note that meanwhile the baseline became four cavities per beam per side of the interaction region, which amounts to a total of 16 crab cavities per beam): 20 to 30 mΩ depending on the option to be compared to ~90 mΩ for the current situation; 10 to 100 kΩ/m at injection energy to be compared to ~2 MΩ/m estimated for the full machine; 300 kΩ/m to 2 MΩ/m at collision energy assuming a beta function at the location of the crab cavities of 4 km, to be compared to ~25 MΩ/m estimated for the full machine.

The impedance of the upgraded experimental beam pipes was also studied, since a reduction of diameter of the inner detector of ATLAS (inner radius from 29 mm to 22.5 mm), CMS (inner radius from 29 mm to 21.7 mm), and ALICE (inner radius from 29 mm to 17.5 mm) as well as of the wakefield suppressor of LHCb (outer radius from 5 mm to 3.5 mm) was proposed to increase their performance [24]. Studies showed an expected increase of ~30% of the power loss for CMS and ATLAS (VI) chambers (from 1.4 W to 1.9 W per meter length), a factor of 4 increase of the transverse inductive impedance at low frequency (from 150 Ω/m to 600 Ω/m), and a 20% increase of the longitudinal inductive impedance at low frequency (from 0.011 mΩ to 0.013 mΩ at injection energy) [25, 26]. The increase of the impedance contributions for the proposed reduction of the ALICE radius was higher (+70% for the power loss, factor 9 for the effective transverse impedance and +50% for the effective longitudinal impedance) but the overall contributions compared to the rest of the machine remained small [27]. Finally, detailed studies of the kick factors, resonant modes and power loss generated by the upgraded vacuum chamber of CMS were performed and showed that differences between the old and new chambers are expected to be small since the trapped fields of the largest resonant modes are not located where the vacuum chamber was planned to be changed [28]. However, this study also confirmed that very large power loss (of the order of 2 kW) could be experienced by the beam pipe if the rms bunch length was reduced to 4 cm as it was planned in some initial HL-LHC scenarios.

More generally, the plan is to have an initial estimate of the machine impedance by November 2013 and an initial estimate of the intensity limitations by May 2014. The final report on beam intensity limitations and specification of machine and beam parameters should be available in November 2014.

Concerning the devices that showed signs of beam-induced RF heating before LS1, the expected situation after LS1 and during the HL-LHC era (with the knowledge at the time of writing) is summarized in Table 1.

Table 1. Summary of the situation for the equipment, which showed sign of beam-induced RF heating before LS1. Black means damaged equipment; red means detrimental impact on operation (dump or delay or reduction of luminosity); yellow indicates need for follow up; green means solved.

Element	Problem	2011	2012	Expected situation after LS1	Expected situation in the HL-LHC era (very preliminary)
Double-bellow VMTSA	Damage	■		All VMTSA should be removed	All VMTSA should be removed
Injection protection collimator TDI	Damage	■	■	Beam screen reinforced; maybe copper coating on the jaws	New design in preparation
Injection kicker MKI	Delay			Beam screen and tank emissivity upgrade	Increased power loss with HL-LHC parameters should be monitored
Primary collimator TCP B6L7.B1	Few dumps			Cooling system; geometry checked, non-conformity should be removed	Depends on the baseline for the HL-LHC collimation
Tertiary collimators TCTVB	Few dumps			All TCTVBs should be removed; situation with new TCTP should be followed up	All TCTVBs should be removed; situation with new TCTP should be followed up
Beam screen standalone Q6R5	Regulation at the limit			Upgrade of the valves; TOTEM check	Upgrade of the valves; forward detectors are not part of the HL-LHC baseline
ALFA roman pot	Risk of damage			New design in preparation	Forward detectors are not part of the HL-LHC baseline
Synchrotron light telescope BSRT	Deformation suspected		■	New design in preparation	

Preliminary computations of heat loads with the HL-LHC beam parameters for several key systems were already performed for the:

(i) Beam screen (see above the application of Eq. (3)).

(ii) New collimator design with integrated BPMs (Beam Position Monitors) and ferrites: 100 W (25 ns with 2.2×10^{11} p/b) to 150 W (50 ns with 3.3×10^{11} p/b) were predicted, of which 5 to 7 W would be dissipated in the ferrites, and 4 to 6 W in the RF fingers [18, 29]; more thorough simulation studies as well as bench measurements are under way to confirm these results.

(iii) Injection kickers (MKIs): simulations for the new design of MKI screen conductors with HL-LHC parameters were performed and predicted 140 W (25 ns with 2.5×10^{11} p/b) to 200 W (50 ns with 3.8×10^{11} p/b) [18]. These values are of the order of heat loads estimated with pre-LS1 parameters for the old MKI design that required significant time to cool down after physics fills. However, the improved understanding of the mechanisms of beam-induced heating to the MKIs, of heat dissipation, and the discovery of non-conforming issues provide a set of solutions that could be used in case of issues.

References

[1] S. Fartoukh and F. Zimmermann, The HL-LHC Accelerator Physics Challenge, in this book: Chapter 4.

[2] O. Brüning, HL-LHC Parameter Space and Scenarios, in *Proc. of the LHC Performance Workshop*, Chamonix, France, 06–10/02/2012.

[3] E. Métral *et al.*, Review of the Instabilities Observed during the 2012 Run and Actions Taken, in *Proc. of the LHC Beam Operation Workshop*, Evian, France, 17–20/12/2012.

[4] B. Salvant *et al.*, Beam Induced RF Heating, in *Proc. of the LHC Beam Operation Workshop*, Evian, France, 17–20/12/2012.

[5] N. Mounet *et al.*, Beam Stability with Separated Beams at 6.5 TeV, in *Proc. of the LHC Beam Operation Workshop*, Evian, France, 17–20/12/2012.

[6] E. Métral, Impedances, Instabilities and Implications for the Future, CMAC meeting #6, CERN, 16–17/08/2012.

[7] A. Burov, Nested Head Tail Vlasov Solver: Impedance, Damper, Radial Modes, Coupled Bunches, Beam-Beam and More…, CERN AP forum, 04/12/2012.

[8] E. Métral, Pushing the Limits: Beam, in *Proc. of the LHC Performance Workshop*, Chamonix, France, 24–28/01/2011.

[9] N. Mounet, The LHC Transverse Coupled-Bunch Instability, EPFL PhD Thesis 5305 (2012).

[10] T. Pieloni *et al.*, Beam Stability with Colliding Beams at 6.5 TeV: in the Betatron Squeeze and Collisions, *Proc. of the LHC Beam Operation Workshop*, Evian, France, 17–20/12/2012.

[11] R. Tomas, Optics and Non-Linear Beam Dynamics at 4 and 6.5 TeV, *Proc. of the LHC Beam Operation Workshop*, Evian, France, 17–20/12/2012.

[12] X. Buffat, Stability Considerations with Beam-Beam and Octupoles, CERN LMC meeting, 29/08/2012.

[13] S. Fartoukh, The Sign of the LHC Octupoles, CERN LMC meeting, 11/07/2012.

[14] S. White, Beam-beam and Impedance, in *Proc. of the ICFA Mini-Workshop on Beam-Beam Effects in Hadron Colliders (BB2013)*, CERN, 18–22/03/2013.

[15] A. Burov, Three-Beam Instability in the LHC, CERN-ATS-Note-2013-012 PERF.

[16] E. Métral *et al.*, Beam-Induced Heating / Bunch Length / RF and Lessons for 2012, in *Proc. of the LHC Performance Workshop*, Chamonix, France, 06–10/02/2012.

[17] E. Métral *et al.*, Beam-Induced RF Heating in the LHC, Diamond Light Source mini-workshop, 30/01/2013.

[18] H. Day, Measurements and Simulations of Impedance Reduction Techniques in Particle Accelerators, Manchester university PhD thesis to be published (2013).

[19] B. Goddard and V. Mertens, Challenges for the Injection and Beam Dump Components, in this book: Chapter 19.

[20] E. Métral *et al.*, Lessons Learnt and Mitigation Measures for the CERN LHC Equipment with RF Fingers, in *Proc. of the IPAC'13 Conference*, Shanghai, China, 12–17 May, 2013.

[21] Web page of the HL-LHC WP2 Task 2.4 on collective effects: https://espace.cern.ch/HiLumi/WP2/task4/SitePages/Home.aspx.

[22] N. Mounet *et al.*, HL LHC: Impedance Considerations for the New Triplet Layout in IR1 & 5, 11th HiLumi WP2 Task Leader Meeting, CERN, 01/07/2013.

[23] B. Salvant *et al.*, The Impedance Model of the HL-LHC with DS Collimators and Crab-Cavities, 2nd Joint HiLumi LHC-LARP Annual Meeting, Frascati, 14/11/2012.

[24] R. Veness *et al.*, Specification of New Vacuum Chambers for the LHC Experimental Interactions, in *Proc. of the IPAC'13 Conference*, Shanghai, China, 12-17 May, 2013.

[25] B. Salvant *et al.*, Impedance Heating, 9th CERN LEB Technical Meeting, 21/01/2011.

[26] N. Mounet *et al.*, Summary of Impedance Heating for the 45 mm ID ATLAS VI, 10th CERN LEB Technical Meeting, 04/02/2011.

[27] B. Salvant *et al.*, RF Analysis of ALICE, LHCb and AFP, 10th CERN LEB Technical Meeting, 15/06/2012.

[28] R. Wanzenberg and O. Zagorodnova, Calculation of Wakefields and Higher Order Modes for the New Design of the Vacuum Chamber of the CMS Experiment for the HL-LHC, CERN-ATS-Note-2013-018 TECH, 10/04/2013.

[29] H. Day *et al.*, TCTP Summary of Power Loss and Heat Load Calculations (with attention paid to the heat load of the ferrite damping material), LHC collimation working group, 01/10/2012.

Chapter 16

Challenges and Plans for the Proton Injectors

R. Garoby

CERN, BE Department, Genève 23, CH-1211, Switzerland

The flexibility of the LHC injectors combined with multiple longitudinal beam gymnastics have significantly contributed to the excellent performance of the LHC during its first run, delivering beam with twice the ultimate brightness with 50 ns bunch spacing. To meet the requirements of the High Luminosity LHC, 25 ns bunch spacing is required, the intensity per bunch at injection has to double and brightness shall almost triple. Extensive hardware modifications or additions are therefore necessary in all accelerators of the injector complex, as well as new beam gymnastics.

1. Introduction

Luminosity in the LHC (L_{LHC}) depends upon beam and machine parameters according to the following formula:

$$L_{\text{LHC}} = \left(\frac{\gamma}{4\pi} \frac{1}{\beta^*} f_{\text{rev}} F \right) \cdot \left(n_b N_b \cdot \frac{N_b}{\varepsilon_n} \right)$$

where γ is the usual relativistic factor, β^* the betatron function at the Interaction Point, f_{rev} the beam revolution frequency, F a form factor depending upon the geometry of the bunch crossing, n_b the number of bunches per ring, N_b the number of protons per bunch and ε_n the normalized transverse emittance of the beam (assumed round).

The second term in brackets in this formula is the product of beam intensity $n_b N_b$ by beam brightness N_b/ε_n which are both established in the injectors and can only degrade in the collider. It clearly shows that LHC luminosity depend crucially upon the injected beam characteristics.

As a typical illustration, the excellent performance of the injector complex ($1.65 \cdot 10^{11}$ p/b with 50 ns bunch spacing within emittances of $1.6\ \mu m$ at ejection from the SPS) has been an essential ingredient to the results obtained during the first run of the LHC which lasted until the end of 2012.

The High Luminosity Upgrade of the LHC is setting an even more challenging goal for the injected beam ($2.3 \cdot 10^{11}$ p/b with 25 ns bunch spacing within emittances of 2.1 μm at ejection from the SPS, assuming 5% beam loss and 20% blow-up in the LHC) [1], corresponding to a factor of three improvement in terms of $n_b N_b (N_b / \varepsilon_n)$.

2. Present LHC Proton Injectors

2.1. *Description*

The CERN accelerators are shown in Fig. 1. The LHC injector complex is composed of six accelerators [Linac2 (50 MeV), PSB (1.4 GeV), PS (25 GeV) and SPS (450 GeV) for protons, plus Linac3 and LEIR for other ions] which were initially commissioned with beam many years ago (in 1959 for the PS). These machines were subjects of extensive consolidations and upgrades during the past decade to meet the nominal specifications of the LHC [2]. For the needs

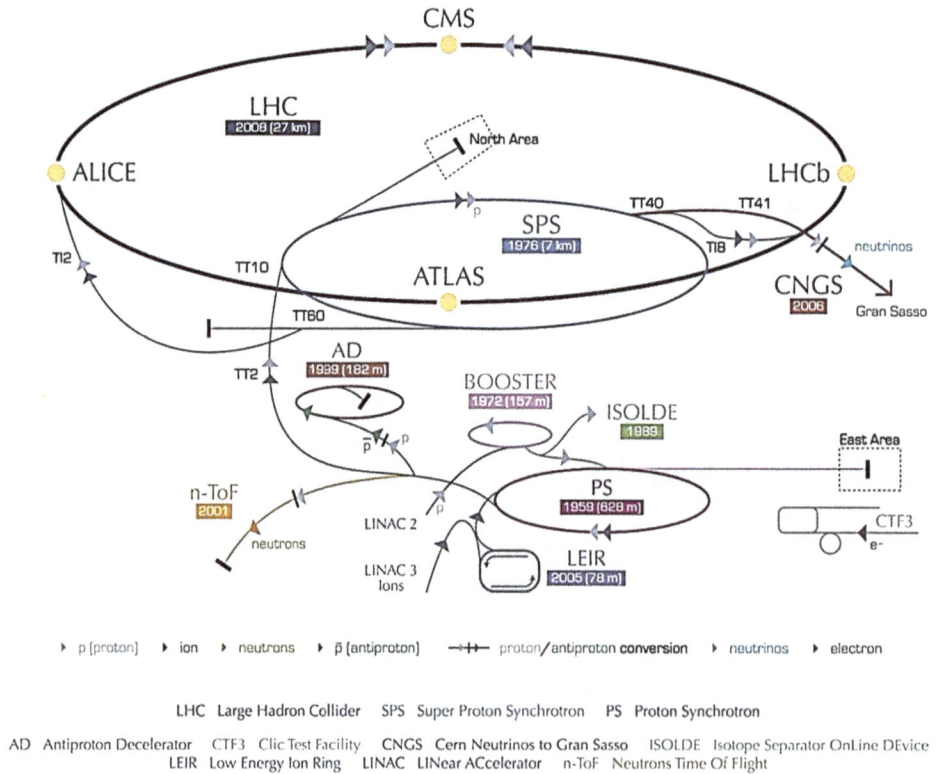

▸ p (proton)   ▸ ion   ▸ neutrons   ▸ p̄ (antiproton)   ⊶ proton/antiproton conversion   ▸ neutrinos   ▸ electron

LHC   Large Hadron Collider   SPS   Super Proton Synchrotron   PS   Proton Synchrotron

AD   Antiproton Decelerator   CTF3   Clic Test Facility   CNGS   Cern Neutrinos to Gran Sasso   ISOLDE   Isotope Separator OnLine DEvice   LEIR   Low Energy Ion Ring   LINAC   LINear ACcelerator   n-ToF   Neutrons Time Of Flight

Fig. 1.   CERN accelerator complex in 2012.

of the High Luminosity project in the LHC, additional measures must be implemented.

In the transverse phase planes, space charge resulting from the high beam brightness is the main concern. The induced tune spread which is directly proportional to $1/\beta\gamma^2 \cdot (N_b/\varepsilon_n)$ [β and γ being the usual relativistic factors] is a basic limitation both in the PSB and in the PS. It is brought to an acceptable level in the PSB by dividing the intensity per pulse N_b by a factor of 2, filling the PS with two batches instead of a single one. This was made possible by operating the PSB on harmonic 1 and hence with a single bunch per ring. Hence more PSB beam pulses could be accumulated in the PS. To reduce the effect of space charge in the PS, where the first batch of bunches stays at injection energy during 1.2 s, the transfer energy from the PSB has been brought up to 1.4 GeV (1.5 times the $\beta\gamma^2$ at 1 GeV).

In the PS, the long and intense bunches delivered by the PSB are transformed into trains of bunches spaced by 25 ns (or 50 ns) before ejection [3, 4], as sketched in Fig. 2. This is obtained with quasi-adiabatic bunch splitting gymnastics which keep the beam bunched and under control of the RF. As a result, the gap without beam corresponding to the empty bucket at injection (six PSB bunches being sent to the PS on $h = 7$) is preserved and used for the rise-time of the ejection kicker, avoiding beam loss at ejection. Moreover, shorter bunch trains can be obtained simply with less bunches from the PSB.

Multiple splitting steps are used:

- Splitting in three takes place at injection energy (1.4 GeV) combining the simultaneous use of three RF systems on harmonics 7, 14 and 21. At the end of the process, the beam is held on $h = 21$ on which it is accelerated up to top energy.
- Splitting in four takes place at 25 GeV, in two successive steps, using RF systems on $h = 21$ and 42 for the first step, and on $h = 42$ and 84 for the second one. Without this last step, bunch spacing is 50 ns.

In addition, the longitudinal emittance is submitted to controlled blow-ups to improve longitudinal stability, pulsing the 200 MHz cavities a few times during the cycle.

Finally, a non-adiabatic bunch length reduction process is used before ejection to the SPS for reducing bunch length to ~ 4 ns which can be captured in a 200 MHz SPS bucket. In total, five families of RF systems are necessary in the PS (3–10 MHz, 20 MHz, 40 MHz, 80 MHz and 200 MHz) to generate the proton beams for LHC.

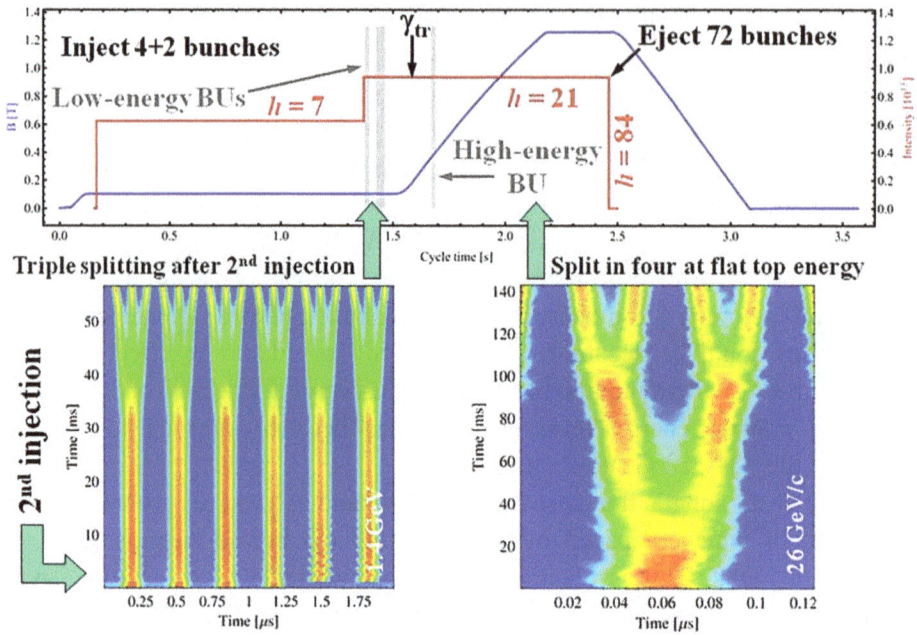

Fig. 2. Longitudinal bunch splitting to generate a 25 ns bunch train in the PS:
- Top graphic: B field (blue) and beam current (red) during a PS cycle. Multiple longitudinal controlled Blow-Ups (BU) are being used for optimizing the gymnastics and avoiding longitudinal beam instabilities.
- Bottom pictures: 2D displays of longitudinal density at 1.4 GeV (left) and at 25 GeV (right). Time evolution during the process is along the vertical axis. Longitudinal density is color-coded from blue (no beam) to red (maximum).

In the SPS, the injected bunches are captured with the main RF system operating at 200 MHz. Up to 4 batches of 25 ns (or 50 ns) bunch trains from the PS are accumulated on a 10.8 s long flat bottom. Longitudinal stability is obtained by adding the 4th harmonic RF (800 MHz) in Bunch Shortening mode (increasing Landau damping) and applying a longitudinal controlled blow-up during acceleration. In the transverse phase plane, the electron clouds that were limiting performance by provoking vertical instability are significantly reduced at present intensities thanks to the scrubbing of the surface of the vacuum chamber.

2.2. *Present performance and future needs*

The beam characteristics delivered at injection in the LHC before its first long shutdown in 2013 are summarized in the first column of Table 1. The corresponding brightness at injection in LHC is 20% higher than the "nominal" value considered in the LHC Design Report [4] for a bunch spacing of 25 ns.

During the first phase of operation, 50 ns spacing has however been preferred, with approximately twice the ultimate brightness and the ultimate intensity per bunch ($\sim 1.7 \cdot 10^{11}$ p/b) at 450 GeV. In spite of transverse blow-up in the LHC (central column in Table 1), it consistently allowed to reach 75% of the nominal peak luminosity ($7.5 \cdot 10^{33}$ instead of 10^{34} cm^{-2}s^{-1}), mostly compensating the effect of the larger physical emittance due to the reduced beam collision energy (4 TeV instead of 7 TeV).

With these beam characteristics, the injector complex is performing as foreseen but without any margin. For the High Luminosity LHC (HL-LHC) project, which aims at accumulating ~ 250 fb^{-1}/year, beam characteristics in collision have to progress to the level described in Table 2. Assuming 20% emittance blow-up and 5% beam loss between injection and collision in LHC [1], the beam intensity required from the injectors (Table 2) has to double in the baseline case (25 ns) and the brightness shall almost triple.

Table 1. Beam characteristics in 2012.

	50 ns bunch trains at LHC injection	50 ns bunch trains at start of collisions	Available 25 ns bunch trains at LHC injection*
Number of bunches (n_b)	1374	1374	2748
Protons/bunch (N_b)	$1.65 \cdot 10^{11}$	$1.6 \cdot 10^{11}$	$1.1 \cdot 10^{11}$
Norm. trans. emittance (ε_n) [μm]	1.6	2.4	2.8

* The number of 25 ns spaced bunches that the LHC could effectively accelerate was limited to 804 because of electron cloud effects [5].

Table 2. Beam characteristics for the High Luminosity LHC project.

	25 ns bunch trains at start of collisions	25 ns bunch trains at injection (estimate)	50 ns** bunch trains at start of collisions	50 ns** bunch trains at injection (estimate)
Number of bunches (n_b)	2748*	2748	1374*	1374
Protons/bunch (N_b)	$2.2 \cdot 10^{11}$	$2.3 \cdot 10^{11}$	$3.5 \cdot 10^{11}$	$3.7 \cdot 10^{11}$
Norm. trans. emittance (ε_n) [μm]	2.5	2.1	3	2.5

* The filling schemes in the accelerator chain, the maximization of the colliding bunches for the four experiments, and the need of non-colliding bunches will slightly reduce the total number of the colliding bunches in the high luminosity interaction points.

** The 50 ns scenario is a back-up, in case fundamental limitations in LHC (e.g. due to electron clouds or total intensity) are encountered with the 25 ns baseline parameters.

The presently identified limitations in the injectors are illustrated in Fig. 3 together with the achieved and expected beam performances. In the coordinate system emittance versus intensity, a constant space charge induced tune spread is represented by a straight line passing through the origin. Below that line, the space charge induced tune spread is excessive. The curve corresponding to the PS is not a straight line because it takes into account the energy spread assuming a constant longitudinal emittance. The other limitations restrict the maximum intensity per bunch, which corresponds to a vertical line parallel to the Y axis. For 25 ns bunch spacing, $1.2 \cdot 10^{11}$ p/b is the maximum intensity in the SPS because of the available RF power and because of longitudinal coupled bunch instabilities. The limit due to electron clouds is nowadays beyond this intensity. For 50 ns, the main limitations result from heat dissipated in the equipment because of the beam image current and from longitudinal instabilities $(N_b < 1.7 \cdot 10^{11}$ p/b).

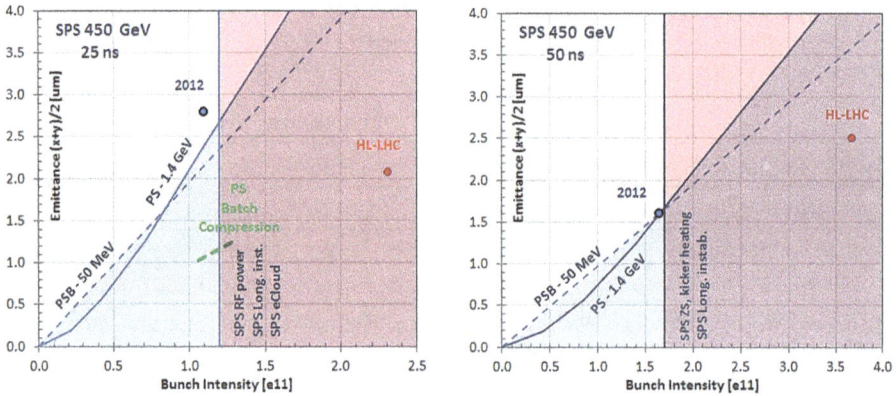

Fig. 3. Performance and limitations at SPS ejection, in 2012, of the LHC proton injector complex for 25 ns (left) and 50 ns (right) bunch spacing.

Before the implementation of the upgrades described in the following part of this document, new sophisticated beam gymnastics have been proposed for generating in the PS 25 ns batches with a brightness similar to 50 ns [6]. The principle is to split the PSB beam in less bunches while keeping spacing at 25 ns. For that purpose, the batch of PSB bunches that fills most of the PS circumference at injection is first accelerated to an intermediate energy (typically 2.5 GeV) where space charge is sufficiently reduced and then compressed into a smaller fraction of the circumference. A typical scenario is illustrated in Fig. 4:

(i) two consecutive batches of four bunches from the PSB are injected in eight buckets of the PS on $h = 9$,

(ii) after acceleration up to an intermediate energy of 2.5 GeV where space charge is smaller and longitudinal acceptance larger, the beam is compressed in 57% of the circumference by adiabatically increasing the harmonic number from $h = 9$ to $h = 14$,

(iii) bunches are merged two by two (reverse process wrt splitting in two) which results in four bunches on $h = 7$,

(iv) triple splitting is finally applied, generating 12 bunches on $h = 21$.

These 12 bunches are then accelerated up to high energy where splitting in four is exercised like nowadays (lower part, bottom right of Fig. 2). Compared to the present process, this "Batch Compression Merging and Splitting" (BCMS) scheme provides only 48 bunches with 25 ns spacing, instead of 72, increasing the filling time of the LHC and decreasing by approximately 10% the maximum number of bunches in the collider because of the gaps required for kickers' rise time in the SPS and LHC.

The corresponding beam characteristics at LHC injection, shown as a dashed green line in Fig. 3 (left), can potentially increase luminosity with respect to the 50 ns scheme while reducing the number of events per crossing and hence easing operation of the detectors in the experiments. The interest of the scheme will however depend upon the LHC capability to preserve the small emittances resulting from the maximum circulating current acceptable in the collider (nominally ~ 0.58 A).

This scheme has already been successfully tested at the end of 2012. A much higher brightness than with the nominal scheme has been obtained, with bunches of $1.15 \cdot 10^{11}$ p/b within emittances of 1.4 μm reproducibly injected in the LHC.

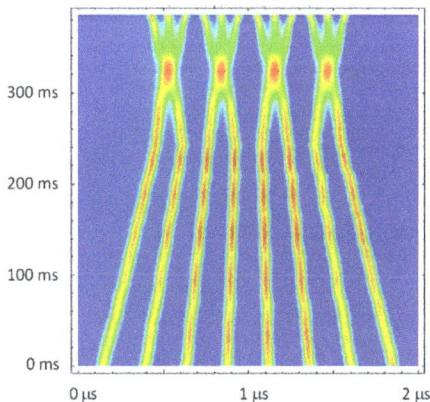

Fig. 4. 2D display (simulated) of longitudinal density in the PS during BCMS at 2.5 GeV. Time evolution during the process is along the vertical axis. Longitudinal density is color-coded from blue (no beam) to red (maximum).

3. Upgrade Plan of the LHC Proton Injector Complex

3.1. *Transverse phase planes*

The primary limitation due to the space charge induced tune spread in the PSB and in the PS will again be addressed by increasing the injection energy.

In the case of the PSB, this will be obtained with a new linac (Linac4 [7]) which will provide beam at a kinetic energy of 160 MeV, doubling $\beta\gamma^2$ with respect to the present 50 MeV Linac2 [8]. Linac4 is the subject of a dedicated chapter in the present book. Its main parameters are summarized in Table 3. Charge exchange injection will replace multi-turn betatron stacking, increasing the efficiency up to ~98% and providing the means to tailor the transverse distribution of protons circulating in the PSB. Painting is also foreseen in the longitudinal phase plane, to maximize capture efficiency and to optimize the longitudinal particle distribution. Operation with the same space charge tune spread $|\Delta Q_y|$ as has been achieved in the current configuration with LINAC2 (0.44), the higher injection energy is expected to allow for a brightness of $1.8 \cdot 10^{12}$ p/μm, twice the present level. With the nominal beam gymnastics and some margin for emittance blow-up, that corresponds to a brightness of 10^{11} p/μm for 25 ns bunch spacing at ejection from the SPS (resp. $2 \cdot 10^{11}$ p/μm for 50 ns).

Table 3. Linac4 beam parameters.

Ion species	H$^-$
Output energy	160 MeV
Bunch frequency	352.2 MHz
Maximum repetition rate	2 Hz
Beam pulse length	400 μs
Mean pulse current	40 mA
Maximum number of particles per pulse	$1.0 \cdot 10^{14}$
Number of particles per bunch	$1.14 \cdot 10^{9}$
Transverse emittance	0.4 $\pi\,\mu$m (rms)

In the case of the PS, the beam injection energy will be increased from 1.4 to 2 GeV kinetic, increasing $\beta\gamma^2$ and decreasing the space charge tune spread by a factor of ~1.6. This energy is attainable in the PSB [8], provided that a number of equipments are upgraded or redesigned, like the power supply for the main dipoles. Likewise, in the PS, important modifications and new equipment must be added for beam injection at 2 GeV [9] and the existing transverse damper will be renovated to avoid transverse instabilities, providing more flexibility in the choice of the tunes at low energy and hopefully stabilizing the beam on the high energy flat top.

Beyond these major changes, an extensive campaign is in progress for optimizing the transverse tunes and improving the compensation of resonances [10]. As a result, operation with larger vertical tune spreads than today is foreseen to be manageable in all synchrotrons, and especially in the PS.

In the SPS, the integer part of the tunes have recently been changed from 26 ("Q26") to 20 ("Q20"), reducing the transition energy and enhancing the slip factor $|\eta| = |1/\gamma_t^2 - 1/\gamma^2|$ to increase the thresholds of longitudinal and Transverse Mode Coupling Instabilities (TMCI) [11, 12]. With this optics, operation with a space charge tune shift in excess of 0.15 is expected to be manageable, corresponding to a brightness of $\sim 10^{11}$ p/μm at SPS ejection, matched to the capability of the upgraded PSB for 25 ns bunch spacing.

3.2. *Longitudinal phase plane*

The PSB is not expected to suffer from limitations in the longitudinal phase plane when providing the high brightness beams for LHC. A major renovation of the main RF systems is however required to guarantee a reliable operation during all the lifetime of the LHC and to let other users [e.g. ISOLDE] benefit from the higher intensity beams allowed with Linac4.

In the PS, the measures presently used to stabilize the beam in the longitudinal phase plane (controlled longitudinal blow-up and coupled bunch instability damper) cannot handle a bunch intensity larger than $\sim 1.7 \cdot 10^{11}$ p/bunch, both for 25 and 50 ns bunch spacing. This limitation will be addressed by a new longitudinal damper using a dedicated "broad band" cavity, aimed at bringing the instability threshold beyond $3 \cdot 10^{11}$ p/bunch. Moreover, transient beam loading in the five families of RF systems will increase with beam intensity, degrading the quality of the multiple beam gymnastics. Fast RF feedback on all high power RF systems will therefore be upgraded, and one-turn delay feedbacks will be renovated on the 3–10 MHz ferrite cavities and implemented on the other cavities. More RF voltage at 40 MHz will be installed to improve longitudinal capture efficiency in the SPS [13]. The combined effect of all these actions is expected to allow for the operational availability of $3 \cdot 10^{11}$ p/bunch at PS ejection.

In the SPS, two new 1.6 MW RF power plants will be installed, doubling the available power at 200 MHz, and the cavities will be reorganized in six assemblies (four today), reducing the beam impedance. This will allow the acceleration of a beam current of up to 3 A, and 10 MV will be available on the high energy flat top, before ejection. Up to $2 \cdot 10^{11}$ p/bunch with 25 ns bunch spacing could then be transferred to the LHC. Longitudinal stability of the beam in the SPS is presently obtained through the combined effects of controlled longitudinal blow-up up to 0.6 eVs and 800 MHz RF voltage used in bunch shortening mode. The instability threshold will increase with the new Q20 optics thanks to the increased slip factor $|\eta|$, although this will be balanced by the smaller longitudinal emittance imposed by the reduced acceptance of the buckets. The lower impedance of the reorganized 200 MHz RF system will also be beneficial, as well as the planned renovation of the low and high power equipment of the 800 MHz system. The present estimate is that $2 \cdot 10^{11}$ p/bunch with 25 ns bunch spacing and $3.5 \cdot 10^{11}$ p/bunch with 50 ns should be attainable. Such intensities might require transferring longer bunches (1.6–1.8 ns) to the LHC where mitigation measures have to be studied [14].

3.3. *Electron clouds*

Electron cloud formation is observed in the PS on the 25 ns beam a few milliseconds before ejection and a transverse instability has repeatedly been diagnosed at the same time. Although not presently affecting performance, it is a subject of theoretical and experimental investigation to determine the risk with the future beam characteristics and to prepare cures or mitigation measures.

In the SPS, electron clouds have been a major concern as soon as an LHC-like beam has been injected [15]. They trigger vacuum pressure rises, instabilities, beam losses and transverse emittance blow-up. Cures and mitigation measures have been developed through modeling/simulation and experimental tests. Scrubbing by the beam for LHC is showing an interesting potential, as demonstrated by the continuous improvement of the SPS since the beginning of operation for LHC. It suffers however from degradation whenever the vacuum chambers are exposed to atmosphere and the minimum obtainable Secondary Electron Yield (SEY) is limited, depending upon the nature and cleanliness of the vacuum chamber. Coating of the vacuum chamber with a low SEY material would be a perfect cure, completely avoiding the appearance of electron clouds. Amorphous carbon is especially efficient in that respect and adequate coating processes for the SPS vacuum chambers have been developed and experimentally demonstrated. The use of clearing electrodes has also been considered, but no satisfying engineering solution has been found which would not reduce the available aperture. In any case, getting rid of the electron cloud limitations in the SPS is considered as feasible, either with scrubbing or with amorphous carbon coating [16]. As a mitigation measure, a wideband (GHz bandwidth) transverse damper is also envisaged as a possible means to counteract electron cloud triggered instabilities [17].

3.4. *Other upgrades*

The equipment in all accelerators must match the increased level of performance and be capable to operate reliably:

- New beam instrumentation has to be developed for measuring with adequate accuracy beams of reduced size and high brightness and intensity. The capability to detect and quantify the intensity in "spurious" bunches in the PS and in the SPS is an important and challenging need.
- New beam intercepting and protection devices have to be built which withstand impact from the higher brightness/higher intensity beam. That concerns beam dumps in all machines, as well as the SPS scraper system for halo shaping and the devices in the SPS to LHC transfer lines protecting the LHC.
- A number of power supplies need to be replaced because of aging and/or because of more demanding specifications.
- Civil engineering and building construction are also necessary for radiation shielding (PS injection and ejection sectors) and to host new large size equipment (PSB new main power supply and new SPS high power RF amplifiers).

Very expensive items like the main dipoles are not planned to be changed, but their ageing will be carefully monitored and spares have to be available.

4. Estimated Performance of the Upgraded LHC Proton Injector Complex

The performance reach of the LHC proton injector complex after the improvements described in the previous section are graphically represented in Fig. 5. Compared, for example, to the present situation with 25 ns (Fig. 3), the intensity per bunch is 70% higher and brightness is more than doubled.

The baseline option preferred by the LHC experiments is 25 ns bunch spacing. It is also preferable for the injectors because the beam characteristics expected by the HL-LHC project (yellow dot) are approximately compatible with all identified limitations, except with the SPS one at $2 \cdot 10^{11}$ p/bunch due to beam loading and longitudinal instabilities.

As a spare solution, in case the 25 ns beam cannot be used in the LHC (e.g. because of electron cloud or beam intensity), 50 ns bunch spacing could be considered. More limitations would then have to be faced in the injectors:

- in the PS, mainly because of longitudinal instability, resulting in an estimated limit of $2.7 \cdot 10^{11}$ p/bunch at injection in LHC, while the HL-LHC specification is at $3.5 \cdot 10^{11}$ p/bunch;
- in the SPS, because of longitudinal instability and because of space charge (the tune spread will reach 0.22 on the injection flat porch).

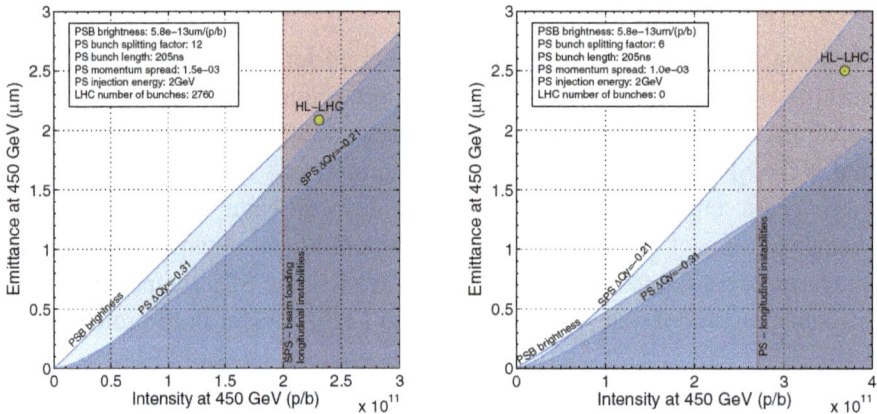

Fig. 5. Performance and limitations at SPS ejection of the upgraded LHC proton injector complex for 25 ns (left) and 50 ns (right) bunch spacing.

The performances shown in Fig. 5 are however only estimates which will have to be regularly revised during the ~10 years duration of the injectors' upgrade program. As past experience with the CERN accelerators has shown, it is not unreasonable to hope that, as a result of the intense effort invested both in theory and in beam experiments [18], beam characteristics will finally exceed the present expectation and meet the present HL-LHC requirements. Similarly, the possibility cannot be discarded that the HL-LHC beam specifications will evolve as experience with the collider progresses.

References

[1] O. S. Brüning, HL-LHC Parameter Space and Scenarios, in *Proc. of Chamonix 2012 Workshop on LHC Performance*, CERN-2012-006, p. 315.

[2] A. Blas *et al.*, Conversion of the PS complex as LHC proton pre-injector, CERN-PS-DI-97-048-DI, http://cdsweb.cern.ch/record/328735/files/ps-97-048.pdf.

[3] R. Garoby, Status of the nominal proton beam for LHC in the PS, CERN-PS-99-013-RF, http://cdsweb.cern.ch/record/382971/files/ps-99-013.pdf.

[4] LHC Design Report, v.3: The LHC Injection Chain, pp. 5–8 and 45–52, CERN-2004-003-V-3.

[5] G. Iadarola *et al.*, Electron Cloud and Scrubbing Studies for the LHC, *IPAC2013*, Shanghai, China, May 2013, TUPFI002, p. 1331.

[6] H. Damerau, Performance Potential of the Injectors after LS1, in *Proc. of LHC Performance Workshop (Chamonix 2012)*, February 2012, p. 268.

[7] M. Vretenar *et al.*, The Linac4 Project at CERN, *IPAC2011*, San Sebastian, Sept. 2011, TUOAA03, p. 900.

[8] K. Hanke *et al.*, PS Booster Energy Upgrade, Feasibility Study, https://edms.cern.ch/document/1082646/3.

[9] S. Gilardoni *et al.*, PS Potential Performance with a Higher Injection Energy, in *Proc. of LHC Performance Workshop (Chamonix 2011)*, January 2011, p. 349.

[10] S. Gilardoni *et al.*, Tune and Space charge Studies for High Brightness and High Intensity at CERN PS, *IPAC2011*, San Sebastian, Spain, Sept. 2011, MOPS014, p. 625.

[11] E. Shaposhnikova, Lessons from SPS studies in 2010, in *Proc. of LHC Performance Workshop (Chamonix 2011)*, January 2011, p. 359.

[12] H. Bartosik *et al.*, Low Gamma Transition Optics for the SPS: Simulation and Experimental Results for High Brightness Beams, in *Proc. of HB2012 Workshop*, Beijing, China, Sept. 2012, WEO1B01.

[13] H. Timko *et al.*, Longitudinal Beam Loss Studies of the CERN PS to SPS Transfer, in *Proc. of HB2012 Workshop*, Beijing, China, Sept. 2012, WEO1C03.

[14] L. Drosdal *et al.*, CERN-ATS-Note-2011-063 MD.

[15] G. Rumolo *et al.*, in *Proc. of CARE-HH-APD Beam'07 Workshop*, CERN 2007.

[16] J. M. Jimenez, SPS: Scrubbing or Coating?, in *Proc. of LHC Performance Workshop (Chamonix 2012)*, February 2012, p. 339.

[17] K. S. B. Li *et al.*, Modelling and Studies for a Wideband Feedback System for Mitigation of Transverse Single Bunch Instabilities, *IPAC2013*, Shanghai, China, May 2013, WEPME042, p. 3019.

[18] G. Rumolo *et al.*, Summary of the LIU Beam Studies Review, CERN-ATS-Note-2012-083 PERF. - 2012.

Chapter 17

New Injectors: The Linac4 Project and the New H⁻ Source

J. Lettry[1] and M. Vretenar[2]

[1]CERN, BE Department, Genève 23, CH-1211, Switzerland
[2]CERN, Accelerator and Technology Sector, Genève 23, CH-1211, Switzerland

Linac4 is a new 160 MeV linear accelerator designed to improve by a factor of 2 the beam brightness out of the LHC injection chain for the needs of the LHC luminosity upgrade. The project started in 2008 and beam commissioning takes place in 2014–2015. The new linac accelerates H⁻ ions that are then stripped at injection into the PS Booster; production of the H⁻ beam takes place in a state-of-the-art ion source of the RF-driven caesiated surface type. Acceleration is provided by four different accelerating sections matched to the increasing beam velocity, including two of novel designs, and focusing is provided by a combination of permanent-magnet and electromagnetic quadrupoles.

1. Introduction

The idea of building a new linear accelerator to make the LHC injectors capable of producing higher brightness beams dates back to the late 90s [1]: at that time the LHC was intended to start at full energy in 2008 and preparing for a high luminosity upgrade after about 10 years of operation was considered an utmost priority. In this perspective, the construction of some new accelerator was unavoidable: although successful, the program for the upgrade of the LHC injectors that took place in 1993–2000 in view of enabling the "nominal" LHC luminosity (10^{34} cm^{-2}s^{-1}) showed inherent limitations in the LHC injection chain that would not allow reaching the "ultimate" luminosity of 2.5×10^{34} cm^{-2}s^{-1} or beyond [2]. Beam brightness from the injectors being one of the main factors limiting the LHC luminosity several upgrades of the injectors were considered; they all had in common a new linac of energy higher than the present 50 MeV of Linac2, the first bottleneck for higher brightness being the limitation to the intensity at injection into the PS Booster (PSB) due to space charge induced tune shift at 50 MeV [3]. Whereas initial upgrade plans concentrated on a high-energy linac, the SPL (Superconducting Proton Linac) that would inject into the PS a beam at 2.2 GeV energy thus overcoming both PSB and PS limitations [4], at a later stage priority was given to a less ambitious program, based on the construction of a new linac injecting into the PSB at higher energy coupled with an increase of the PSB energy. The new linac was called Linac4, being the fourth hadron linac to be built at CERN [5]; after analyzing different options, the Linac4 energy was eventually fixed to 160 MeV, corresponding to a factor of 2 in $\beta\gamma^2$ with respect to the present 50 MeV injection (with β, γ relativistic beam parameters). Space charge induced tune shift scaling as $1/\beta\gamma^2$, the same tune shift as with Linac2 could be expected with twice the intensity in the same PSB transverse emittance, making possible a gain in brightness by a factor of 2. The Linac4 Technical Design Report was presented in 2006 [6] and construction of the new linac was approved by the CERN Council in June 2007; the project officially started in January 2008. After completing the civil engineering, the infrastructure and the construction of the machine components, beam commissioning has started in 2013 and will continue until end of 2015. Connection to the PSB will be possible from end of 2016.

Together with the increase in brightness for the LHC, the new linac is expected to bring other advantages to the CERN injector complex. The use of charge exchange injection in the PSB made possible by the acceleration in Linac4 of H⁻ ions instead of protons is expected to simplify the injection process reducing at the same time beam loss and activation; the modern construction technology of Linac4 will remove the concerns for long-term reliability related to the aging vacuum structure of Linac2; finally, a higher beam intensity will be available for non-LHC users of the PSB. As an additional feature, the Linac4 accelerator is designed to be able to operate at higher duty cycle than what was required by the PSB, following an upgrade of the power converters and of electrical and cooling infrastructure. This would make possible a future upgrade to become the injector for the proton driver accelerator of a neutrino facility.

2. Parameters, General Design and Layout

The main Linac4 parameters are reported in Table 1.

The new linac is dimensioned to double the maximum intensity from the PSB with the same transverse emittances, providing up to 10^{14} protons per pulse; it is expected to supply this charge with 400 μs long pulses at 40 mA current. The pulse repetition frequency is limited to a maximum of about 1 Hz by the PSB magnetic cycle, giving a beam duty cycle of less than 1%. In case Linac4 would be used in a future high-intensity facility for neutrino physics, the accelerating structures have been designed for a maximum duty cycle of 10%; infrastructure and power supplies are dimensioned only for the low duty cycle. About 35% of

Table 1. Main Linac4 design parameters.

Output energy	160	MeV
Bunch frequency	352.2	MHz
Repetition frequency	1.1 (max. 2)	Hz
Beam pulse length	0.4	ms
Beam duty cycle	0.08	%
Chopper beam-on rate	62	%
Linac pulse current	40	mA
N. of particles per pulse	1.0	$\times 10^{14}$
Transverse emittance	0.4	π mm mrad
Maximum RF duty cycle	10	%

the beam is chopped at 3 MeV and sent to a collimator/dump to allow low-loss capture of the beam in the PSB: this requirement together with some beam loss expected at low energy brings the required current out of the ion source to 80 mA. The RF frequency has been fixed at 352.2 MHz as in the old LEP accelerator, in order to recuperate klystrons and other RF equipment while remaining in the optimum frequency range for modern linear accelerators. The linac is composed of a 3 MeV injector followed by three different accelerating structures bringing the energy up to 160 MeV, for a total length of 75 m (Fig. 1).

The 3 MeV section is made of the ion source, of a Low Energy Beam Transport (LEBT), of a 3 m long Radio Frequency Quadrupole (RFQ) and of a 3.6 m long chopping and matching line that prepares the beam to be injected into the main accelerating structures. The three accelerating structures are a Drift Tube Linac (DTL) that reaches 50 MeV energy with three tanks, followed by a sequence of 7 Cell-Coupled Drift Tube Linac (CCDTL) modules going up to 104 MeV, and finally by 12 Pi-Mode Structure (PIMS) cavities reaching the final energy. All accelerating structures are normal conducting, the investment for superconducting accelerating structures not being justified in this range of energy and duty cycle, and have the same RF frequency to reduce cost and simplify

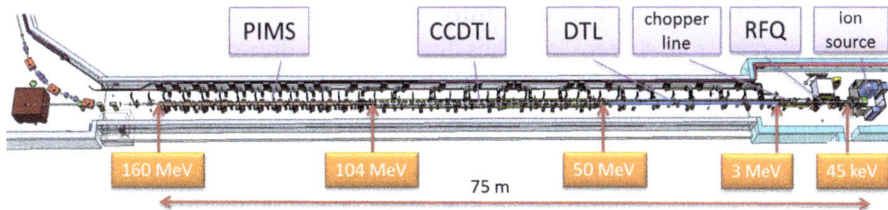

Fig. 1. Linac4 layout.

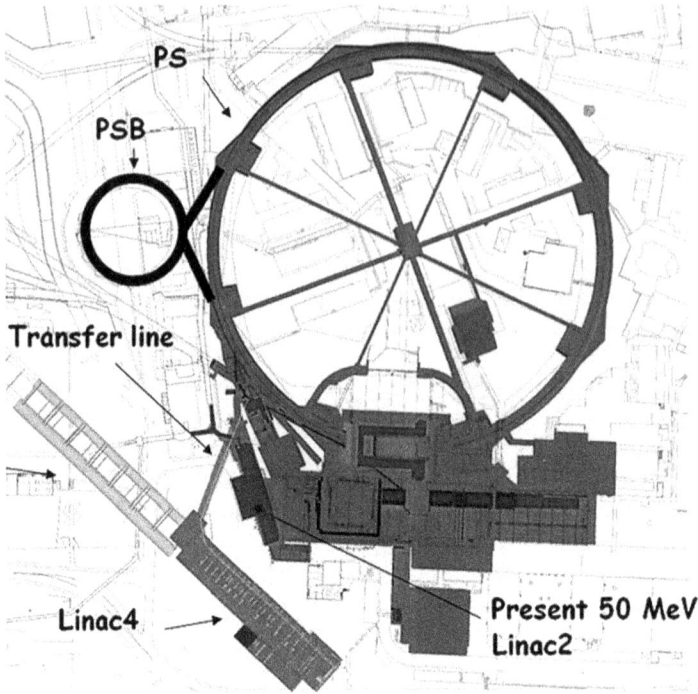

Fig. 2.　Side and 2D view of tunnel and surface building.

maintenance thanks to standardised RF systems. A bending magnet at the end of the linac can send the beam to a transfer line connected to the existing Linac2 to PSB transfer line; beam diagnostics elements and a beam dump are placed in a straight line after the bending. The basic Linac4 layout is presented in Fig. 1, while Fig. 2 shows the location of Linac4 with respect to PSB, PS and Linac2.

3.　Challenges of the Ion Source

The Linac4 requires the development of a completely new H⁻ source or at least a serious upgrade of the nearest existing design. None of the four H⁻ ion production mechanisms that were successfully demonstrated so far and operated in linear accelerator ion sources matches the specified duty-factor, emittance and ion beam current criteria at the 45 kV voltage required to inject into the RFQ. Additionally, for reliable operation for the LHC, Mean Time Between Maintenance (MTBM) driving factors such as sparking, material sputtering or contaminants should be identified and modeled and their detrimental effects minimized. Common feature to all H⁻ ion sources, the electrons co-extracted with the H⁻ beam must be properly dumped.

The H^- production mechanism taking place in the volume of the hydrogen plasma relies on the dissociation of an excited H_2^v molecule by a low energy electron in the vicinity of the extraction region. This ion source does not require cesium to operate; this is a particularly valuable asset during test periods that may be more prone to non-nominal operation conditions. A challenging feature of volume sources is their large amount of co-extracted electrons (typ. e/H^- ratio of 30).

Cesium is a highly reactive alkali, requiring particularly low levels of oxygen or water vapor levels, it is an outstanding electron donor and in the seventies, its admixture to hydrogen discharge sources opened the path to larger H^- ion currents [7] while sizably reducing co-extracted electrons. In penning or mag-netron discharge ion sources 0.3 to 3 g of metallic cesium is injected monthly; sputtering induced by the presence of hydrogen and Cs in the discharge electrical field is a relevant wear mechanism [8, 9].

An alternative design has been demonstrated by the SNS ion source team; in their RF sustained plasma ion source, a mono-layer of cesium deposited on their molybdenum extraction surface was providing a stable 60 mA beam of H^- for a period of five weeks at a record duty factor of 6% [10]. This caesiated surface mode of production reduces the average Cs consumption by 2 orders of mag-nitude and keeps a low e/H^- ratio. The hydrogen plasma of the SNS ion source is sustained by an internal solenoid RF system but a prototype with an external solenoid was also successfully operated for four weeks at SNS [11].

On the basis of today's state of the art H^- source designs/knowledge, the strategy decided for Linac4 has been to build an H^- volume source dedicated to low current tests and RFQ/Chopper commissioning while an H^- caesiated surface source driven by an external RF antenna solenoid is used for Linac4

Table 2. Summary of ion source parameters for the most representative ion sources operated at accelerator facilities for each of the H⁻ ion production mechanism; Volume, cesiated surface, Penning and Magnetron discharge [12, 13, 14].

Parameter	Unit	L4 Spec.	DESY	ISIS	SNS	BNL
Beam energy	keV	*45*	35	17–35	65	35, 40
Pulse duration	ms	*0.4*	0.1	0.5	1.	0.8
Repetition rate	Hz	*2*	6	50	60	6.6
H⁻ current	mA	*40/80*	30	35	60	65, 100
H⁻ production mode			Volume	Cs-Arc	Cs-surface	Cs-surface
Plasma heating:		*RF*	RF	Penning	RF	Magnetron
Emittance$_{Norm RMS}$	mm mrad	*0.25*	0.25	0.2	0.25	0.4, 0.56
Cs-consumption	mg/day		0	100	<1	12
Operation MTBM	weeks	*50*	50	5	6	36

operation. A 100 kW, 2 MHz RF amplifier developed at CERN will be used for both sources. The only system producing routinely 80 mA H^- current is BNL's Magnetron [12], however within a larger emittance than specified by Linac4; therefore, this production mode is the backup option in case the SNS-type source could not provide the design intensity. A summary of the main parameters of existing H^- ion sources is given in Table 2. The Cs consumption normalized to the H^- beam current is lowest for the SNS caesiated surface production systems.

The challenge of the Linac4 ion source is to design and produce a 45 kV volume H^- source for the tests of the 3MeV section and a caesiated surface ion source prototype operating at the 0.1% duty factor required for Linac4 operation [15]. Operation and measurement of ion sources performance is foreseen in stages (20 mA, 40 mA and maximum H^- current). The layout of the ion source is shown in Fig. 3 and the details of the plasma generators of a volume and a caesiated surface source are illustrated in Fig. 4.

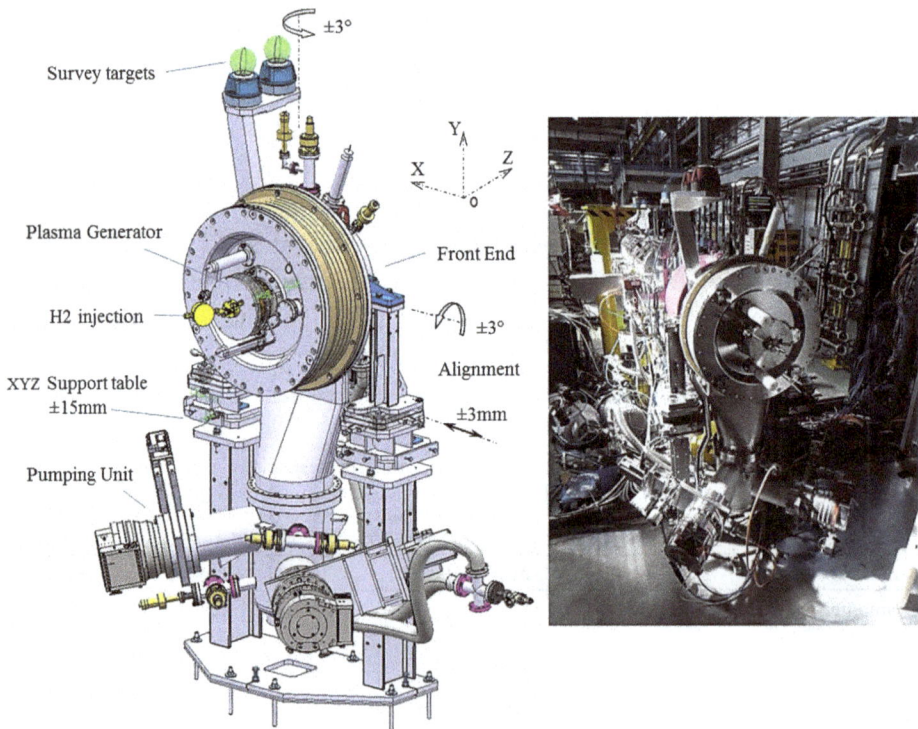

Fig. 3. Layout of the Linac4 ion source (volume production version) and its dedicated turbo-molecular pumping unit; the front end is installed on an adjustable support table. Once aligned, beam based minor offset of the horizontal position and angular orientations can be corrected. The plasma generator and beam extraction electrodes are mounted on an exchangeable flange.

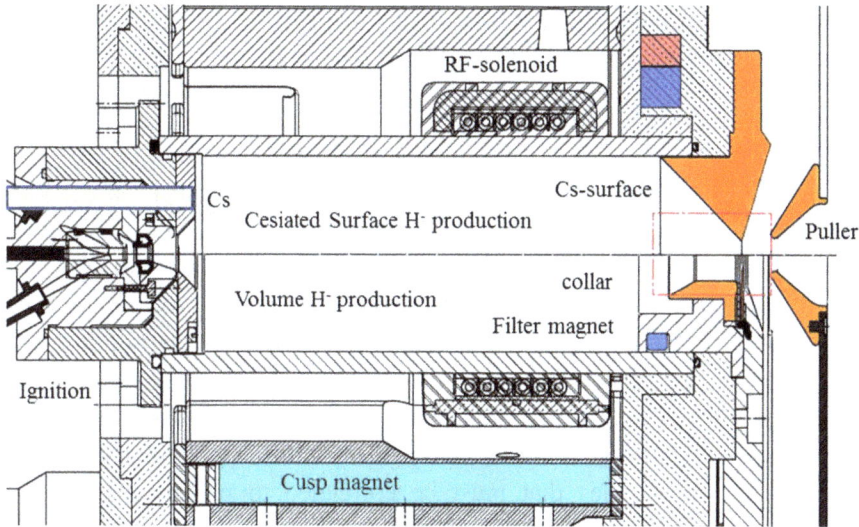

Fig. 4. Comparative schemes of plasma generators dedicated to cesiated surface (top) and volume production of H⁻ ions. The extraction geometry (diameter of the extraction hole and geometry of the puller) is similar to compare easily simulation and measurement conditions. The plasma-beam-formation region simulated in [24] is illustrated by a red dotted line.

After commissioning of the prototypes, the priority will be to characterize the plasma parameters best suited for H⁻ production, to measures the figures of merit of reliability and stability, identify the driving factors of the Mean Time Between Maintenance MTBM and minimize down time or ion source exchange/restart time. Depending on these findings, minimization of the emittance growth will motivate a 2nd order iteration of the design.

Modeling and simulation of the various physical and chemical processes involved in the H⁻ production must be complemented by dedicated experimental techniques and measurement campaigns. Observables are mandatory to validate the design and to further optimize the design; on-going simulation, control and experimental tasks are summarized in the next sections.

H_2-*injection*: The base line selected is a piezo valve driven pulsed injection into the plasma chamber. The dynamics of the flow from the injection line (0.3 to 2.5 bar, stabilized to 0.1%) down to below 10^{-6} mbar in the pumping system was simulated by electrical equivalent circuits [16] and calibrated [17].

RF-coupling: At plasma ignition, the electron/proton density ramps from 0 to typically 10^{18} m^{-3} within few tens of micro seconds. The plasma is ignited either via arc discharge or capacitive coupling; it is then sustained by inductive coupling of the RF-injected power into the plasma. The RF amplifier and matching circuit are engineered to follow in real time major changes of the load while being constantly matched via constant adaptation of the frequency [18]. The RF power and frequency are driven by arbitrary functions and plasma ignition is time tagged via an intensity threshold of the optical light emitted from the plasma.

Plasma: The hydrogen plasma is confined in a ceramic chamber surrounded by a permanent octupole magnet cusp in Halbach configuration. In the plasma expensing region, a magnetic dipole field reduces the electron energy in the vicinity of the beam extraction hole. The plasma is characterized by its neutrals and H^-/proton/electron densities, electron temperature and distribution of the excitation levels of the hydrogen atoms and molecules. Langmuir gauge measurements of the electron energy distribution function and plasma density [19] are complementary to non-invasive photometry and spectrometry [20] that provide on-line observables that must be correlated to plasma modeling (i.e. collision-radiation model) via Particle in Cell simulation. Preliminary simulation results at low plasma density [21] illustrate the pulsing of the plasma and the effect of the cusp field on the average electron energy.

Pulsed High Voltage: Minimization of the HV-sparks induced damages observed on cw high voltages stabilized by capacitor banks motivated the choice of pulsed power supplies. The high voltage system includes a low voltage electron dump (0–10 kV), a puller electrode (20–30 kV) extracting the beam and a high voltage (40–50 kV) acceleration station (or ground electrode?) defining the beam energy. The pulsed HV-system is based on standard power converters feeding dedicated HV-transformers located in the vicinity of the ion source [22]. It is specified to provide fast ramps and a 0.7 ms flat top held at nominal voltage under the pulsed H^- and electron beam loads. Furthermore, current limits allow identifying overcurrent and switching off the HV within a few micro seconds.

Beam formation: The beam formation region is composed of puller, electron dump and ground electrodes. Electrodes geometries and voltages must be optimized for the H^- beam and co-extracted electron currents [23]. The electron to H^- ratio is driven by the plasma parameters at the extraction hole and the extraction field; rather difficult to predict it must be measured to fine tune the space charge driven beam formation region. Particle in cell simulation of the beam formation region based on expected plasma parameters and puller geometries were achieved [24] for volume and surface production modes. An electrostatic accelerating Einzel lens matches the beam into the two solenoid Low Energy Beam Transport (LEBT) section.

4. Challenges of the 3 MeV Injector

The injector is the most critical part of any linear accelerator: this is where the particles are produced, where the beam emittance is generated, where the Coulomb repulsion between particles is the highest requiring particular focusing solutions, where the relativistic beta of the particles increases drastically imposing the use of complex mechanical structures, and where safety and collimation system have to be placed, before the energy becomes too high forbidding any particle loss.

The Linac4 injector section is about 9 m long; made of the ion source, of a Low Energy Beam Transport (LEBT), of a Radio Frequency Quadrupole (RFQ) and of a chopping/matching transport line it brings the beam to the energy of 3 MeV, a value still below the activation limits for commonly used metals. The LEBT is of the 2-solenoid type, and includes beam stoppers, some diagnostics and a gas injection system to control beam neutralization. The RFQ (Fig. 5), a combined focusing-bunching-accelerating device of 3-m length is of the brazed-copper 4-vane type and has been entirely built in the CERN Workshop; the RFQ electrodes ("vanes") have been accurately brazed in position after a complex sequence of machining steps and thermal treatments intended to keep vane position errors after brazing within the $30 \, \mu$m tolerance required by beam dynamics and adjustment of the RF field. The overall quadrupole field error is below $\pm 1\%$ [25].

The 3.6 m long chopping line that follows the RFQ houses two chopping structures, fast electrostatic deflectors capable of deflecting selected trains of bunches into a dump placed inside the line. The chopper will thus create beam free intervals within the linac current pulse at positions corresponding to the

Fig. 5.   The Linac4 RFQ.

transitions between the 2 MHz buckets in the PSB, where particles would be lost during the RF capture process. The choppers are 40 cm long meander-line deflectors mounted on a ceramic substrate and inserted into a quadrupole. Pulses of > 650 V with rise times of 1.5 ns (10 to 90%) have been produced during testing, a value that allows the chopper voltage to rise between two 352 MHz bunches thus avoiding partially deflected bunches [26]. The deflected beam is sent on a conical dump that acts like a collimator for the non-deflected beam. Three rebunching cavities, 11 quadrupoles and diagnostic equipment complete the chopper line and allow transporting, measuring and matching of the beam before injection into the DTL.

5. Challenges of the Accelerating Structures

In the energy range between 3 MeV and 160 MeV the beam velocity and thus the basic geometrical period (proportional to $\beta\lambda$) increase by a factor 6.5; at the same time space charge defocusing decreases, allowing to progressively lengthening the focusing periods including an increasing number of geometrical periods. This broad range of mechanical and focusing parameters has to be covered with a series of different accelerating structures, each of them providing optimized mechanical configuration, RF efficiency (shunt impedance), and focusing for the beam energy range in which it operates. An additional constraint is the requirement to keep the same RF frequency up to the final energy, to standardize the RF system reducing cost and simplifying maintenance. A thorough optimization of the Linac4 accelerating section led to a design based on a sequence of three accelerating structures: a Drift Tube Linac (DTL), a Cell-Coupled Drift Tube Linac (CCDTL) and a Pi-Mode Structure (PIMS) [27].

The DTL is divided into three tanks and accelerates the beam up to 50 MeV. Focusing is provided by 108 Permanent Magnet Quadrupoles (PMQ) placed in vacuum inside the DTL drift tubes. A special mechanical design has been developed for the DTL, providing a precise positioning of the drift tubes inside the tank without using bellows or flexible vacuum joints to align the tubes after assembly [28]. Figure 6 shows the first segment of DTL Tank1 during low-power RF tests.

Although fully successful, the design of the DTL has confirmed that this structure remains complex and expensive to build; moreover, the fact that the quadrupole magnets are welded inside the drift tubes makes access for repair extremely difficult. For these reasons, as soon as the increase in energy of the beam permits longer distances between quadrupoles the Linac4 design adopts another type of structure, the Cell-Coupled DTL. In a CCDTL short DTL-like

Fig. 6.   DTL.

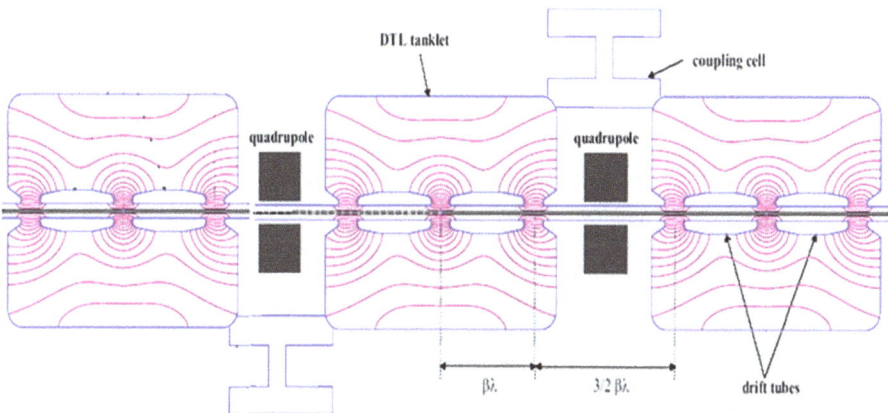

Fig. 7.   CCDTL scheme (with electric field lines); module 2 during tests.

tanks are connected by coupling cells to form a single RF structure; the coupling cells leave space for placing the quadrupoles in an accessible position between the tanks and the fact that the drift tubes do not contain quadrupoles allows for smaller diameter tubes providing higher RF shunt impedance. The Linac4 CCDTL is made of 7 accelerating modules for a maximum RF power of 1 MW each, accelerating the beam to 104 MeV; a module is composed of 3 DTL-type tanks housing two drift tubes each plus two coupling cells [29]. Focusing is provided by a combination of PMQs (between tanks) and electromagnetic quadrupoles (between modules). Figure 7 shows the CCDTL principle and one of the modules produced for Linac4.

The third accelerating structure, called Pi-Mode Structure (PIMS), covers the energy range 104–160 MeV where focusing is required only between modules; here a module is composed of a single resonator made of seven cells coupled via slots and operating in π-mode [30]. In the PIMS section there are 12 modules of this type. The cells are made of copper and electron-beam welded. The first PIMS module is shown in Fig. 8.

Fig. 8.   The first PIMS module under high-power tests.

6. Infrastructure and Operational Challenges

Linac4 is a modern linear accelerator that requires a complex infrastructure, housed in a large surface building above the machine tunnel. Particularly complex is the high-power Radio-Frequency system (Fig. 9), composed of thirteen 1.3-MW-klystrons, previously used in the CERN LEP accelerator, and six new klystrons of 2.8 MW, all operating in pulsed mode. The power out of the

2.8 MW klystrons is divided in two, allowing to either feed two couplers on the DTL tanks 2 and 3 or at high energy to feed two PIMS cavities in parallel. It is foreseen that as the stock of LEP klystrons runs out, pairs of LEP klystrons will be replaced by new klystrons equipped with a power splitter, leaving the wave-guide network unchanged and progressively increasing the number of RF stations feeding two cavities [31]. The 110 kV modulators, equipped with a HV pulse transformer and a droop compensation bouncer, have a modular design that allows feeding of either one high-power klystron or two LEP-type units. A digital Low-Level RF derived from the LHC system and equipped with feed-forward capability is used to precisely set the cavity voltages.

Fig. 9. Linac4 klystron gallery.

Linac4 will have to face a number of challenges related to the operation as injector of the PSB and of the entire LHC injection chain. First of all, the reliability of the linac injector must be the highest of all accelerators in the chain; achieving beam availability levels around 98%, as is the case with Linac2, is a challenge in itself and requires a careful design of all components taking sufficient safety margins, with particular care in the design and margins of high voltage components, a preference for proven technical solutions, the selection of component architectures less prone to faults, a thorough testing of all compo-nents before installation in the linac, and finally a year-long reliability run before connection to the PSB.

Another important requirement for Linac4 is providing a beam perfectly adapted to the PSB in terms of beam quality, matched beam parameters, synchronization, and safe operation. The beam parameters in Linac4 are constantly monitored via a large number of beam diagnostics devices, often derived from standard units used in other machines at CERN and adapted for the reduced longitudinal space and the specific beam parameters of a linac. There are in Linac4 27 beam position monitors based on strip-line detectors. Beam profile along the machine is measured by 16 SEM-grids and 6 wire scanners and beam intensity by 15 beam current transformers [32]. The beam parameters before injection into the PSB are measured in two dedicated measurement lines, for transverse and longitudinal beam parameters. A complex interlock system protects the machine from damage coming from beam loss in the long multi-branch transport lines between Linac4 and the PSB.

References

[1] R. Garoby and M. Vretenar, Proposal for a 2 GeV Linac Injector for the CERN PS, PS/RF/Note 96-27.

[2] M. Benedikt *et al.*, The PS complex produces the nominal LHC beam, in *Proc. of EPAC 2000*, Vienna.

[3] M. Benedikt, K. Cornelis, R. Garoby, E. Métral, F. Ruggiero and M. Vretenar, Report of the High Intensity Protons Working Group, CERN-AB-2004-022-OP-RF.

[4] M. Vretenar (ed.), Conceptual Design of the SPL, a high-power superconducting Linac at CERN, CERN 2000-012.

[5] M. Vretenar, R. Garoby, K. Hanke, A.M. Lombardi, C. Rossi and F. Gerigk, Design of Linac4 a new injector for the CERN Booster, in *Proc. of 2004 Linac Conf.*, Lübeck.

[6] F. Gerigk, M. Vretenar (eds.), Linac4 Technical Design Report, CERN-AB-2006-084.

[7] Yu. I. Bel'chenko, G. I. Dimov and V. G. Dudnikov, *Doklady Akademii Nauk SSSR*, **213**, 1283 (1973).

[8] J. Lettry, J. Alessi, D. Faircloth, A. Gerardin, T. Kalvas, H. Pereira and S. Sgobba, Investigation of ISIS and BNL ion source electrodes after extended operation, *Review of Scientific Instruments* **83**, 02A728 (2012).

[9] H. Pereira, J. Lettry, J. Alessi and T. Kalvas, Estimation of Sputtering Damages on a Magnetron H^- Ion Source Induced by Cs^+ and H^+ Ions, *AIP Conf. Proc.* **1515**, 81–88 (2013).

[10] M. P. Stockli, B. X. Han, S. N. Murray, T. R. Pennisi, M. Santana and R. Welton, Recent performance of the SNS H^- source for 1-MW neutron production, *AIP Conf. Proc.* **1515**, 292 (2013).

[11] R. F. Welton, N. J. Desai, B. X. Han, E. A. Kenik, S. N. Murray Jr., T. R. Pennisi, K. G. Potter, B. R. Lang, M. Santana and M. P. Stockli, Ion source development at the SNS, *AIP Conf. Proc.* **1390**, 226–234 (2013).

[12] J. Alessi, in *Proc. of 20th ICFA Advanced Beam Dynamics Workshop on High Intensity and High Brightness Hadron Beams*, 2002.

[13] J. Peters, The HERA Volume H⁻ Source, in *Proc. of PAC05*, TPPE001, p. 788 (2005).

[14] D. C. Faircloth, J. W. G. Thomason, M. O. Whitehead, W. Lau and S. Yang, *Review of Scientific Instruments* **75**(5), 1735 (2004).

[15] J. Lettry, D. Aguglia, Y. Coutron, A. Dallochio, H. Perreira, E. Chaudet, J. Hansen, E. Mahner, S. Mathot, S. Mattei, O. Midttun, P. Moyret, D. Nisbet, M. O'Neil, M. Paoluzzi, C. Pasquino, J. Sanchez Arias, C. Schmitzer, R. Scrivens, D. Steyaert and J. Gil Flores, H⁻ ion sources for CERN's Linac4, *AIP Conf. Proc.* **1515**, 302–311 (2013).

[16] C. Pasquino, P. Chiggiato, A. Michet, J. Hansen and J. Lettry, Vacuum simulation and characterisation for the LINAC4 H⁻ source, *AIP Conf. Proc.* **1515**, 401–408 (2013).

[17] E. Mahner, J. Lettry, S. Mattei, M. O'Neil, C. Pasquino and C. Schmitzer, Gas injection and fast-pressure-rise measurements for the Linac 4 H⁻ source, *AIP Conf. Proc.* **1515**, 425–432 (2013).

[18] M. Paoluzzi, M. Haase, J. Marques Balula and D. Nisbet, CERN LINAC4 H⁻ Source and SPL plasma generator RF systems, RF power coupling and impedance measurements, *AIP Conf. Proc.* **1390**, 265–271 (2011).

[19] C. Schmitzer, M. Kronberger, J. Lettry, J. Sanchez-Arias and H. Störi, Plasma characterization of the SPL plasma chamber using a 2 MHz compensated Langmuir Probe, *Review of Scientific Instruments* **83**, 02A715 (2012).

[20] J. Lettry, U. Fantz, M. Kronberger, T. Kalvas, H. Koivisto, J. Komppula, E. Mahner, C. Schmitzer, J. Sanchez, R. Scrivens, O. Midttun, P. Myllyperkiö, M. O'Neil, H. Pereira, M. Paoluzzi, O. Tarvainen and D. Wünderlich, Optical Emission Spectroscopy of the Linac4 and SPL Plasma Generators, *Review of Scientific Instruments* **83**, 02A729 (2012).

[21] S. Mattei, M. Ohta, A. Hatayama, J. Lettry, Y. Kawamura, M. Yasumoto and C. Schmitzer, Plasma modeling of the Linac4 H⁻ ion source, *AIP Conf. Proc.* **1515**, 386–393 (2013).

[22] D. Aguglia, Design of a System of High Voltage Pulsed Power Converters for CERN's Linac4 H⁻ Ion Source, to be presented at the IEEE Pulsed Power & Plasma Science International conference, San Francisco (USA), 12–16 June 2013.

[23] Ø. Midttun, T. Kalvas, M. Kronberger, J. Lettry and R. Scrivens, A magnetized Einzel-lens electron dump for the Linac4 H⁻ ion source, *AIP Conf. Proc.* **1515**, 481–490 (2013).

[24] S. Mochalskyy, J. Lettry, T. Minea, A. F. Lifschitz, C. Schmitzer, O. Midttun and D. Steyaert, Numerical modeling of the Linac4 negative ion source extraction region by 3D PIC-MCC code ONIX, *AIP Conf. Proc.* **1515**, 31–40 (2013).

[25] C. Rossi, L. Arnaudon, G. Bellodi, J. Broere, O. Brunner, A. M. Lombardi, J. Marques Balula, P. Martinez Yanez, J. Noirjean, C. Pasquino, U. Raich, F. Roncarolo, M. Vretenar, M. Desmons, A. France and O. Piquet, Commissioning of the Linac4 RFQ at the 3 MeV Test Stand, in *Proc. of IPAC 2013*, Shanghai.

[26] M. Paoluzzi, A fast 650V chopper driver, in *Proc. of IPAC 2011*, San Sebastian.

[27] F. Gerigk, N. Alharbi, M. Pasini, S. Ramberger, M. Vretenar and R. Wegner, RF Structures for Linac4, in *Proc. of 2007 PAC*, Albuquerque.

[28] S. Ramberger, G. De Michele, F. Gerigk, J. M. Giguet, J. B. Lallement, A. M. Lombardi and M. Vretenar, Production design of the Drift Tube Linac for the CERN Linac4, in *Proc. of 2010 Linac Conf.*, Tsukuba.

[29] M. Vretenar, Y. Cuvet, G. De Michele, F. Gerigk, M. Pasini, S. Ramberger, R. Wegner, E. Kenjebulatov, A. Kryuchkov, E. Rotov and A. Tribendis, Development of a Cell-Coupled Drift Tube Linac for Linac4, in *Proc. of 2008 Linac Conf.*, Victoria.

[30] F. Gerigk, P. Ugena Tirado, J. M. Giguet and R. Wegner, High Power Tests of the first PIMS Cavity for Linac4, in *Proc. of IPAC 2011*, San Sebastian.

[31] O. Brunner, N. Schwerg and E. Ciapala, RF Power Generation in Linac4, in *Proc. of 2010 Linac Conf.*, Tsukuba.

[32] F. Roncarolo *et al.*, Overview of the CERN Linac4 Beam Instrumentation, in *Proc. of 2010 Linac Conf.*, Tsukuba.

Chapter 18

Challenges and Plans for the Ion Injectors

D. Manglunki

CERN, TE Department, Genève 23, CH-1211, Switzerland

We review the performance of the ion injector chain in the light of the improvements which will take place in the near future, and we derive the expected luminosity gain for Pb–Pb collisions in the collider during the HL-LHC era.

1. Introduction

The Pb^{82+} ion beam brightness, as delivered at 177 GeV/nucleon from the injectors in February 2013 at the end of the first LHC p–Pb run [1], exceeded the design value [2] by a factor of 3, as shown in Table 1.

Table 1. Pb^{82+} ion beam properties at SPS extraction in 2013, compared to design values.

	Pb^{82+} 2013	Pb^{82+} design	
N_B	22	9	10^7 ions/bunch
n_B	24	52	bunches
ε_H	1.1	1.2	μm (norm. RMS)
ε_V	0.9	1.2	μm (norm. RMS)
N_B/ε	22.1	7.5	10^7 ions/μm

With this high brightness beam, a Pb–Pb run at $7Z$ TeV performed with the current filling scheme, described in Section 2.2 below, would deliver a peak luminosity of 2.3×10^{27} cm^{-2}s^{-1}. Since the peak luminosity requested by the ALICE experiment for the HL-LHC era [3] is of the order of 7×10^{27} cm^{-2}s^{-1}, a missing factor of 3 is still to be found. The retained solution is to increase the number of bunches in the collider, as the bunch brightness is already a limiting factor on the flat bottom of the SPS, due to space-charge and intra-beam scattering [4]. Increasing the bunch intensity would also decrease the luminosity lifetime due to a resulting larger burnoff rate [5].

2. The Current Scheme

2.1. *The ion accelerator chain*

CERN's heavy ion production complex (Fig. 1) was first designed in the 1990s for the needs of the SPS fixed target programme [6, 7]. It was rejuvenated at the beginning of the 2000s in order to cope with the LHC's stringent demands for high brightness ion beams. It currently consists of:

- The Electron Cyclotron Resonance (ECR) ion source,
- The ion Radio Frequency Quadrupole (RFQ),
- The Interdigital H Linear Accelerator (LINAC3),
- The Low Energy Ion Ring (LEIR), an accumulation and cooling storage ring,
- The 25 GeV Proton Synchrotron (PS), and
- The 450 GeV Super Proton Synchrotron (SPS).

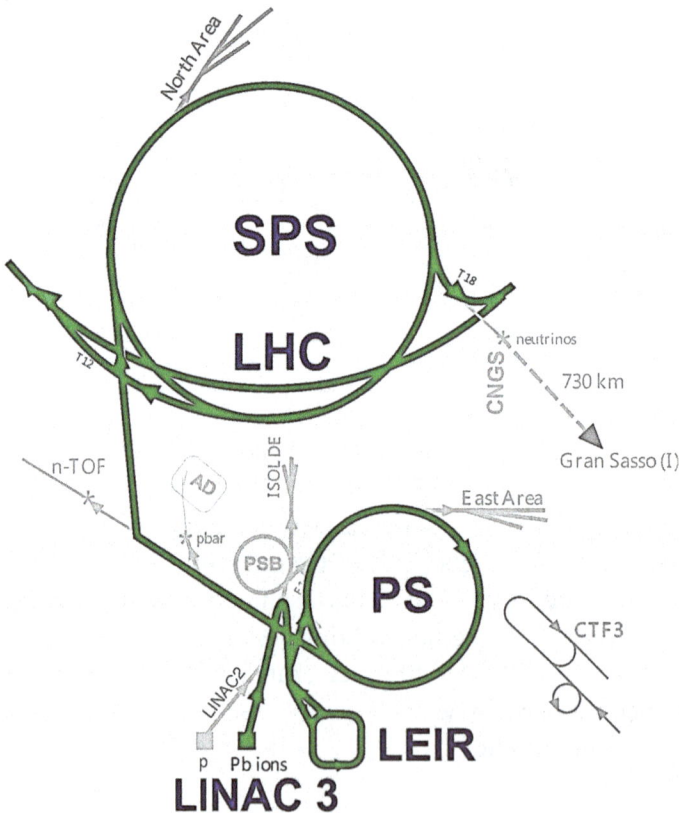

Fig. 1. The ion injector part of the CERN accelerator complex (not to scale).

2.2. *Production of the bunch trains for the LHC*

A sample of isotopically pure ^{208}Pb is inserted in a filament-heated crucible at the rear of the ECR source, whilst oxygen is injected as support gas. With a 10 Hz repetition rate, a 50 ms long pulse of 14.5 GHz microwaves accelerates electrons to form an oxygen plasma, which in turn ionizes the lead vapor at the surface of the crucible. The source operates in the so-called afterglow mode: after the microwave pulse is switched off, the intensity of the high charge state ions from the source increases dramatically. The ions are electrostatically extracted from the source with an energy of 2.5 keV/nucleon.

Out of the different ion species and charge states, a 135° spectrometer situated at the exit of the source selects those lead ions which have been ionized 29 times (Pb^{29+}). The beam is then accelerated to 250 keV/nucleon by the 2.66 m long RFQ which operates at 101.28 MHz. The RFQ is followed by a four-gap RF cavity which adapts the longitudinal bunch parameters to the rest of the linear accelerator: a three-cavity interdigital H (IH) structure, Linac3, which brings the beam energy up to 4.2 MeV/nucleon, requiring about 30 MV of accelerating voltage, for a total acceleration length of 8.13 m. The first cavity operates at the same frequency as the RFQ (101.28 MHz), while the second and third cavities operate at 202.56 MHz. Finally, a 250 kV "ramping cavity", also operating at 101.28 MHz, distributes the beam momentum over a range of ±1%, according to the time along the 200 μs pulse. The linac currently only operates at 5 Hz.

A 0.3 μm thick carbon foil provides the first stripping stage at the exit of the linac, followed by a spectrometer which selects the Pb^{54+} charge state.

A current of about 22 μA of Pb^{54+} is injected over 70 turns in the Low Energy Ion Ring LEIR, whose unique injection system fills the 6-dimension phase space. This is achieved by a regular multi-turn injection with a decreasing horizontal bump, supplemented by an electrostatic septum tilted at 45°, and the time dependence of the momentum distribution coupled with the large value (10 m) of the dispersion function in the injection region. The injection process is repeated up to six times, every 200 ms. The whole process is performed under electron cooling. At the end of the seven injections, the beam is bunched on harmonic 2, accelerated to 72 MeV/nucleon, and fast-extracted towards the PS. At this point the bunch population is of the order of 5×10^8 ions/bunch.

In the PS, the two bunches are injected in two adjacent buckets of harmonic 16 and accelerated by the 3–10 MHz system to 5.9 GeV/nucleon, with an intermediate flat-top for batch expansion. This process consists of a series of harmonic changes: $h = 16$, 14, 12, 24, 21, in order to finally reach a bunch spacing of 200 ns at top energy. Before the single turn extraction, the bunches are finally rebucketed into $h = 169$, using one of the PS's three 80 MHz cavities. This rebucketing only affects the bunch length, not the inter-bunch spacing.

At the exit of the PS, the beam traverses a final stripping stage to Pb^{82+}, through a 1 mm thick Aluminum foil, where the FODO lattice of the transfer line is modified to provide low beta functions in both planes, so as to minimize the transverse emittance blowup. As the stripping foil is located in the same transfer line as the high intensity proton beams for other users (the Antiproton Decelerator, the neutron Time-of-Flight facility, CERN Neutrinos to Gran Sasso, etc.), it needs to move in or out of its position in less than a hundred milliseconds, before or after a series of ion cycles.

The two bunches, now about 3×10^8 ions each, are injected in the SPS, with a bunch-to-bucket transfer into the 200 MHz system. This process can be repeated up to 12 times every 3.6 seconds; the repetition rate is imposed by the duration of the LEIR cycle, while the total number of injections is currently limited by the SPS control hardware.

At a fixed harmonic, the heavy mass of the ions would yield too large a frequency swing (198.51–200.39 MHz) for the range of the travelling wave cavities of the SPS (199.5–200.4 MHz). Hence, instead of using a fixed harmonic, the ions are accelerated using the "fixed frequency" method in the SPS: a non-integer harmonic number is used, by turning the RF on at the cavity center frequency during the beam passage and switching it off during its absence, to correct the RF phase and be ready for its next passage in the cavities. The phase is adjusted by an appropriate modulation of the frequency. Aside from its complexity, one drawback of the method is that the beam has to be constrained in a relatively short portion ($\sim 40\%$) of the circumference of the machine.

The SPS ion beam, now consisting of 24 bunches, is then accelerated to 177 GeV/nucleon before its single turn extraction towards the LHC via one of the transfer lines TI2 or TI8. This operation is repeated 30 times, fillings each ring of the LHC with 360 bunches.

3. Planned Upgrades

3.1. *Doubling the repetition rate of Linac3*

The possibility of accelerating up to three charge states in Linac3 to double the intensity as originally planned at the time of design [8] is now abandoned, as it is likely that the off-central charge states would not transmitted into the LEIR machine, due to their large momentum error.

Hence, aside from new developments on the ECR source, the most promising path consists in doubling the repetition rate from 5 to 10 Hz [9]. The investment will be relatively modest as the source itself is already pulsing at 10 Hz, and the Linac3 has been designed for operating at 10 Hz as well. The LEIR injection system has also been tested at 10 Hz, so only a few magnets and power supplies in the Linac-to-LEIR transfer lines will have to be replaced or upgraded.

3.2. Overcoming the intensity limitation in LEIR

Due to a loss that occurs at the beginning of the acceleration ramp (Fig. 2), the intensity extracted out of LEIR is currently limited to about 6×10^{10} proton charges, corresponding to 5.5×10^{8} Pb^{54+} ions per bunch. A thorough machine development program has started in order to try and understand, then overcome or mitigate this limitation [10].

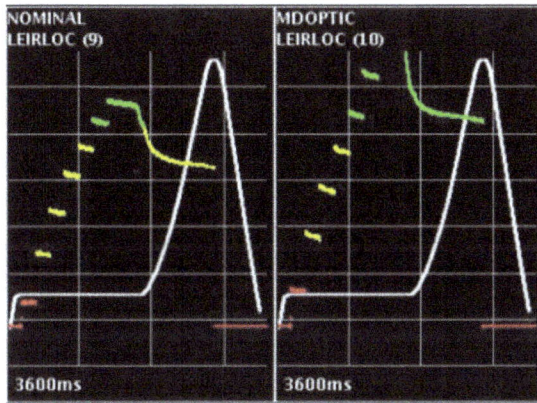

Fig. 2. Loss at acceleration on the LEIR nominal and development cycles. The white curve represents the magnetic field, while the coloured one represents the number of ions circulating in the machine. One can see the 7 injections on the flat bottom, followed by the acceleration ramp, at the beginning of which the loss occurs. The bunch population is presently limited to 5.5×10^{8} Pb^{54+} ions per bunch.

3.3. Bunch splitting in the PS

In order to further increase the bunch number in the LHC, one would have to split the bunches in the PS, whose RF system cannot accelerate on 20 MHz. Hence, splitting down to a 50 ns bunch spacing would have to be done at high energy, close to transition. So, not only new cavities would have to be installed in the PS machine for the RF gymnastics themselves, but also a new gamma-jump scheme, specific to ions, would have to be designed and implemented. For the same reason, a batch compression down to 50 ns bunch spacing at high energy would be impractical. Hence, only splitting to 100 ns bunch spacing in the PS is considered, using the same bunch structure as already planned for the nominal beam in the design report [11]: the RF gymnastics are similar to the ones currently used, and described in Section 2.2 above ($h = 16$, 14, 12, 24, 21). The only difference is the phase of harmonic 24 which is shifted by 180°, resulting in bunch splitting. The final spacing between bunches at the PS flat top is 100 ns. These gymnastics have been tried and tested during machine development sessions in parallel to the early LHC runs.

3.4. *New injection scheme into the SPS*

In order to maximize the number of bunches, the 4-bunch batches from the SPS will be injected next to each other, with a batch spacing of 100 ns, thanks to a new ion injection system [12]. A faster, 50 ns injection system had been considered [13] but had to be discarded due to impedance and resources considerations. The new system consists of:

- New pulsers for the fast kicker magnets, allowing a rise time of 100 ns. These fast pulsers had already been foreseen at the time of design [14].
- A new injection septum, which will be recuperated from the PS Booster extraction line, after its upgrade to 2 GeV.
- An improved injection damper to mitigate the large oscillations of the bunches situated at the limit of the kickers rise time.

3.5. *Momentum slip-stacking in the SPS*

As bunch-splitting or batch compression to reach a bunch spacing of 50 ns both prove difficult, if not impossible, to perform in the PS, elegant RF gymnastics in the SPS have been proposed, which will be made possible by the planned upgrade of the SPS radio frequency systems. Momentum slip-stacking [15] consists of capturing two trains of bunches with independent beam controls, and by detuning them in momentum, bringing them closer together. Those gymnastics, originally proposed to increase bunch densities [16], can be applied in our case to interleave the bunches, as sketched in Fig. 3:

- The PS transfers two times six batches of four bunches to the SPS, with a batch spacing and a bunch spacing of 100 ns. In between the two trains of 24 bunches, is a 1 μs gap, large enough for two independent RF systems to capture each of them on the same frequency.
- The two trains are detuned in momentum by opposite amounts.

- Due to the frequency difference, the two trains start slipping towards each other.
- Once the bunches are interleaved, they are recaptured at average frequency and filament in a large bucket.

One issue is the larger resulting longitudinal emittance, but early simulations indicate the bunch length would still be within the accepted limits of the LHC RF at injection [17]. This issue would be completely dismissed in case of the addition of 200 MHz cavities in the LHC.

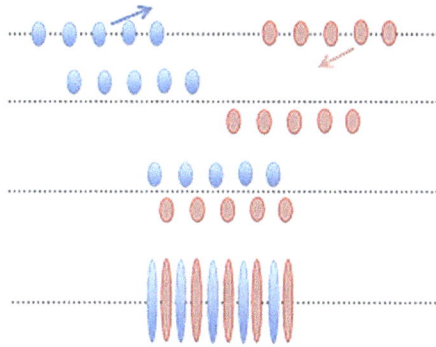

Fig. 3. Schematic for halving the bunch spacing using momentum slip-stacking.

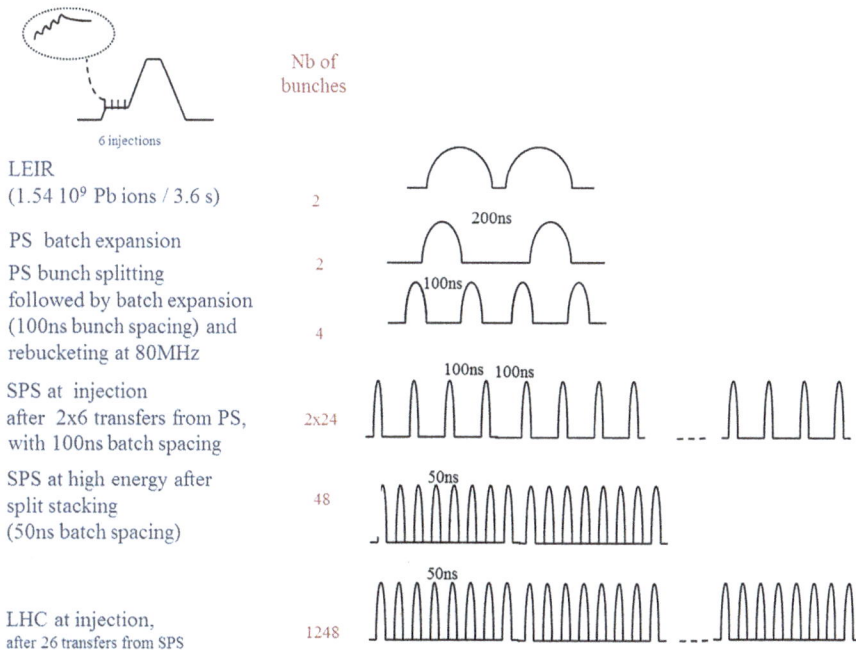

Fig. 4. Upgraded filling scheme.

4. The Upgraded Filling Scheme

The upgraded filling scheme assumes all upgrades listed above have been implemented, namely:

- Doubling the Linac3 repetition rate to 10 Hz.
- A reasonable increase (~ 50%) of the bunch density in LEIR.
- RF gymnastics splitting the two bunches in the PS, in order to send 4-bunch trains into the SPS with 100 ns bunch spacing.
- A new 100 ns rise time injection scheme into the SPS.
- Momentum slip-stacking in the SPS.

The new proposed filling scheme, sketched in Fig. 4, should deliver up to 1248 bunches in each one the LHC rings, bringing the peak Pb–Pb luminosity at 7Z TeV to $\mathcal{L}_{peak} > 7.0 \times 10^{27}$ cm^{-2}s^{-1}. Table 2 summarizes the projected beam parameters in the injectors.

Table 2.   Projected beam parameters in the circular machines

	LEIR	PS	SPS
Charge state [1]	54+	54+/82+ [1]	82+
Output Energy [GeV/u]	0.0722	5.9	176.4
Output Bρ [T.m]	4.8	86.7/57.1 [1]	1500
Injections into the next machine	1	6+6	26
Bunches/ring	2	4	48
Number of ions/pulse	1.6×10^9	7.2×10^8	5.2×10^9 [3]
Total extracted charge	8.6×10^{10}	3.9×10^{10} / 5.9×10^{10} [1]	4.3×10^{11} [3]
Ions/bunch at injection	10^9 [4]	8×10^8	1.8×10^8 [2]
Ions/LHC bunch at extraction	4×10^8	1.8×10^8	1.5×10^8 [2] 1.1×10^8 [3]
Bunch spacing at extraction [ns]	350	100	50
Normalized transverse rms emittance [μm]	0.70	1.0	1.2
Longitudinal emittance [eVs/u]	0.025	0.045	0.24
4σ Bunch length [ns]	200	3.9	1.8 [5]
2σ Momentum spread	1.2×10^{-3}	1.1×10^{-3}	6.4×10^{-4} [5]
Repetition time [s]	3.6	3.6	49.2
Space charge ΔQ on flat bottom	0.13 [4]	0.23	0.13 [2]

[1] Stripper stage between PS & SPS; parameters before/after stripping.
[2] Corresponding to the maximum bunch intensity.
[3] Average intensity taking into account the distribution of bunch intensities along the train.
[4] At the end of the flat bottom, just before acceleration.
[5] Assuming 7.5MV RF voltage.

References

[1] D. Manglunki *et al.*, The first LHC p-Pb run: performance of the heavy ion production complex, WEPEA061, IPAC'13, Shanghai, China.

[2] LHC design report, Vol. I, Chap. 21, The LHC as a Lead Ion Collider.

[3] The ALICE Collaboration, Upgrade of the Inner Tracking System Conceptual Design Report, CERN-LHCC-2012-013 (LHCC-P-005), September 2012.

[4] D. Manglunki *et al.*, Plans for the Upgrade of CERN's Heavy Ion Complex, WEPEA060, IPAC'13, Shanghai, China.

[5] J. M. Jowett *et al.*, Future heavy-ion performance of the LHC, in *Proc. RLIUP Workshop*, Archamps, France, October 2013.

[6] H. Haseroth (ed.), Concept for a lead-ion accelerating facility at CERN, CERN-90-01, Geneva, 1990.

[7] D. J. Warner (ed.), CERN heavy-ion facility design report, CERN-93-01, Geneva, 1993.

[8] LHC design report, Vol. III, Part 4, Chap. 34, Source and Linac3.

[9] D. Küchler, D. Manglunki and R. Scrivens, How to run ions in the future?, in *Proc. RLIUP Workshop*, Archamps, France, October 2013.

[10] M. Bodendorfer, Chromaticity in LEIR performance, CERN-ACC-NOTE-2013-0032.

[11] LHC design report, Vol. III, Part 4, Chap. 37, The LHC Ion Injector Chain: PS and transfer line to SPS.

[12] M. Benedikt (ed.), LIU-SPS 50ns Injection review, EDMS 1331860, https://edms. cern.ch/file/1331860/1.0/LIU-SPS_50ns_Injection_review_ExecutiveSummary_ 22Nov2013.pdf, November 2013.

[13] B. Goddard *et al.*, A New Lead Ion Injection System for the CERN SPS with 50 ns Rise Time, MOPFI052, IPAC'13, Shanghai, China.

[14] LHC design report, Vol. III, Part 4, Chap. 38, The LHC Ion Injector Chain: SPS.

[15] R. Garoby, RF Gymnastics in a synchrotron, in *Handbook of Accelerator Physics and Engineering*, World Scientific ed., 1999, pp. 286–287.

[16] D. Boussard and Y. Mizumachi, Production of Beams with High Line-Density by Azimuthal Combination of Bunches in a Synchrotron, CERN-SPS/ARF/79-11.

[17] T. Argyropoulos, Th. Bohl and E. Shaposhnikova, private communication.

Chapter 19

Challenges and Plans for Injection and Beam Dump

M. Barnes, B. Goddard, V. Mertens and J. Uythoven

CERN, TE Department, Genève 23, CH-1211, Switzerland

The injection and beam dumping systems of the LHC will need to be upgraded to comply with the requirements of operation with the HL-LHC beams. The elements of the injection system concerned are the fixed and movable absorbers which protect the LHC in case of an injection kicker error and the injection kickers themselves. The beam dumping system elements under study are the absorbers which protect the aperture in case of an asynchronous beam dump and the beam absorber block. The operational limits of these elements and the new developments in the context of the HL-LHC project are described.

1. Introduction

The beam transfer into the LHC is achieved by the two long transfer lines TI 2 and TI 8, together with the septum and injection kicker systems, plus associated machine protection systems to ensure protection of the LHC elements in case of mis-steered beam. The LHC is filled by ~ 10 injections per beam. The MKI kicker pulse length is 8 μs, with a rise time of 0.9 μs and a fall time of 2.5 μs. Filling each ring takes 8 minutes with the SPS supplying interleaved beams to other facilities. The foreseen increase in injected intensity and brightness for the HL-LHC means that the protection functionality of the beam-intercepting devices needs upgrading. In addition the higher beam current significantly increases the beam induced power deposited in many elements, including the injection kicker magnets in the LHC ring.

The beam dumping system is also based on DC septa and fast kickers, with various beam intercepting protection devices including the beam dump block. Again, the significant change in the beam parameters for the HL-LHC implies redesign of several of the dump system devices, because of the increased energy deposition in the case of direct impact, but also because of increased radiation background which could affect the reliability of this key machine protection system.

In the following sections the function and required changes planned for the different LHC beam transfer systems are described.

2. Protection Against Injection Errors

The high injected beam intensity and energy mean that precautions must be taken against damage and quenches, by means of collimators placed close to the beam in the injection regions. The layout of the injection and associated protection devices is shown schematically in Fig. 1. The beam to be injected passes through five horizontally deflecting steel septum magnets (MSI) with a total deflection of 12 mrad, and four vertically deflecting kickers (MKI) with a nominal total kick strength of 0.85 mrad. Uncontrolled beam loss resulting from errors (missing, partial, badly synchronised or wrong kick strength) in the MKI could result in serious damage of the equipment in the LHC injection regions, in particular the superconducting separation dipole D1, the triplet quadrupole magnets near the ALICE or LHCb experiments, or in the arcs of the LHC machine itself. Damaging detector components, in particular close to the beam pipe, by excessive stray radiation is also possible.

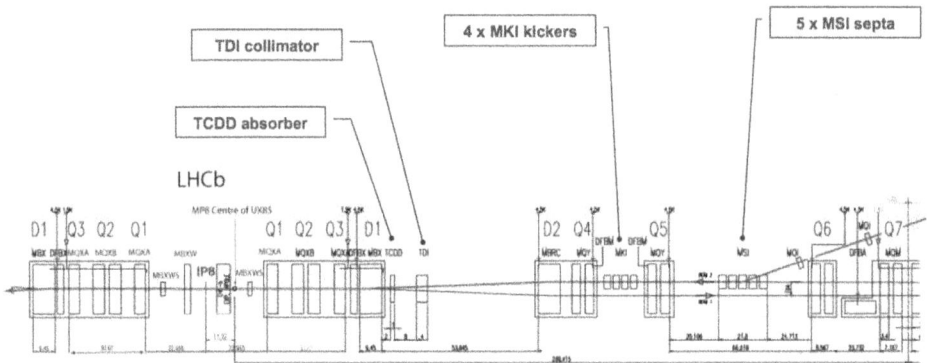

Fig. 1. Overview of injection into LHC (Beam 2, P8). The beam is injected from the right hand side.

To protect against this failure, a movable 2-sided absorber TDI is installed about 70 m from the MKI, at a 90° phase advance, in order to protect the LHC equipment. Each TDI consists of two absorber jaws. The upper jaw intercepts injected beam which is not (or not sufficiently) deflected by the injection kickers. The lower jaw intercepts the deflected circulating beam in the event of a kicker synchronisation error. Other failure scenarios have been explored in depth [3]. The most critical MKI failure is a flashover, or breakdown, in the kicker magnet which could lead to a grazing incidence beam on the TDI absorber. An accurate setup of the TDI jaws is extremely important for machine protection, in the light of the occurrence of several injection kicker failures per year with high injected intensity in LHC Run 1.

A fault on the MKI could result in the whole injected batch being mis-steered. The damage level for LHC equipment for fast and localised losses is estimated to be around 20 J/g [4]. Secondary and scattered particles must not cause damage to local equipment, in particular the D1 magnet (Fig. 1). For that purpose the TDI is complemented by a fixed mask TCDD, just in front of the D1. During such events the TDI should not itself be damaged, either in terms of the jaw material, coating, vacuum system, positioning system or in other functional ways such as integrity of the impedance shielding.

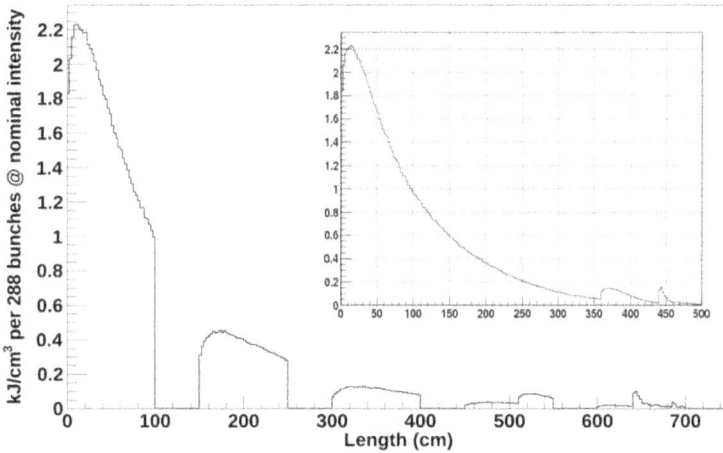

Fig. 2.   Energy deposition in segmented CfC/Mo TDI absorber, with inset showing the energy deposition for the existing device.

The increased injected intensity and brightness with the HL-LHC beam parameters [1] mean that the present TDI needs redesigning, since the energy deposition and resulting thermal stresses are above the design limits of the present system. In addition, the experience from LHC Run 1 revealed several issues with the vacuum level, beam induced heating and mechanical design of the present TDI device. As a result, for HL-LHC, a completely new TDI design is being developed. The design considerations remain the same — to protect the downstream elements from damage in the event of a kicker failure — but the device concept will be revised to consist of a sequence of several shorter ~1 m long collimators with jaw materials designed to survive the impact, Fig. 2, while providing enough dilution to protect the downstream D1 and triplet. In addition, the vacuum performance of the TDI needs to improve.

The TCDD masks are 1 m long Cu blocks, which in P2 are required to open for data-taking of the ALICE ZDC [2]. For the injection protection functionality a reduction of aperture of the TCDD masks is under study.

Finally, the TDI is complemented by two auxiliary collimators TCLI, which are located on the outgoing side of the triplet, at phase advances which are designed to be ±20° (modulo 180°). The present TCLIA collimators are 1.0 m long graphite jaws, based on the TCS design, and will possibly need to be replaced by more robust and more absorbing jaws. The design issues here are, besides the robustness of the TCLI jaws, the impedance heating of the two-in-one design for the TCLIA, the protection of the downstream superconducting elements and the risk of quench from injection beam losses on the TCLI jaws.

Overall, the TDI and TCLI elements need to protect the LHC arc aperture from mis-steered beams. The criteria for the protection are based on the assumed safe beam limit, which is taken as 10^{12} protons. Tracking studies showed that settings of 6.8σ for the TDI and TCLI systems adequately protect the LHC arc aperture against MKI flashovers [3]. As this depends on the injected intensity, it may be necessary to reduce the 6.8σ settings slightly for injection of higher intensity beams, which needs to be analysed in the context of the beam cleaning collimation system settings for injection.

3. Injection Kicker MKI Performances

The injection kicker (MKI) systems, at Points 2 (MKI2) and 8 (MKI8), deflect the incoming particle beams onto the accelerators equilibrium orbits. The four MKI magnets per injection are named D, C, B and A: D is the first to see injected beam. The total vertical deflection by the four MKI magnets is 0.85 mrad, requiring an integrated field strength of 1.3 Tm. Reflections and flat top ripple of the field pulse must be less than ±0.5%, a demanding requirement, to limit the beam emittance blow-up due to injection oscillations.

Each MKI system consists of a Pulse Forming Network (PFN) and a multi-cell travelling wave kicker magnet, connected by matched transmission lines and terminated by a matched termination resistor. Each MKI magnet has 33 cells. A cell consists of a U-core ferrite sandwiched between HV conducting plates, and two ceramic capacitors sandwiched between an HV plate and a plate connected to ground (Fig. 3, right). Either 8C11 or CMD5005 ferrite is used for the MKI yoke: the data sheets for these ferrites show that the initial permeability starts to reduce for temperatures above $\sim 100\,°C$.

After an MKI magnet is installed in a vacuum tank (Fig. 3, left), and prior to mounting vacuum valves, the complete magnet is baked out, in an oven, to $300\,°C$ for at least 48 hours: the bake-out permits the MKIs to achieve a vacuum of around 10^{-11} mbar. Following cool-down the MKI is HV pre-conditioned to a PFN voltage of 56.6 kV. Subsequently the MKI is returned to the clean-room, vacuum valve actuators are installed, and bake-out jackets are mounted on each vacuum

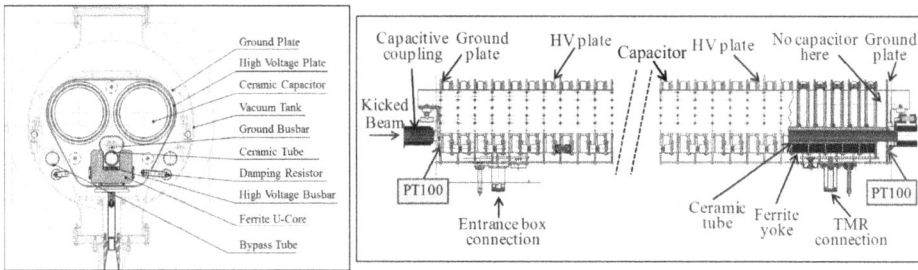

Fig. 3.   MKI kicker magnet.

tank. The jackets are used to re-bake the MKI before a final HV conditioning. The jackets remain on the tank such that a bake-out of the MKI can be carried out in the LHC tunnel, if required.

The 3414 mm long, 540 mm diameter, tanks housing the MKI magnets are reused from the LEP accelerator: they were used for 300 kV electrostatic separators and thus were electro-polished. Each vacuum tank requires a bypass tube for the counter-rotating beam: high conductivity copper is used. Figure 3 shows two bypass tubes — this allows an MKI to be used at either LHC injection point. Small longitudinal slots, in the copper, connect the vacuum of the tank and the bypass tube.

A number of effects have been observed for the MKI magnets during operation in the LHC: these include beam induced heating of the ferrite yoke, inefficient cooling of the ferrite yoke, occasional beam induced heating of ferrite toroids outside of the magnet yoke, electrical flashovers, unidentified falling objects (UFOs) and electron-cloud [6].

3.1. *Beam induced heating of ferrite yoke*

A beam screen (electromagnetic shield) is placed in the injected beam aperture of each magnet, to provide a path for the image current of the LHC beam and screen the ferrite yoke against wake fields. The screen consists of a ceramic tube with conductors lodged in grooves in the inner wall. To limit longitudinal beam coupling impedance, while allowing a fast magnetic field rise-time, the conductors are connected to the standard LHC vacuum chamber at one end and are capacitively coupled to it at the other end [7]. In the original design the extruded ceramic tube had 24 nickel-chrome (80/20) conductors, each 0.7 mm $\times$ 2.7 mm with rounded corners, inserted into slots. In the version installed in the LHC prior to the 2013–2014 long shutdown one (LS1), nine conductors closest to the HV busbar were removed to reduce the maximum electric field. This beam screen ensured a low rate of flashover on the inner surface of the ceramic tube and, initially, an adequately low beam coupling impedance. However with high LHC beam currents, integrated over the several hours of a good physics fill, the real component of beam coupling impedance of the magnet ferrite yoke can lead to significant beam induced heating.

When the temperature of the MKI ferrite yoke approaches the Curie point the strength of the kick reduces and the mis-kicked injected beam could result in quenches of several superconducting magnets. Hence there is an interlock to inhibit injection if the measured yoke temperature is above specified thresholds. As a result of low emissivity of the inside of the MKI vacuum tanks the time-constants for the measured ferrite cool-down are relatively long: on about ten occasions, during 2012, after a series of long fills, it was required to wait longer than one hour before injecting to allow the cool-down of the MKI8D yoke, thus limiting the running efficiency of the LHC.

Extensive simulations have been carried out to understand the cause of, and to significantly reduce, the beam induced heating: the beam screen implemented during LS1 (2013–2014) will have a full complement of 24 screen conductors [8]. The cause of heating of the hottest MKI magnet, during 2012, has been identified to be caused by a non-conformity of the ceramic tube, leaving part of the ferrite yoke unscreened: this led to about double the power deposition. Going from 15 to 24 screen conductors would have reduced the power deposition by a factor of 4, relative to a conforming ceramic tube, for pre-LS1 beam conditions (50 ns bunch spacing, 1.2 ns 4σ bunch length, 1380 bunches and 1.6×10^{11} ppb). For post LS1 beam conditions (25 ns bunch spacing, 1.0 ns 4σ bunch length, 2808 bunches and 1.15×10^{11} ppb) the change from 15 to 24 screen conductors even lead to a reduction of beam induced power by a factor seven, due to the difference in the beam power spectrum.

The HL-LHC beam option with 50 ns bunch spacing and 3.5×10^{11} ppb leads to the highest power deposition in the MKI magnets up to 240 W/m [9], always assuming 24 screen conductors. This is significantly more than the power deposited in the magnet yoke associated with the non-conform ceramic tube which was installed pre-LS1, estimated to be 160 W/m and leading to temperatures which limited operation. For 25 ns bunch spacing and 2.5×10^{11} ppb the expected deposited power has been calculated to be between 125 W/m and 190 W/m. Thus, in addition to the increased number of screen conductors, the cooling of the ferrite yoke is studied as an additional measure.

3.2. *Cooling of ferrite yoke*

The MKI magnet is operated in a vacuum of $\sim 10^{-11}$ mbar thus convection does not contribute to cooling of the ferrite yoke. In addition, many of the materials used in the kicker magnet have relatively poor thermal conductivity hence cooling

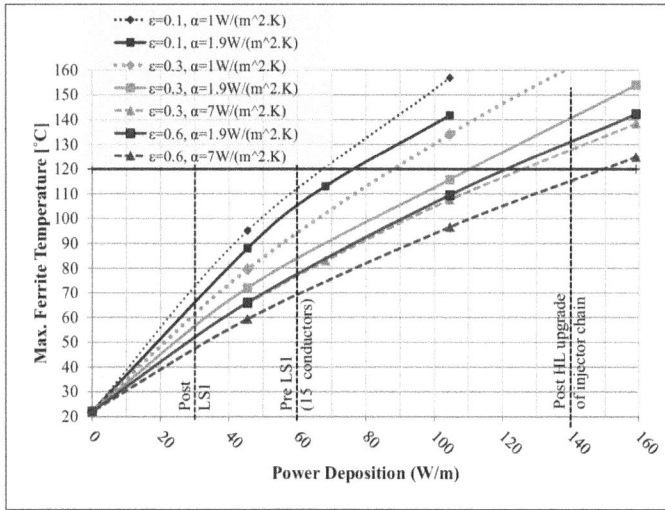

Fig. 4. Predicted steady-state ferrite temperature versus total power deposition per meter length of ferrite yoke, for various emissivities of the inside of the MKI tank and convection (α). 15 and 24 straight screen conductors are considered for pre and post LS1, respectively.

is mainly due to thermal radiation. Thermal simulations, confirmed by measurements, show that the inside of an electropolished MKI tank has an emissivity (ϵ) of ~0.1. Such a low emissivity results in a relatively high steady-state temperature of the ferrite yoke. Thermal simulations predict that the steady-state yoke temperature is proportional to $\epsilon^{-0.16}$. In addition transient thermal simulations predict that the time constant for cool-down of the ferrite yoke, in hours, is given by $(10/\sqrt{\epsilon})$ [10]. Thus an increased emissivity has the benefit of both reducing the maximum ferrite temperature and the cool-down time-constant.

Figure 4 shows a plot of predicted ferrite temperature versus total power deposition per metre length of ferrite yoke, for various emissivities of the tank and heat transfer coefficients by convection (α). A standard bake-out jacket for an MKI has an α in the range 1.9–2.9 W/(m^{2}K). An MKI without a bake-out jacket is represented by $\alpha = 7$ W/(m^{2}K). Figure 4 shows that for ϵ of 0.1 the expected ferrite temperature will be well below the Curie temperature for beam parameters to be used after LS1, but not for HL-upgrade parameters. A tank emissivity of at least 0.6, without bake-out jackets, is required to maintain the ferrite yoke temperature below 120 °C. Water cooling of the outside of the vacuum tank is studied to reduce the ferrite temperature. In addition water cooling of the internal side plates, which are nominally at ground potential, is being considered.

The pre-treatment of the MKI vacuum tanks by an ion bombardment technique, in an atmosphere of argon and oxygen [10] is being studied during LS1. Results are not conclusive at the moment and other means of increasing the emissivity of the inside of the tank are being studied.

3.3. Ferrite toroid heating

There has also been occasional unexpected heating of some toroidal ferrites, nine of which are mounted at each end of the ceramic tube, whose purpose is to damp low-frequency resonances. Each set of nine toroids has two types of Ferroxcube NiZn ferrite, namely 4M2 and 4B3, with a Curie point of 200 °C and 250 °C, respectively. One set of nine toroids, at the capacitively coupled end of the beam screen occasionally reached 193 °C measured; all others remained below 100 °C measured. The source of the heating is still under investigation.

3.4. Surface flashover of ceramic tube

A voltage is induced on the screen conductors, mainly by mutual coupling with the MKI magnet cell inductance. Hence the voltages, at the open end of the screen conductors, show a positive peak (max.) during field rise and a negative peak during field fall: the height of the maximum is about twice the magnitude of the minimum. Figure 5 shows the predicted maximum voltage versus conductor number. The predicted maximum voltage between adjacent screen conductors is also shown, the highest value is ∼2.7 kV.

Extensive 3D electromagnetic simulations have been carried out to study electric fields on the surface of the ceramic tube. The predictions show that, as a result of the high permittivity (10) of the alumina, equipotential lines penetrate into the ceramic between adjacent screen conductors. The predictions, for 15 and 19 conductor versions, together with previous observations of the surface flashover

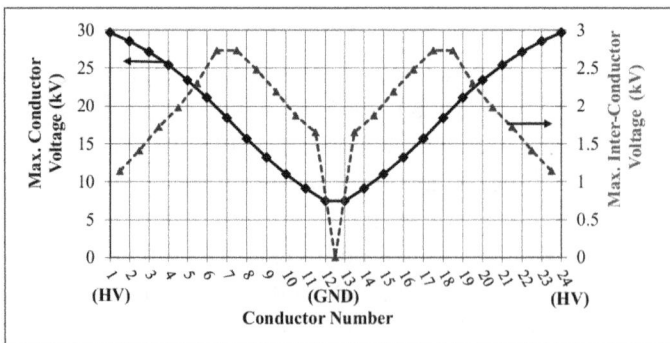

Fig. 5.   Conductor (solid) and inter-conductor (dashed) voltages, for 24 screen conductors, for 60 kV PFN voltage.

inception voltage, have been used to determine the upper limit for the predicted electric field (~ 12 kV/mm). Removing some metallisation and replacing it with an offset cylinder, reduces the maximum electric field by a factor of more than 2, providing a safety margin for operation at the PFN design voltage of 60 kV. Tests in the laboratory, with all screen conductors pulsed to the same voltage, confirm the aforementioned predictions.

3.5. *Electron cloud*

Prior to LS1 significant pressure rises, due to electron-cloud, occurred in and nearby the MKIs: the predominant gas desorbed from surfaces is H_2. A pressure rise also increases the number of unidentified falling objects (UFOs) [11] and may augment the probability of electrical breakdown in the magnet and surface flashover on the ceramic tube. Conditioning of surfaces reduces electron-cloud, and thus pressure rise, but further conditioning is often required when beam parameters (e.g. bunch spacing, bunch length and bunch intensity) are pushed.

The ceramic tube of an MKI magnet replaced during Technical Stop 3 (TS3) in September 2012 had a high secondary electron yield when installed and required conditioning with beam, together with metallic surfaces facing the beam (e.g. screen conductors): Fig. 6 shows pressure, measured both in the upgraded MKI8D tank and nearby interconnects, normalized to the number of protons (p). The highest normalized pressure occurred in interconnect MKI8D-Q5, followed by interconnect MKI8C-MKI8D. Initially the beam current was kept low to maintain

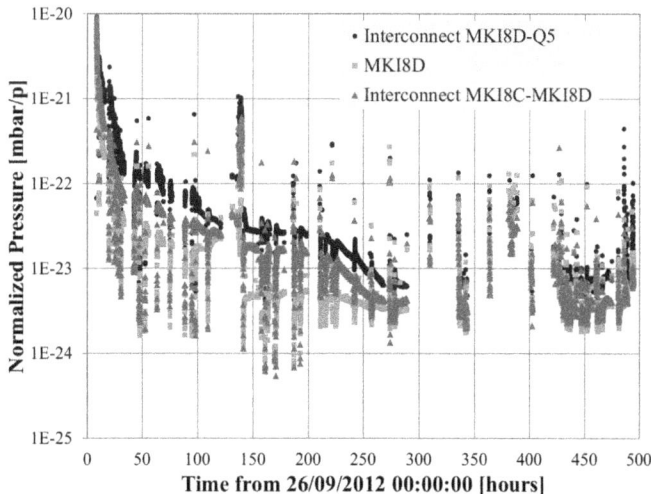

Fig. 6. Pressure after TS3, measured both in and nearby tank MKI8D, normalized to the number of protons.

the pressure below the interlock thresholds. The ceramic tube conditioned relatively quickly but required 250 hours, with beam, to achieve a normalized pressure similar to the pre-replacement ($\sim 4 \times 10^{-24}$ mbar/p) level.

In order to mitigate electron multipacting the MKI interconnects and bypass tubes (Fig. 3) have been NEG coated during LS1. The adjacent beam instrumentation (BTVSI and BPTX) have also been NEG coated during LS1. In addition NEG cartridges have been installed, on the cold-warm transition, to supplement existing ion pumps. On the MKI interconnects the ion pump has been exchanged for a version which also includes a NEG cartridge.

3.6. Fast transient beamlosses (UFOs)

There have been a total 21 protective beam dumps due to fast transient beamlosses (termed UFOs) at the MKIs. This is about half the number of UFO related beam dumps which occurred in the LHC. The UFO activity around the injection kicker in P2 generally exceeded that around the MKI8s [11]. After a comprehensive study program in 2011, the MKI UFOs were identified as macro particles originating from the ceramic tube inside the MKI magnets [14]. Thus the MKI8D installed during TS3 had improved cleaning of the ceramic tube, which included iterations of flushing of the inside of the tube with N_2 at 10 bar and dust sampling, until no significant further reduction of macro particles was noted. Before TS3, MKI8D had the highest UFO activity of the MKIs in P8; the replacement MKI8D had the lowest UFO activity [11]. Extensive additional cleaning has been carried out on the ceramic tubes which have been put in place during LS1.

With 15 screen conductors installed, macro particles could be detached from the inside of the ceramic tube and accelerated towards the beam by the high electric field when the kicker is pulsed [14]. The 24 screen conductors reduce the electrical field during most of the MKI pulse hence decreasing the probability of a particle being detached from the ceramic.

3.7. Possible future upgrades and ongoing R&D

Ferrite such as CMD10, which has a higher Curie temperature than the CMD5005 or 8C11 presently used for the MKI yoke, would permit high-intensity beam operation with better availability. However, operating at higher yoke temperatures will result in higher pressure in the vacuum tank which may result in an increased electrical breakdown and surface flashover rate [15].

As mentioned above, as a result of multipacting in the copper bypass tube, significant pressure rise can occur in an MKI tank, hence the copper bypass tubes have been NEG coated during LS1. However, if the counter-rotating beam still has a detrimental effect upon the MKI operation, the vacuum of the bypass tube could be separated from the vacuum of the MKI tank.

A series of high voltage tests are planned for the laboratory in which different gases are injected into a test tank: this will allow a careful and systematic study of the effect of gas pressure upon surface flashover of the ceramic tube. These studies will permit the MKI vacuum interlock thresholds to be optimized and the possibility, from the MKI vacuum perspective, of using a higher Curie temperature ferrite to be evaluated.

4. Beam Loss Control at Injection

Losses at injection into the superconducting LHC can adversely affect the collider performance in several important ways. Injection related losses can produce spurious signals on the sensitive beam loss monitoring system which will trigger beam dumps. In addition, the use of the two injection insertions to house downstream high energy physics experiments brings constraints on permitted beam loss levels, to avoid trips or even damage to the sensitive detector sub-assemblies.

The apertures in the injection region are small, in particular at the MSI septum and the MKI kicker, to be able to achieve the required deflections with achievable septum and kicker currents and to minimise the stray field in the field free septum holes. The physical radius of the protective mask immediately upstream of the MSI is 10 mm, from which orbit and alignment tolerances need to be subtracted. For the circulating beam the MSI provides 7.3σ aperture in n1 notation [16]. The MKI has a ceramic chamber with 39 mm ID. For HL-LHC beams the energy in each injected batch will increase to ~4.5 MJ, which is about a factor 25 above that required to damage accelerator components in the event of direct impact. To protect the small apertures in the injection regions, and also to prevent injection into the LHC of beams with large oscillations, protection devices TCDI are set around the injected beam trajectory in the transfer lines TI 2 and TI 8. There are three devices per plane, spaced at 60 degrees in betatron phases, Fig. 7, with settings of $\pm 4.5\sigma$. The TCDIs are two-jawed movable devices, with 1.2 m of Carbon to intercept the beam.

The beam loss monitoring BLM system for the LHC ring is designed to protect the machine against quenches and damage from beam loss. The location near the superconducting elements in the LHC tunnel of some of the TCDI protection devices, together with the TDI and TCLIs, means that there is very strong crosstalk between injection losses and the beam loss signals from circulating beam. Although the loss limits are well below quench levels, for beams with large transverse tails, large emittance or poor trajectories, the BLMs can frequently trigger, resulting in a beam abort and reducing machine availability.

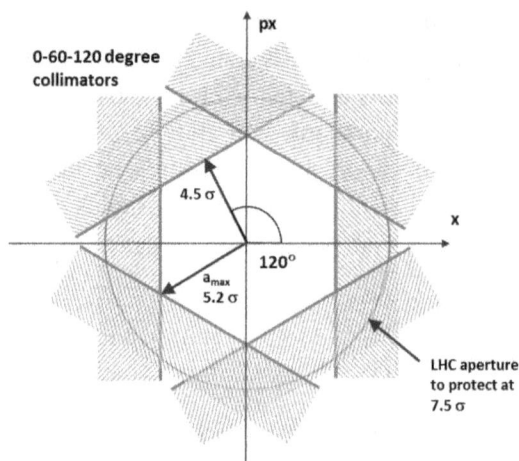

Fig. 7.   Phase space coverage of TCDI collimators in TI 2 and TI 8 transfer lines, courtesy V. Kain.

The nominal emittance for the LHC beam transfer and injection is 3.5 mm mrad, normalized emittance. The emittances injected into the LHC in the years 2010 to 2012 were slowly reduced from about 3.0 mm mrad to below 2.0 mm mrad, the latter made possible by the change to double-batch injection with the 50 ns beam. Despite this reduction in emittance, the increasing beam intensity and push for better availability meant that many mitigations against beam loss in the injection regions were deployed through Run 1. First was an increase in the thresholds of the LHC BLMs. Then local shielding was added between the critical TCDIs and the LHC elements, optimized after a series of FLUKA simulations. The reduction factor in beam loss per proton at the affected BLMs was expected to be a factor of 3, and was measured to be slightly less, at around a factor of 2–2.5 depending on location.

The key factor for losses on TCDIs are scraping of tails in the SPS, which has been shown to provide a factor of ~10 reduction [17]. The same studies also demonstrated that injection of 3.5 mm mrad emittance beams is possible with similar loss levels to 2 mm mrad, an important result for future LHC operation and HL-LHC parameters.

In addition to the beam size and tails, trajectory instability in the transfer lines can increase losses on TCDIs. The stability of the lines has been analyzed in detail [18] and is the subject of ongoing studies. The main source of errors are shot-to-shot variations in the horizontal plane, likely due to variations in the field of the extraction septa in the SPS. The lines need correcting about once per week in order to keep the trajectories within the allocated tolerances, which gives enough operational margin for injection beam losses. Importantly, this is sufficient to run without repeated setup of the TCDI collimators, despite their close settings of $\pm 4.5\sigma$, and it was possible to run a full year without setting up the TCDIs again. For the future an automated procedure is being considered.

In the LHC, uncaptured circulating beam leads to beam loss on the TDI and TCLIs when the injection kicker pulses, with high loss rates recorded on the TCTVBs (which are not part of the injection protection system); this was a major problem in 2010 operation and was the subject of several mitigation measures. The first mitigation was to increase BLM thresholds to avoid dumping, especially on TCTVB collimators. FLUKA simulations were performed to understand energy deposition and to optimize shielding design. The RF capture voltage in the LHC was increased, and finally the use in regular operation of a gated excitation of the transverse damper was deployed for both abort gap cleaning AGC [19] and injection gap cleaning IGC [20]. In these modes the damper excitation is gated such as to clean away uncaptured beam which would otherwise drift around the ring and be swept onto the TDI by the MKI pulse. The combination of AGC and IGC is very effective, and reduces the beam loss of uncaptured beam on the TDI by a factor of ~ 10.

For the HL-LHC era, the foreseen improvements to overcome injection losses are to remove the most troublesome TCDI collimators from the LHC tunnel, to new locations in the transfer line with suitable phase and betas. The scraping in the SPS and the stability of the transfer lines are also areas where improvements are being studied. For the uncaptured beam, the AGC and IGC functionality will need to be maintained and possibly improved. Further developments on the beam loss measurement system are ongoing to cope with injection losses. Smaller ionization chambers have been developed, which have a larger dynamic range than the standard beam loss monitors used across the LHC and avoid saturation during injection. A second development is to temporarily disable the interlocking ability of a selected set of beam loss monitors in the injection region, for some ms at the moment of injection [21].

5. Beam Dump System Performance Reach

The beam in the LHC is aborted or dumped by a dedicated system based on pulsed extraction kickers and DC septum magnets located in the dedicated insertion in Point 6 (P6), followed by a dilution kicker system MKB, long drift chamber and graphite beam dump absorber block TDE, kept at atmospheric pressure of N_2. The $3\ \mu s$ rise time of the kicker field is synchronized by a highly reliable timing system [24] to a beam-free abort gap in the circulating bunch pattern. The horizontal and vertical dilution kickers are powered with anti-phase sinusoidal currents so as to paint the bunches onto the TBE with an elliptical shape, Fig. 8.

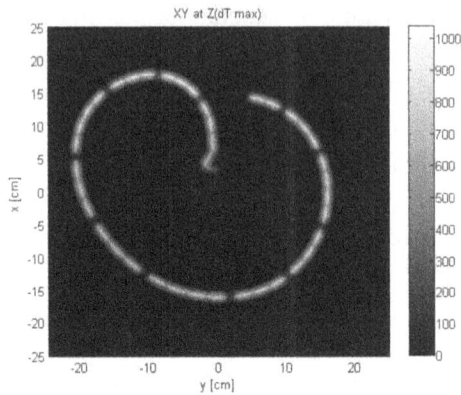

Fig. 8.   Sweep form of 25 ns spacing LHC beam on TDE dump block, with 1.7×10^{11} p+ per bunch.

Various instruments are also used to set up and monitor the quality of the beam dump action, and comprehensive internal and external quality checks are made using sophisticated pattern recognition algorithms after each dump action to verify the dump quality, both in terms of losses and trajectory and in terms of hardware performance and detection of any failures in redundant pathways. The dump system is designed to have a very high reliability, SIL4, which corresponds to an expected beyond-design failure rate of fewer than once per 10,000 years [25].

The beam dump block TDE is designed to withstand the impact of the so-called 'ultimate' LHC beam, which contains 2808 bunches of 1.7×10^{11} protons at 25 ns spacing, with acceptable thermal stresses in the TDE core [26]. The peak energy deposition and temperature rise depend on the details of the sweep shape. With the present sweep the maximum temperature increase with the HL-LHC beam increases from 1000 to 1350 °C, Fig. 9. It remains to be seen whether this is still acceptable in terms of the TDE robustness — further simulations are needed, and possibly test in the HiRatMat beam facility can be used to confirm the models used.

In case the TDE temperature rise after the impact of the HL-LHC beam is not acceptable, possible solutions could be to replace the TDE cores with new blocks, with lower density and higher robustness carbon fibre composite, or to modify the MKB dilution kicker system to increase the sweep frequency by about 50%. Both approaches need more detailed study.

Another concern will be the total energy deposited in the dump blocks, which have a thermal time constant of about 4.5 hours and reached a maximum temperature of about 20 °C above ambient during 2012, when the maximum dumped energy was 140 MJ, a factor of 5 lower than the 690 MJ for HL-LHC. The block cooling and N_2 gas handling may also need upgrading to cope with the increase in the average and peak beam power.

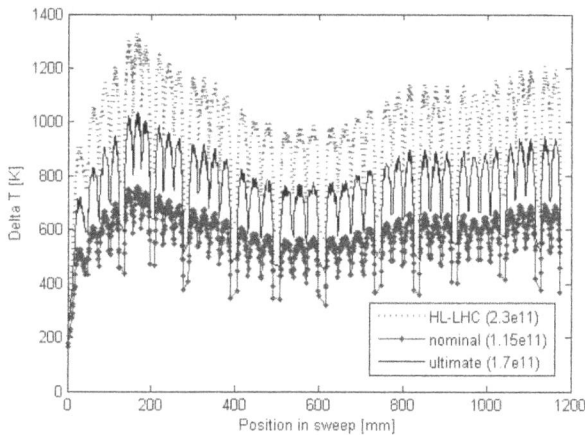

Fig. 9. Temperature rise at shower maximum in the TDE dump block, as a function of position along the sweep length, for different 25 ns beams.

6. Protection Against Beam Dumping Errors

Several failure modes exist in the synchronization system and in the kicker switches which could lead to an asynchronous dump, in which the full beam intensity would be swept across the LHC aperture by the rising kicker field. Without dedicated protection devices this would lead to massive damage of the LHC magnets in LSS6 and the downstream arcs 5-6 and 6-7, and depending on the operational configuration a number of collimators and possibly experimental triplet magnets.

A large amount of redundancy and fault tolerance has been built into the LHC dump system to avoid producing these asynchronous dumps; nevertheless, during Run 1, two such events happened, although without beam loss as one was at injection energy with a single pilot bunch and the second event occurred without beam. To avoid the machine damage, a series of passive protection devices have been designed to intercept the swept particles. The first such element TCDS is a fixed absorber block located directly in front of the extraction septum, which intercepts about 50 bunches at 25 ns spacing. The second system is the TCDQ which is located just upstream of the next superconducting quadrupole Q4, and this is a movable device which has to be adjusted as a function of beam size and orbit. It intercepts about 30 bunches, and is complemented by a two-sided collimator TCSG set $\sim 0.5\sigma$ closer to the beam and a 2 m long fixed mask TCDQM which is slightly smaller than the aperture of the Q4.

The increased beam intensity and brightness for HL-LHC will require a redesign of the TCDS and TCDQ absorbers. The TCDQ re-design has been analyzed in detail [27] and the new design will already be installed for LHC Run 2.

This extends the absorber length from 6 to 9 m, and replaces the higher density graphite absorber material by different densities (1.4 and 1.8 g/cm^3) of carbon fibre composite (CfC). The energy deposition and induced thermal stresses then remain acceptable [22] during an asynchronous abort, Fig. 10, and the protection of Q4 and downstream elements remains sufficient with energy deposition in the magnet coils of around 20 J/cm^3 [23]. Operational experience and further simulations will show if it will be necessary to improve the TCDQ positioning accuracy, which is presently around 50 μm. If required, changes of the electronics and control system but also of the mechanical system of the TCDQ need to be foreseen.

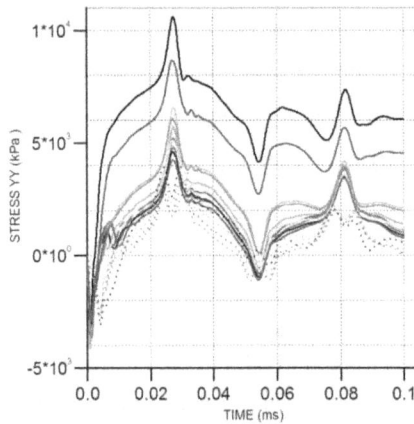

Fig. 10.   Time evolution of vertical (YY) stresses in the new CfC TCDQ block for an asynchronous dump with HL-LHC beam, courtesy T. Antonakakis.

A similar level of redesign will be needed for TCDS. Here any additional length will reduce slightly the aperture for the circulating or extracted beams, by a small fraction of a sigma, which should be acceptable. The TCSG collimators will be upgraded with integrated button BPMs in the jaws, which should allow faster and more accurate setup.

7. Beam Dump Kicker Performance Upgrades

The spontaneous triggering of the LHC beam dumping system switches, resulting in a dump asynchronous to the abort gap, cannot completely be avoided. The expected rate of asynchronous dumps is about one per beam dumping system per year. Asynchronous beam dumps have a negative effect on the machine availability, as one can expect the quench of a number of magnets in case that the asynchronous dump occurs at full beam energy during a normal physics run. The asynchronous beam dump also puts a strain on the protection elements described above, as their correct positioning must be guaranteed to avoid damage to machine elements at the moment of an asynchronous beam dump.

One of the sources of spontaneous triggering of the beam dump kicker is the Single Event Burnout (SEB) of the Fast High Current Thyristor (FHCT) switches when exposed to radiation. The SEB failure rate depends primarily on the voltage over the switch, which in the case of the LHC beam dumping system is tracking the beam energy, the energy of the radiation to which the switch is exposed and its charge. Measurements of the effect of radiation on the MKD and MKB FHCT switches have been performed at the H4IRRAD irradiation facility [28]. As a result the switches of some of the MKD generators are exchanged during LS1 so all switches installed will be of the one manufacturer which showed a significantly lower SEB during the tests. The shielding of the cable ducts between the generators and the LHC tunnel is also improved during LS1. When operation at almost double the beam energy will start in 2015 the performance of the system in the light of erratic triggering due to radiation will need to be closely surveyed. If necessary, further shielding against radiation will be required.

To follow the recommendations of the FHCT switch manufacturer it is foreseen to increase the power trigger dI/dt from the present 400 A/μs to 1 kA/μs and increase the switching trigger peak current from 400 A to about 800 A. This requires the upgrade of the trigger transformer system which is presently under development. At the same time the IGBT switches of the power triggers have already been upgraded during LS1 to allow for an increased trigger voltages and the larger dI/dt. These changes have the aim to increase the switch lifetime. The end of lifetime is expected to manifest itself by an increased leakage current which will result in an internally triggered synchronous beam dump. However, it is also possible that at the end of the lifetime of a switch the likelihood of erratic triggering increases, which will lead to asynchronous beam dumps. The expected lifetime of a switch is around 1 million pulses. With approximately 20,000 pulses per system per year, a large scale exchange of the FHCT switches during the HL-LHC era should normally not be required. The same applies to the capacitors used in the pulse generators.

Other possible changes of the generators include the addition of saturable cores at the top of the FHCT stack. The cores are foreseen to avoid resonances in the generator at the moment of switching and by this reduce the spark rate between the switch and the insulators of the generators. The saturable cores can at the same time give a gain in rise time, which is very favourable. Operation at full voltage of the MKD generators with the newly inserted insulators during LS1 will need to be carefully monitored.

The delay between the erratic switching of one power switch and the re-triggering of the complete beam dumping system (the asynchronous dump) is presently around 800 ns. If this delay is proven to be critical, based on operational experience and simulations, the delay can possibly be shortened. This will require the development of new electronic systems.

Acknowledgments

All systems described in this chapter have been designed, built, commissioned and successfully operated as a result of the diligent work and efforts of many colleagues and collaborators from various groups, who are far too numerous to mention individually. We thank them all.

References

[1] G. Rumolo, HL-LHC beam parameters, CERN EDMS document 1296306, https://edms.cern.ch/document/1296306/1.

[2] ALICE Collaboration, The Zero Degree Calorimeters for the ALICE Experiment, CERN-ALICE-PUB-99-17, 1999.

[3] V. Kain *et al.*, The expected performance of the LHC injection protection system, presented at EPAC'04; 9th European Particle Accelerator Conference 2004.

[4] O. Brüning and J. B. Jeanneret, Optics constrains imposed by the injection in IR2 and IR8, CERN/LHC Project Note 141 1998.

[5] LHC Design Report, http://ab-div.web.cern.ch/ab-div/Publications/LHC-Design Report.html.

[6] M. J. Barnes *et al.*, Upgrade of the LHC Injection Kicker Magnets, in *Proc. of IPAC'13*, MOPWA030; http://www.JACoW.org.

[7] M. J. Barnes, F. Caspers, L. Ducimetiére, N. Garrel and T. Kroyer, The Beam Screen for the LHC Injection Kicker Magnets, in *Proc. EPAC'06*, TUPLS011; http://www.JACoW.org.

[8] H. Day, Measurements and Simulations of Impedance Reduction Techniques in Particle Accelerators, CERN-THESIS-2013-083.

[9] H. Day *et al.*, Beam Coupling Impedance of the New Beam Screen of the LHC Injection Kicker Magnets, in *Proc. IPAC'14*, TUPR1030; http://www.JACoW.org.

[10] M. J. Barnes *et al.*, Beam Induced Ferrite Heating of the LHC Injection Kickers and Proposals for Improved Cooling, in *Proc. IPAC'13*, MOPWA031; http://www.JACoW.org.

[11] T. Baer *et al.*, UFOs: Observations, Statistics and Extrapolations, in *Proc. LHC Beam Operation Workshop*, Evian, 17–20 December 2012.

[12] G. Rumolo and G. Iadarola, private communication, 16 April 2013.

[13] T. S. Sudarshan and J. D. Cross, *The Effect of Chromium Oxide Coatings on Surface Flashover of Alumina Spacers in Vacuum*, IEEE Trans. on Electrical Insulation, Vol. EI-11, No. 1, March 1976, pp. 32–35.

[14] B. Goddard *et al.*, Transient Beam Losses in the LHC Injection Kickers from Micron Scale Dust Particles, in *Proc. IPAC'12*, TUPPR092; http://www.JACoW.org.

[15] M. J. Barnes, L. Ducimetiére, N. Garrel, B. Goddard, V. Mertens and W. Weterings, Analysis of Ferrite Heating of the LHC Injection Kickers and Proposals for Future Reduction of Temperature, in *Proc. IPAC'12*, TUPPR090; http://www.JACoW.org.

[16] J. B. Jeanneret and R. Ostojic, Geometrical acceptance in LHC version 5.0, CERN-LHC-Note-111, 1997.

[17] B. Goddard *et al.*, Beamloss Control at Injection into the LHC, CERN-ATS-2011-274, 2011.

[18] L. Drosdal *et al.*, Analysis of LHC Transfer Line Trajectory Drifts, in *Proc. IPAC'13*, MOPWO033; http://www.JACoW.org.

[19] W. Bartmann *et al.*, LHC abort gap cleaning studies during luminosity operation, in *Proc. IPAC'12*, MOPPD058; http://www.JACoW.org.

[20] A. Boccardi *et al.*, Summary of Injection pre-cleaning tests performed on October 27, 2010, CERN-ATS-Note-2010-050 PERF. 2010.

[21] C. Bracco *et al.*, Injection and lessons for 2012, in *Proc. Chamonix 2012 Workshop on LHC Performance*, pp. 65–68.

[22] T. Antonakakis and C. Maglioni, Upgrade of the TCDQ: A dumping protection system for the LHC, sLHC Project Note 0041, 2013.

[23] R. Versaci, B. Goddard and V. Vlachoudis, LHC Asynchronous Beam Dump: Study of new TCDQ model and effects on downstream magnets, CERN-ATS-Note-2012-084 MD, 2012.

[24] A. Antoine, E. Carlier and N. Voumard, The LHC Beam Dumping System Trigger Synchronisation and Distribution System, *ICALEPCS2005*, Geneva, 2005.

[25] R. Filippini, Dependability analysis of a safety critical system: the LHC beam dumping system at CERN, CERN-THESIS-2006-054 - Pisa U., 2006.

[26] J. Zazula and S. Péraire, LHC beam dump design study; 1, simulation of energy deposition by particle cascades; implications for the dump core and beam sweeping system, CERN-LHC-Project-Report-80, 1996.

[27] R. Versaci, B. Goddard and V. Vlachoudis, LHC Asynchronous Beam Dump: Study of new TCDQ model and effects on downstream magnets, CERN-ATS-Note-2012-084 MD, 2012.

[28] V. Senaj and L. Ducimetiére, Attempt to a non-destructive single event burnout test of high current thryristors, in *Pulsed Power Conference, 2011 IEEE* (IEEE, 2011), pp. 797–796.

Chapter 20

Beam Instrumentation and Diagnostics for the LHC Upgrade

E. Bravin, B. Dehning, R. Jones and T. Lefevre

CERN, BE Department, Genève 23, CH-1211, Switzerland

The extensive array of beam instrumentation with which the LHC is equipped, has played a major role in its commissioning, rapid intensity ramp-up and safe and reliable operation. High Luminosity LHC (HL-LHC) brings with it a number of new challenges in terms of beam instrumentation that will be discussed in this chapter. The beam loss system will need significant upgrades in order to be able to cope with the demands of HL-LHC, with cryogenic beam loss monitors under investigation for deployment in the new inner triplet magnets to distinguish between primary beam losses and collision debris. Radiation tolerant integrated circuits are also being developed to allow the front-end electronics to sit much closer to the detector. Upgrades to other existing systems are also envisaged; including the beam position measurement system in the interaction regions and the addition of a halo measurement capability to synchrotron light diagnostics. Additionally, several new diagnostic systems are under investigation, such as very high bandwidth pick-ups and a streak camera installation, both able to perform intra-bunch measurements of transverse position on a turn by turn basis.

1. Introduction

The extensive array of beam instrumentation with which the LHC is equipped, has played a major role in its commissioning, rapid intensity ramp-up and safe and reliable operation. In addition to all of this existing diagnostics HL-LHC brings a number of new challenges in terms of beam instrumentation that are currently being addressed.

The beam loss system, designed to protect the LHC from losses that could cause damage or quench a superconducting magnet, will need a significant upgrade in order to be able to cope with the new demands of HL-LHC. In particular, cryogenic beam loss monitors are under investigation for deployment in the new inner triplet magnets to distinguish between collision debris and primary beam losses. Radiation tolerant integrated circuits are also under development to allow the front-end electronics to sit much closer to the detector, so minimizing the cable length required and reducing the influence of noise.

The proposed use of crab cavities implies new instrumentation in order to allow for optimization of their performance. Several additional diagnostic systems will therefore be investigated, including very high bandwidth pick-ups and a streak camera installation. These would be able to perform intra-bunch measurements of transverse position on a turn-by-turn basis.

An upgrade to several existing systems is also envisaged, including the beam position measurement system in the interaction regions and the addition of a halo measurement capability to synchrotron light diagnostics.

2. Beam Loss Measurement for HL-LHC

Monitoring of beam losses is essential for the safe and reliable operation of the LHC. The beam loss monitoring (BLM) system provides knowledge of the location and intensity of such losses, allowing an estimation to be made of the energy dissipated in the equipment along the accelerator. The information is used for machine protection, to qualify the collimation hierarchy, to optimize beam conditions and to track the radiation dose to which equipment has been exposed. This is currently done using nearly 4000 ionization monitors distributed around the machine and located at all probable loss locations, with the majority mounted on the outside of the quadrupole magnets, including those in the inner triplet regions. Around one third of the arc monitors have recently been relocated in order to optimize the system for protection against fast beam losses believed to be caused by dust particles falling into the vacuum pipe. While the existing system is expected to meet the needs of the HL-LHC for the arcs, this will no longer be the case for the high luminosity interaction points.

In the HL-LHC high luminosity insertions the magnets will be subjected to a greatly enhanced continuous radiation level due to the increase in collision debris resulting from the higher luminosity. With the presently installed configuration of ionization chambers in this region the additional signal from any dangerous accidental losses would be completely masked by that coming from collision debris. This is a critical issue for LHC machine protection and therefore R&D has started to investigate possible options for placing radiation detectors inside the cryostat of the triplet magnets as close as possible to the superconducting coils. The dose measured by such detectors would then correspond much more precisely to the dose deposited in the coils, allowing the system to be used once again to prevent a quench or damage.

The quench level signals estimated for 7 TeV running are, for some detectors, very close to the noise level of the acquisition system. This is mainly determined by the length of cable required to bring the signal from the radiation hard detector to the more radiation sensitive front-end electronics. Although qualified for use in the low radiation environments of the LHC arcs the current electronics cannot be located close to the detectors in the higher radiation insertion regions. Development has therefore also started to implement this electronics in a radiation hard Application Specific Integrated Circuit (ASIC).

2.1. *Beam loss monitors for the HL-LHC triplet magnets*

Three detectors are currently under investigation as candidates for operation at cryogenic temperatures inside the cryostat of the triplet magnets [1]:

- Single crystal chemical vapor deposition (CVD) diamond with a thickness of 500 μm, an active area of 22 mm^2 and gold as the electrode material.
- p^+-n-n$^+$ silicon wafers with a thickness of 280 μm, an active area of 23 mm^2 and aluminium as the electrode material.
- Liquid helium ionization chambers.

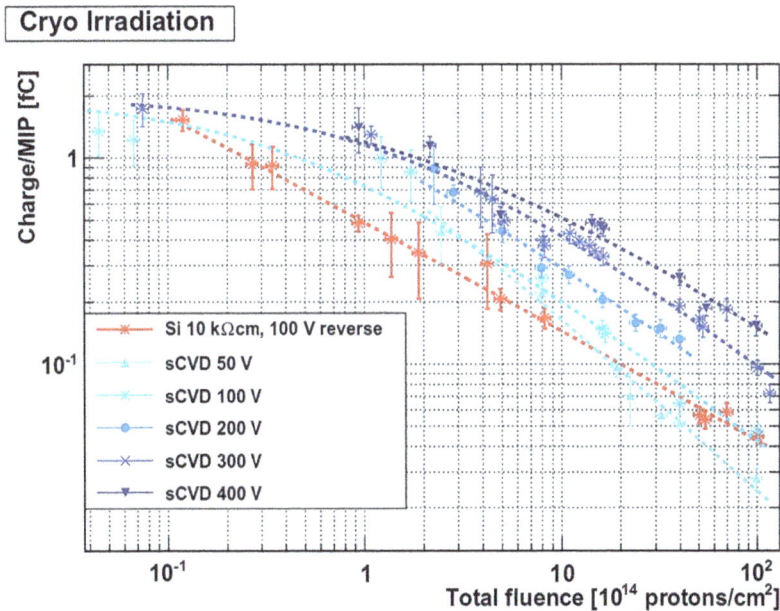

Fig. 1. Charge collection efficiency for silicon and diamond detectors with increasing radiation fluence in a cryogenic environment.

Experiments have already been performed to observe the behavior of such detectors in a cryogenic environment and on the radiation effects at such temperatures for silicon and single crystal diamond. Irradiation up to several Mega-Gray showed a degradation of the charge collection efficiency in both CVD diamond and Si by a factor of 15 (see Fig. 1). The major downside of silicon compared to diamond, its much higher leakage current when irradiated, has been observed to disappear at liquid helium temperatures, with the leakage current remaining below 100 pA at 400 V, even under forward bias for an irradiated diode. Further experiments combining irradiation with cryogenic temperatures will be necessary to optimise the final detector design. These experiments will be accompanied by tests of detectors mounted inside the cryostats of existing LHC magnets with the aim of gaining experience with the long term performance of such detectors under operational conditions.

2.2. A radiation tolerant ASIC for the HL-LHC beam loss monitoring system

The front-end electronics design for both solid state and ionization chamber beam loss detectors gives rise to several challenges, namely the requirement for a wide dynamic range, a relatively short conversion time and a low offset current. Additionally, particularly in the case of the HL-LHC, this electronics needs to be as close as possible to the detector, meaning that it is exposed to the same ionizing radiation as the detector itself.

The current front-end electronics for the LHC BLM system, while providing a 40 μs integration time, is limited in the dynamic range it can handle and is not radiation tolerant. The latter implies the use of long cables in the higher radiation LSS regions, which further limits the dynamic range and in some cases brings the noise floor close to the quench level signal at 7 TeV. Instead of the discrete component currently used, an optimized ASIC is therefore under development. This is still based on the current to frequency conversion used in the existing system, but packaged in a compact, radiation-tolerant form with an increased dynamic range. The technique employed allows the digitization of bipolar charge over a 120 dB dynamic range (corresponding to an electric charge range of 40 fC–42 nC) with a 40 μs integration time and a conversion reference provided by an adjustable, temperature-compensated current reference [2].

Figure 2 shows the block diagram of the integrated circuit. It is composed of a bipolar, fully-differential integrator that converts the charge received from the detector into a voltage input for a synchronous comparator system. A three-level Digital-to-Analogue Converter (DAC) drives the discharge current for the

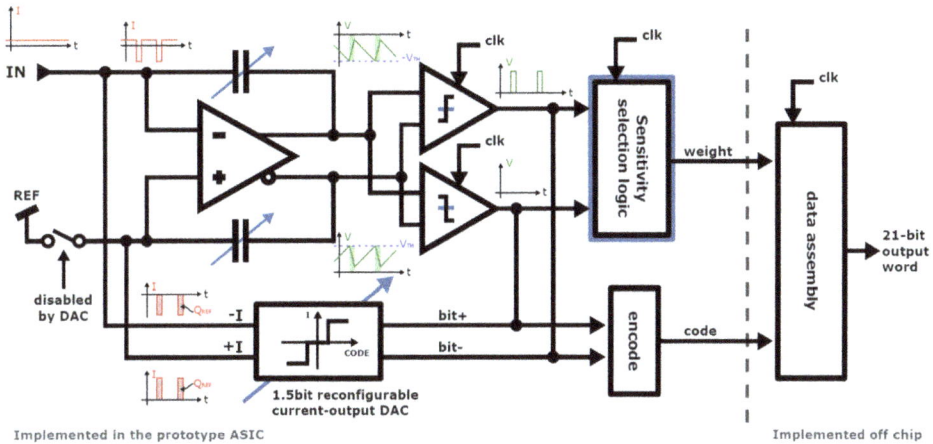

Fig. 2.   Schematic representation of the ASIC implementation currently under development.

Table 1.   Characteristics of the prototype HL-LHC BLM system ASIC.

A/D converter

Integration time	40 μs	Linearity error	< ±5%
Input current range	−1.05 mA to 1.05 mA	Peak S/N ratio	53 dB
Input charge range	−42 nC to 42 nC	SFDR at 999 Hz, 1 mA	50 dB
Offset	< 40 aC, 40 μs integration	Total ionizing dose	10 Mrad(Si)
	< 1 pA	Supply voltage	2.5 V
Default LSB step	50 fC ±20%, adjustable	Clock	12.8 MHz
Dynamic range	120 dB	Power	40 mW

Reference charge

Drift with TID	3% at 10 Mrad(Si)	Drift with temperature	< 600 ppm/°C

integrator and is connected in a feedback loop to the comparator output. A first logic block is used to select the gain in the integrator, the current step in the DAC and the threshold in the comparators, while a second logic block encodes the output signal from the comparators. The results of both of these are used by a third block to assemble the final, correctly weighted output word.

The prototype ASIC is designed in commercial CMOS technology and has two Analogue-to-Digital (A/D) channels and a sensitivity selection logic that can be disabled to implement the circuitry externally. This strategy has been useful for testing the device and improving the algorithm. Its measured characteristics are listed in Table 1.

The measured linearity is limited by transistor matching imperfections in the DAC, introducing an error at the transition between the sensitivities. However, overall the error is less than 5% and well inside specification (< 10%).

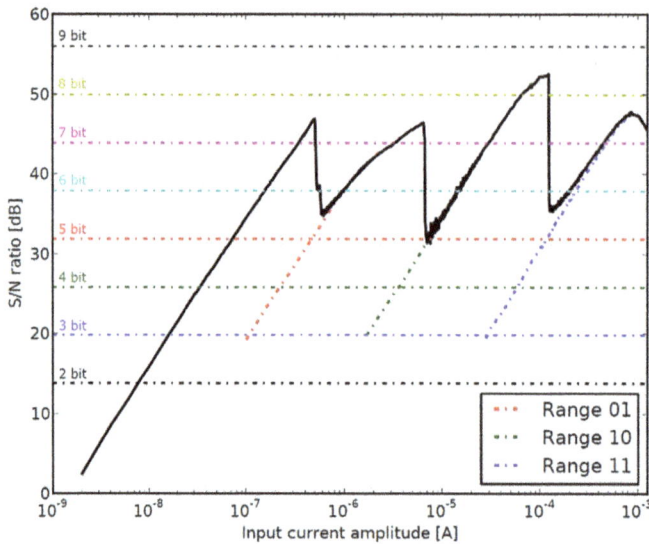

Fig. 3. Measurements of the Signal-to-Noise ratio as a function of the input current.

Figure 3 shows the measured Signal-to-Noise (S/N) ratio versus input amplitude for a sinusoidal current stimulus at 91 Hz. In the figure, the measurement is
compared with the theoretical peak S/N ratio for fixed resolution A/D converters
(horizontal lines) and with the measured S/N ratio of each sensitivity setting
(dashed lines extrapolated from the saw-tooth shape of the output). The S/N ratio
for this conversion principle results in a saw-tooth waveform: the S/N ratio
increases with the amplitude as the input current increases until saturation occurs
and the interface switches to the next range, increasing the full-scale of the
conversion.

Total Ionizing Dose (TID) effects on the ASIC have been investigated using
an X-ray beam with 20 keV peak energy. The characteristics of the device were
measured up to 100 kGy (Si), followed by a 1-week annealing cycle at 100 °C.
From the beginning to the end of the irradiation cycles, the functionality was
always preserved, with the conversion offset remaining below 1 Least Significant
Bit (LSB) and the value of the full scale charge drifting by less than 3%.

Development will now continue to address the issues found using the prototype and to implement more advanced logic blocks within the ASIC.

3. Beam Position Monitoring for the HL-LHC

With its 1070 monitors, the LHC Beam Position Monitor (BPM) system is the
largest BPM system in the world [3]. Based on the Wide Band Time Normalizer

(WBTN) principle [4], it provides bunch-by-bunch beam position over a wide dynamic range (~ 50 dB). Despite its size and complexity (3820 electronic cards in the accelerator tunnel and 1070 digital post-processing cards in surface buildings) the performance of the system during the last three years has been excellent, with greater than 97% overall availability.

3.1. *Current performance and limitations*

The position resolution of the LHC arc BPMs is better than $150 \, \mu m$ when measuring a single bunch on a single turn and better than $10 \, \mu m$ for the average position (orbit) of all bunches [5]. The main limitation on the accuracy of the BPM system is linked to temperature dependent effects in the acquisition electronics, which can generate offsets of up to a millimeter over a timescale of hours. On-line compensation was introduced to limit this effect during operation and temperature controlled racks are currently being installed with the hope of eliminating this limitation from the Run II start-up in 2015.

The non-linearity of the BPMs located near the interaction points has also proven to be problematic, in particular for accurate measurements during the beta-squeeze and during machine development periods. A new correction algorithm has therefore been developed, based on exhaustive electromagnetic simulations, with the aim of bringing down the residual error to below $20 \, \mu m$ over most of the useable BPM area [6]. Developed to be able to distinguish between the positions of two counter propagating beams in the same beam pipe, these BPMs also suffer from non-optimal decoupling between the beams, which is something that will need to be addressed for HL-LHC.

3.2. *A high resolution orbit measurement system for HL-LHC*

Originally developed to process signals from BPM buttons embedded in LHC collimator jaws, orbit measurement using a compensated diode detector scheme [7], has already been demonstrated to be simple and robust, and to provide a position resolution down to the nanometer level. A comparison of the orbit measured on a single BPM during a van der Meer scan by the current orbit system and the new diode orbit system is presented in Fig. 4, where the resolution of the new system can be seen to be over 50 times better. All new LHC collimators will have BPMs using this acquisition system installed with them, with plans to also equip the BPMs in all four LHC interaction regions. It is important to note, however, that the new system does not provide the bunch-by-bunch measurement capability of the existing system.

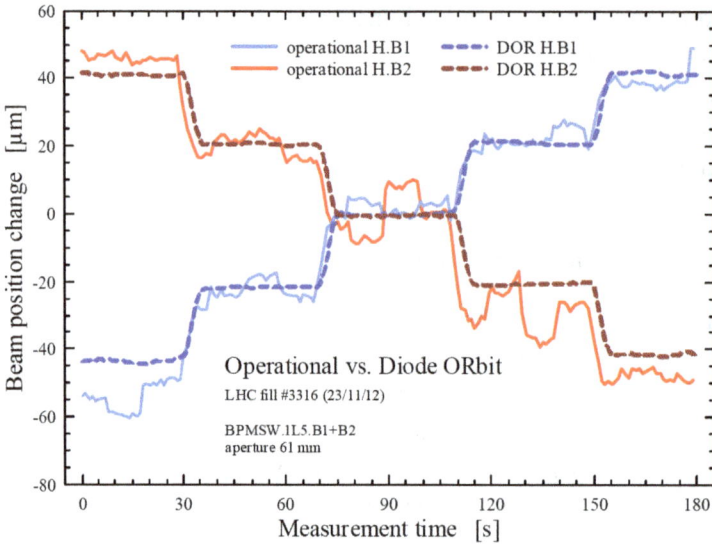

Fig. 4. Comparison of the new LHC orbit system electronics with the existing system during a van der Meer scan.

At the start of the HL-LHC era the existing BPM system will have been operational for over 15 years, using components which are over 20 years old. It is therefore likely that a completely new system will need to be installed during HL-LHC running. One candidate would be to extend the new diode orbit system to the whole machine for accurate global orbit measurements, and complement this with a system capable of providing the high-resolution bunch-by-bunch, turn-by-turn measurements required in particular for optics studies and the many other accelerator physics experiments that will be needed to understand and optimize HL-LHC performance.

With the higher bunch intensities foreseen, the dynamic range of the BPM system for HL-LHC would need to be increased accordingly. The present system implements two sensitivity ranges, optimized for pilot and nominal bunch intensities. Issues have been observed in the first three years of operation with BPMs providing large errors when reaching the limit of their dynamic range. For the interlock BPMs located in Point 6, this can trigger false beam dumps, which clearly has an impact on machine availability. Although improvements have already been made to this interlock system for Run II, any consolidation of the LHC BPM system will need to include developing dedicated electronics for this system, optimized for both high reliability and availability.

3.3. *High directivity strip-line pick-ups for the HL-LHC insertion regions*

In the BPMs close to the interaction regions, the two beams propagate in the same vacuum chamber. Directional strip-line pick-ups are therefore used to distinguish between the positions of both beams. When the two beams pass through the BPM at nearly the same time, the two signals interfere due to the limited directivity of the strip-line, which in the present design, only gives a factor of 10 isolation between the two incoming signals for the same bunch intensity. This effect can be minimized by installing the BPMs at a location where the two beams do not temporally overlap, which is a constraint included in both the present and future layout, but which cannot be maintained for all BPM locations. In addition, for the HL-LHC BPMs in front of the Q2a, Q3 magnets and the triplet corrector magnet package, there is the additional constraint that tungsten shielding is required at the level of the cold bore to minimize the heat deposition in these magnets. A mechanical re-design coupled with extensive electromagnetic simulations is therefore necessary to optimize the directivity under these constraints.

3.4. *Collimator BPMs*

All next generation collimators in the LHC will have button electrodes embedded in their jaws for on-line measurement of the jaw to beam position. This is expected to provide a fast and direct way of positioning the collimator-jaws and subsequently allow constant verification of the beam position at the collimator location, improving the reliability of the collimation system as a whole. The design of such a BPM was intensively simulated using both electromagnetic (EM) and thermomechanical simulation codes. In order to provide the best accuracy, the BPM readings must be corrected for the non-linearity coming from the varying geometry of the collimator jaws as they are closed and opened, for which a 2D polynomial correction has been obtained from EM-simulations and qualified with beam tests in a prototype system installed in the CERN-SPS [8].

The collimator BPM hardware, i.e. the button electrode located in the jaw, the cable connecting the electrode to the electrical feed-through mounted on the vacuum enclosure and the feed-through itself, has been chosen to withstand the radiation dose of 20 MGy expected during the lifetime of the collimator.

4. Emittance Measurement for the HL-LHC

The LHC is currently fitted with a host of beam size measurement systems used to determine the beam emittance. These different monitors are required in order to overcome the specific limitation of each individual system. Wire-scanners are used as the absolute calibration reference, but can only be operated with a low number of bunches in the machine due to intensity limitations linked to wire breakage at injection and the quenching of downstream magnets at 7 TeV. A synchrotron light monitor is therefore used to provide beam size measurements, both average and bunch-by-bunch, during nominal operation. However, the small beam sizes achieved at 7 TeV, the multiple sources of synchrotron radiation (undulator, D3 edge radiation, central D3 radiation), and the long optical path required to extract the light mean that the correction needed to be applied to extract an absolute value is of the same order of magnitude as the value itself. This implies an excellent knowledge of the error sources to obtain meaningful results. The third system installed is an ionization profile monitor. Originally foreseen to provide beam size information for lead ions at injection, where there is insufficient synchrotron light, this monitor has also been used for protons. However, the system can currently only measure the average size of all bunches and suffers from space charge effects at high energy.

Whilst efforts are ongoing to improve the performance of all the above systems, alternative techniques to measure the bunch-by-bunch transverse beam size and profile are also under study for the HL-LHC.

4.1. *A beam gas vertex emittance monitor for the HL-LHC*

The vertex detector (VELO) of the LHCb experiment has shown how beam gas interactions can be used to reconstruct the transverse beam profile of the circulating beams in the LHC [9]. The new concept under study is to see whether a simplified version of such a particle physics tracking detector can be used to monitor the beams throughout the LHC acceleration cycle. This concept has up to now never been applied to the field of beam instrumentation, mainly because of the high level of data treatment required. However, the advantages compared to the standard beam profile measurement methods listed above are impressive: high resolution profile reconstruction, single bunch measurements in three dimensions, quasi non-destructive, no detector equipment required in the beam vacuum, high radiation tolerance of the particle detectors and accompanying acquisition electronics.

Such a beam shape measurement technique is based on the reconstruction of beam gas interaction vertices, where the charged particles produced in inelastic beam gas interactions are detected with high-precision tracking detectors (Fig. 5). Using the tracks left in the detectors, the vertex of the particle-gas interaction can be reconstructed so, with enough statistics, building up a complete 2-dimensional

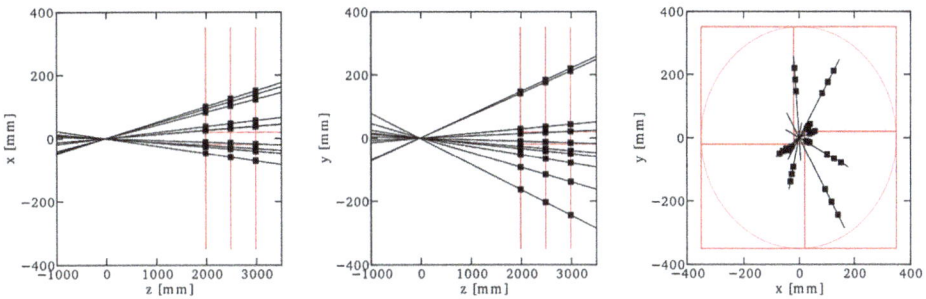

Fig. 5. A sketch demonstrating the beam gas vertex measurement concept.

transverse beam profile. The longitudinal profile could also be re-constructed in this way if relative arrival time information is additionally acquired by the system, something which is not currently planned.

In order to obtain a reasonable time for reconstruction of the profile a dedicated gas-injection system is required to provide a local pressure bump in the vicinity of the detectors. The pressure and type of gas used are of principal importance for the statistical and systematic uncertainties of the measured beam profiles.

The prototyping of such a detector is currently underway in collaboration with the LHCb experiment, EPFL (Lausanne, Switzerland) and RWTH (Aachen, Germany), with the aim of developing a system capable of providing:

- Transverse bunch size measurements with a 5% resolution within 1 minute.
- Average transverse beam size measurements with an absolute accuracy of 2% within 1 minute.

The envisaged measurement times and accuracy will allow meaningful measurements to be performed during the LHC ramp and provide a direct calibration for other beam size measurement instruments.

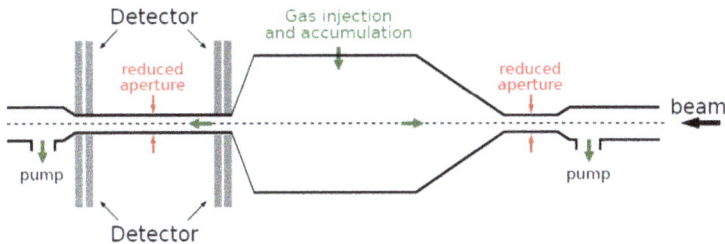

Fig. 6. Conceptual layout of the beam gas vertex emittance monitor. The two major sub-systems are the gas target (vacuum chamber, gas injection and pumping systems), and the tracking detectors.

A proof-of-principle demonstrator is foreseen for installation on the left side of LHC IP4 within the next few years. The basic layout is shown in Fig. 6. This relies on a gas target with a length of 1–2 m and tracking detectors external to the vacuum chamber. There are no moving parts and the aperture is defined to ensure that the chamber does not become a local aperture restriction. The chamber impedance is taken into account in the chamber design.

5. Halo diagnostics for the HL-LHC

One of the major issues for high intensity accelerators is the control of beam losses. In the case of HL-LHC the stored energy per beam is of the order of 700 MJ while the collimation system can sustain a maximum of 1 MW continuous power deposition. For this reason it is very important to study and understand loss dynamics. An important mechanism for slow losses consists of populating the beam "halo", i.e. populating the periphery of the phase-space with particles at large amplitudes (by IBS, beam-gas collisions, resonances etc.). These halo particles gradually increase their amplitude until they hit a collimator. Measurement of the beam halo distribution is important for understanding this mechanism to allow a minimization of its effects. Moreover, in the HL-LHC, crab cavities will be used to counter the geometric loss factor introduced by the increased crossing angle, which reduces the luminosity. In case of failure of a crab cavity module the whole halo can be lost in a single turn. If the halo population is too high this can cause serious damage to the collimation system or to other components of the machine. The total halo population that can be absorbed by the collimation system in case of a fast loss is of the order of 10^{-5} of the nominal beam intensity. The halo monitor for HL-LHC should thus have a dynamic range of at least 10^{5}.

There are two main ways of measuring the beam halo: either measuring the whole transverse space with a high dynamic range monitor or sampling only the tails using a monitor with a standard dynamic range. Both methods have already been attempted at other machines offering a good example of what can be achieved. A third technique often used to measure the halo consists in removing it by scraping the beam and recording the loss rate during the process. This technique is, however, not suitable for the intense, nominal HL-LHC beams and can only be used in dedicated low intensity experiments.

Most diagnostics used for transverse beam profile measurement can be adapted for halo measurements. For the LHC this consists of beam imaging using synchrotron radiation, wire-scanners and ionization profile monitors and the new technique based on beam-gas vertex reconstruction.

5.1. *Halo measurement using synchrotron radiation imaging*

This technique seems the most promising, as it is non-invasive and allows the continuous monitoring of the beams. Halo measurement could be achieved by using, as examples, one of the following techniques:

- High dynamic range cameras (like the SpectraCam from ThermoFisher) [10].
- Core masking and standard cameras (possibly with adaptive masks based on micro mirror arrays) [11].
- Performing an X-Y scan of the image plane with a photo-detector located behind a pinhole.
- Single photon counting with a pixelated photo-detector.

A variety of options are therefore available, but the limiting factor in all cases is likely to be the unavoidable presence of diffused synchrotron light coming from reflections in the vacuum chamber or optics, diffusion by dust particles and diffraction. The first two can, in principle, be mitigated with an appropriate surface treatment and a clean and hermetic setup, although diffusion by scratches and defects on the optical components cannot be entirely removed, while the third source is a physical limitation and its effects have to be studied carefully.

5.2. *Halo measurements using wire-scanners*

Halo measurement using wire scanners is possible by measuring the induced signal as the wire is moved into the tails of the beam. For very high sensitivity the downstream shower detector could be used in counting mode instead of the usual pulse height mode if needed. The problem associated with this method is that the wire has to be brought quite close to a high intensity beam with inherent safety issues. Another problem is the possibility of burning the wire by just exposing it to the electromagnetic field of the beam (RF heating). A modified version of the wire scanner, the so-called vibrating wire scanner, can also be used to monitor the faint halo around the core [12]. The advantage of this instrument is the high sensitivity, but it again involves moving the wire to a very small distance from a very intense beam.

5.3. *Halo measurements using ionization profile monitors*

Ionization profile monitors use the electrons or ions generated by beam collisions with residual gas to image the transverse beam profile. This technique again has the advantage of being quasi non-invasive and allowing continuous monitoring. Solutions for a high dynamic range readout could possibly be found by, for example, using a mask on the multi-channel plate (MCP) input. In reality in order to obtain a measurable signal for the tails the gas pressure has to be much larger than for just a core measurement. This would lead to the generation of large losses and hence enhanced radiation levels in these locations. Another problem arises from the fact that many ions and electrons generated by the core of the beam will drift outward under the influence of the beam space charge creating a background difficult to remove or correct for.

5.4. *Halo measurement summary*

Looking at the diagnostics that have been used to date for the measurement of beam halo, synchrotron radiation imaging seems to be the best candidate for halo monitoring for the HL-LHC. Work will start soon to study this in more detail in particular to identify the final potential of such techniques and any related problems.

6. Diagnostics for Crab Cavities

The crab cavities for the HL-LHC are proposed to counter the geometric reduction factor and so to enhance luminosity. These cavities will be installed around the high luminosity interaction points (IP1 and IP5) and used to create a transverse bunch rotation at the IP. The head and tail of each bunch is kicked in opposite directions by the crab cavities such that the incoming bunches will cross parallel to each other at the interaction point. These intra-bunch bumps are closed by crab cavities acting in the other direction on the outgoing side of the interaction region. If the bumps are not perfectly closed the head and tail of the bunch will follow different orbits along the ring. Monitors capable of measuring the closure of the head-tail bump and any head-tail rotation/oscillation outside of the interaction regions are therefore required.

6.1. *Bunch shape monitoring using electromagnetic pick-ups*

Electromagnetic monitors for intra-bunch diagnostics are already installed in the LHC. These so-called "Head-Tail" monitors mainly provide information on instabilities and have a bandwidth of some 2 GHz. To go to a higher resolution within the bunch a bandwidth of 10 GHz or more is desirable. This will be important to better understand instabilities in HL-LHC and to help with the tuning of the crab cavities. Several of these systems are therefore foreseen for installation around the interaction points. In addition to studies aimed at improving the existing electromagnetic pick-ups, which include optimization of the pick-up design and the testing of faster acquisition systems, pick-ups based on electro-optical crystals in combination with laser pulses are also being considered [13]. Such pick-ups have already demonstrated fast time response in the picosecond range [14]. Developed mainly for linear accelerators, this technology is now also being considered for circular machines, with a design for a prototype to be tested on the CERN-SPS recently initiated.

6.2. Bunch shape monitoring using streak cameras

The use of synchrotron light combined with a streak camera may be an easier alternative to electromagnetic or electro-optical pick-ups for high resolution temporal imaging. Using an optical system to re-image the synchrotron light at the entrance of a streak camera allows the transverse profile of the beam to be captured in one direction (X or Y) with a very fast time resolution (below the picosecond level). Only one transverse axis can be acquired with a given setup, while the other is used for the streaking. Using a sophisticated optical setup it is however possible to monitor both axes at the same time, as was performed in the CERN-LEP accelerator [15].

Streak cameras can be used to observe a number of parameters simultaneously: bunch length, transverse profiles along the bunch, longitudinal coherent motion, head-tail motion etc. The main limitation of the streak camera is the repetition rate of the acquisition, typically < 50 Hz, and the limited length of the recorded sample, given by the CCD size. Double scan streak cameras also exist that allow an increase in the record length. By using a CCD with 1000×1000 pixels working at 50 Hz and adjusting the optical magnification and scan speed such that the image of each bunch covers an area of about 100×100 pixels one can record a maximum of 100 bunch images per 20 ms, i.e. 5000 bunches per second. This is clearly just an optimistic upper limit with other factors likely to reduce this value.

The longitudinal resolution of around 50 ps required for HL-LHC is rather easy to achieve using streak cameras, where measurements down to the sub-picosecond are now possible. In terms of transverse resolution two distinctions have to be made:

(1) The beam width is affected by diffraction due to the large relativistic gamma of the beam, with the diffraction disk of the same order as the beam size. Measurement of the absolute transverse beam size will therefore not be very precise.

(2) The centroid motion (i.e. the center of gravity) is not directly affected by the diffraction, which produces a symmetrical blur, and therefore the resolution for this type of measurement will be much better.

As only the average position along the bunch is of importance in this measurement and not changes to the beam size, the streak camera should therefore be able to achieve the resolution of a few percent of the beam sigma necessary to quantify any residual non-closure of the crab cavity bumps.

Streak cameras are expensive and delicate devices not designed for the harsh environment inside an accelerator. Radiation dose studies are therefore required in order to verify if a streak camera can be installed directly in the tunnel or if, which seems more likely, it has to be housed in a dedicated, shielded, hutch. The latter would imply an optical line to transport the synchrotron light from the machine to the camera, something for which an integration study is required.

Another point to consider is the synchrotron light source. At the moment two synchrotron light telescopes are installed in the LHC, one per beam. These telescopes already share their light amongst three different instruments, the synchrotron light monitor, the abort gap monitor and the longitudinal density monitor, and in the future will also have to accommodate the halo monitor. It will therefore be difficult to integrate yet another optical beam line for the streak camera. The installation of additional light extraction mirrors may therefore be necessary to provide the light for the streak cameras. Since the crab cavities are only needed at high energy, dipole magnets can be used as the source of this visible synchrotron radiation, with no need for the installation of additional undulators which are only required at injection energy, where the dipole radiation is in the infra-red.

7. Luminosity Measurement for HL-LHC

The measurement of the collision rate at the luminous interaction points is very important for the regular tuning of the machine. The LHC experiments can certainly provide accurate information about the instantaneous luminosity, but this information is often not available until stable collisions have been established, and is often missing altogether during machine development periods. For this reason simple and reliable collision rate monitors, similar to those now used in LHC, are also needed for HL-LHC. The collision rate is currently obtained by measuring the flux of forward neutral particles generated in the collisions using fast ionization chambers installed at the point where the two beams are separated into individual vacuum chambers. The detectors (BRAN) are installed inside the neutral shower absorber (TAN) whose role is to avoid that the neutral collision debris, and the secondary showers these induce, reach and damage downstream machine components. The luminosity monitors therefore already operate in a very high radiation area, which for HL-LHC is anticipated be a further ten times the nominal LHC value. For this reason the technology chosen for HL-LHC is likely to be based on the radiation hard LHC design [16], with the geometry adapted to the new TAN design. In order to further increase the radiation resistance some current features, such as bunch to bunch measurement capability, may need to be sacrificed and redundancy added.

8. Summary

The LHC was constructed with a comprehensive array of beam instrumentation. This played an important role in the early commissioning of the accelerator and is essential for its safe and reliable operation. In addition, a full understanding of this complex machine cannot be achieved without having dedicated, specialized instruments available for the measurement of nearly every conceivable accelerator parameters. HL-LHC will push the performance of the LHC even further, requiring an even deeper understanding of beam related phenomena, which can only be delivered through its beam instrumentation. The upgrade of many of the existing systems in conjunction with the development of new diagnostics will therefore be mandatory to address the specific needs of the HL-LHC.

References

[1] C. Kurfuerst *et al.*, Radiation Tolerance of Cryogenic Beam Loss Monitor Detectors, 4th International Particle Accelerator Conference, Shanghai, China, 12–17 May 2013.

[2] G. Venturini, F. Anghinolfi, B. Dehning and M. Kayal, Characterization of a wide dynamic-range, radiation-tolerant charge-digitizer ASIC for monitoring of Beam losses, International Beam Instrumentation Conference, Tsukuba, Japan, 1–4 Oct 2012.

[3] E. Calvo *et al.*, The LHC Orbit and Trajectory System, in *Proc. of DIPAC*, Mainz, Germany, 2003, p. 187.

[4] D. Cocq, The Wide Band Normaliser: a New Circuit to Measure Transverse Bunch Position in Accelerators and Colliders, *Nucl. Instrum. Methods Phys. Res. A* **416**, 1 (1998).

[5] R. J. Steinhagen, Real-time Beam Control at the LHC, in *Proc. of the PAC Conference*, New York, USA, 2011, p. 1399.

[6] A. Nosych and M. Wendt, Analysis of Geometric Non-Linearities of LHC BPMs by 2D and 3D Electromagnetic Simulations, to appear in *Phys. Rev. ST Accel. Beams* (2015).

[7] M. Gasior, J. Olexa and R. J. Steinhagen, BPM electronics based on compensated diode detectors – results from development systems, in *Proc. of BIW*, Newport News, Virginia, USA, 2012, CERN-ATS-2012-247.

[8] A. Nosych *et al.*, Electromagnetic Simulations of an Embedded BPM in Collimator Jaws", in *Proc. of DIPAC*, Hamburg, Germany, 2011, p. 71.

[9] LHCb Collaboration, Absolute luminosity measurements with the LHCb detector at the LHC, *J. Instrum.* **7**, P01010 (2012).

[10] C. P. Welsch *et al.*, Alternative Techniques for Beam Halo Measurements, *Meas. Sci. Technol.* **17**, 2035c (2006) and CERN-AB-2006-23.

[11] H. D. Zhang *et al.*, Beam halo imaging with a digital optical mask, *Phys. Rev. Spec. Top. Accel. Beams* **15**, 072803(2012).

[12] S. G. Arutunian *et al.*, Vibrating wire scanner: First experimental results on the injector beam of the Yerevan synchrotron, *Phys. Rev. Spec. Top. Accel. Beams* **6**, 042801 (2003).

[13] M. A. Brubaker *et al.*, Electro-Optic Beam Position and Pulsed Power Monitors for the Second Axis of DARHT, in *Proc. of PAC*, Chicago, USA, 2001, p. 534.

[14] Y. Okayasu *et al.*, The First Electron Bunch Measurement by Means of Organic EO Crystals, in *Proc. of IBIC*, Tsukuba, Japan, 2012.

[15] E. Rossa, Real time single shot three-dimensional measurement of picosecond photon bunches, *6th Beam Instrumentation Workshop*, Vancouver, British Columbia, Canada, 3–6 Oct 1994.

[16] J. F. Beche *et al.*, Rad hard luminosity monitoring for the LHC, in *Proc. EPAC06*, Edinburgh, Scotland, June 2006.

Chapter 21

Heavy-Ion Operation of HL-LHC

J. M. Jowett[1], M. Schaumann[2] and R. Versteegen[1]

[1]*CERN, BE Department, Genève 23, CH-1211, Switzerland*
[2]*CERN and RWTH Aachen University, D-52056 Aachen, Germany*

The heavy-ion physics programme of the LHC will continue during the HL-LHC period with upgraded detectors capable of exploiting several times the design luminosity for nucleus–nucleus (Pb–Pb) collisions. For proton–nucleus (p–Pb) collisions, unforeseen in the original design of the LHC, a comparable increase beyond the 2013 luminosity should be attainable. We present performance projections and describe the operational strategies and relatively modest upgrades to the collider hardware that will be needed to achieve these very significant extensions to the physics potential of the High Luminosity LHC.

1. Introduction

The 2003 Design Report [1] of the LHC set out a "Nominal" design luminosity goal for Pb–Pb collisions in *two* experiments of

$$L_{AA} = 1. \times 10^{27} \text{ cm}^{-2}\text{s}^{-1} \text{ at } E_b = 7Z \text{ TeV} = 2.76A \text{ TeV or } \sqrt{s_{NN}} = 5.5 \text{ TeV}, \quad (1)$$

where E_b is the beam energy, $\sqrt{s_{NN}}$, the average center-of-mass energy per colliding nucleon pair and $Z = 82, A = 208$ for the lead nuclei accelerated in the CERN complex. These parameters corresponded closely to the capabilities of the specialized ALICE detector and the anticipated performance of the injectors.

About that time, there emerged a number of potential performance limits from beam physics and instrumentation [1–6], some of which were unprecedented, reflecting the projected factor 27 increase in collision energy beyond RHIC, the only previous heavy-ion collider. Within intrinsic uncertainties, most of these limits were expected to occur in a band of about an order of magnitude spanning the design luminosity.

Since then, the performance limits have been clarified considerably as the LHC has delivered two Pb–Pb physics runs, to *three* experiments, in 2010 and 2011. The peak Pb–Pb luminosity has reached

$$L_{AA} \simeq 0.5 \times 10^{27} \text{ cm}^{-2}\text{s}^{-1} \text{ at } E_b = 3.5Z \text{ TeV} = 1.38A \text{ TeV or } \sqrt{s_{NN}} = 2.76 \text{ TeV}. \quad (2)$$

With the natural scaling $L_{AA} \propto E_b^2$, this translates to roughly twice the design luminosity, Eq. (1), at full energy. The first experience of heavy-ion collisions in the new regime has also given opportunities to test some measures proposed to overcome the performance limits. Single-bunch intensities from the injectors are, with $N_b > 16 \times 10^7$ (for some bunches), already far higher than the design value, $N_b = 7 \times 10^7$. The greatest remaining uncertainties stem from present knowledge of the quench limits of superconducting magnets operating at full field [7].

Hybrid p–Pb collisions were not considered in the design [1] but a luminosity goal of

$$L_{pA} = 1.15 \times 10^{29} \text{ cm}^{-2}\text{s}^{-1} \text{ at } E_b = 7Z \text{ TeV or } \sqrt{s_{NN}} = 8.8 \text{ TeV}. \tag{3}$$

was set later [8, 9]. This was almost reached — or exceeded by about a factor of 3 if energy scaling is taken into account — with two of the four experiments that took data in early 2013 recording [10]

$$L_{pA} \simeq 1.1 \times 10^{29} \text{ cm}^{-2}\text{s}^{-1} \text{ at } E_b = 4Z \text{ TeV or } \sqrt{s_{NN}} = 5.0 \text{ TeV}. \tag{4}$$

2. Performance with Pb–Pb Collisions

2.1. *Experiments' requirements*

The ALICE experiment has requested a peak Pb–Pb luminosity of up to $L_{AA} \simeq 7. \times 10^{27}$ cm^{-2}s^{-1} after its upgrade with a view to integrating 10 nb^{-1} over several years operation (cf. the original request of 1 nb^{-1}). Similar requirements come from ATLAS and CMS.

Broadly speaking, the accumulation of comparison data in p–Pb and p–p collisions should keep pace with the Pb–Pb data in terms of integrated nucleon–nucleon luminosity:

$$A^2 \int L_{AA} \, dt \sim A \int L_{pA} \, dt \sim \int L_{pp} \, dt, \tag{5}$$

at equivalent $\sqrt{s_{NN}}$. In practice, p–Pb collisions usually take place at maximum energy for the sake of higher luminosity and faster commissioning. However Run 1 included p–p running at $\sqrt{s_{NN}} = 2.76$ TeV to fulfil Eq. (5) for all experiments.

2.2. *Optical configuration*

In heavy-ion, as opposed to proton, operation, a low value of β^*, is required at three, rather than two, experiments. The triplet quadrupoles around the ALICE experiment are not being upgraded within the HL-LHC project as are those of ATLAS and CMS. As a baseline, we envisage heavy-ion operation with $\beta^* = 0.5$ m in IP1, IP2 and IP5 using a conventional LHC optics, i.e., without the achromatic telescopic squeeze (ATS). The ATS optics for p–p operation is being designed to maintain this functionality for Pb–Pb or p–Pb physics [11]. The changes in phase advance with respect to the previous optics may provide some additional tunability for $\beta^* < 0.5$ m which can be explored although the beam–beam separation around ALICE may become marginal at small values of β^* (see below). As in 2010 and 2011, the beams will remain separated at LHCb in an unsqueezed optics for Pb–Pb operation.

When the beams are colliding, the vertical half-crossing angle at the ALICE experiment is [12]

$$\theta_{y_c} = \frac{\pm 490 \, \mu\text{rad}}{E_b/(7Z \text{ TeV})} + \theta_{yext} \tag{6}$$

where the first term is the angle created by the orbit bump (entirely inside the innermost quadrupoles) required to compensate the detector's muon spectrometer magnet (whose field can vary in polarity, less often in magnitude) and the second term is the contribution of an "external" bump created by orbit correction dipoles further out.

In order to provide an unimpeded path for "spectator" neutrons emerging from the collisions to the Zero-Degree Calorimeter (ZDC) [13,14], the condition

$$|\theta_{y_c}| < 60 \, \mu\text{rad} \tag{7}$$

has been imposed in heavy-ion operation up till now. With bunch spacings, $S_b/c = 100$ ns, as foreseen in the original LHC design [1], this provides adequate separation at the parasitic beam–beam encounters around the IP.

Some developments in the injectors [15] are aimed at increasing the number of bunches by achieving $S_b/c = 50$ ns — half of the original design — for at least some of the spacings between bunches in the LHC. Figure 1 shows that, for this value, it is no longer possible to satisfy the usual separation requirements in AL-ICE together with Eq. (7) at the closest parasitic encounters to the IP. However, experience [12] suggests that, given the relatively low charge of the Pb bunches (compared to p bunches), it may be possible to operate with more relaxed conditions, say, $r_{12}/\max(\sigma_x, \sigma_y) \gtrsim 3$ where r_{12} is the transverse distance between the closed orbits of the two beams. This is sufficient to reduce the parasitic luminosity to acceptable levels [14]. Thus the efforts to achieve shorter bunch spacing remain well-motivated; it is unlikely, although not strictly excluded, that the data quality of the ZDC may be somewhat compromised by a need for a larger crossing angle.

The ATLAS and CMS experiments do not have a muon spectrometer and separation requirements for Pb beams are less demanding than those established for protons.

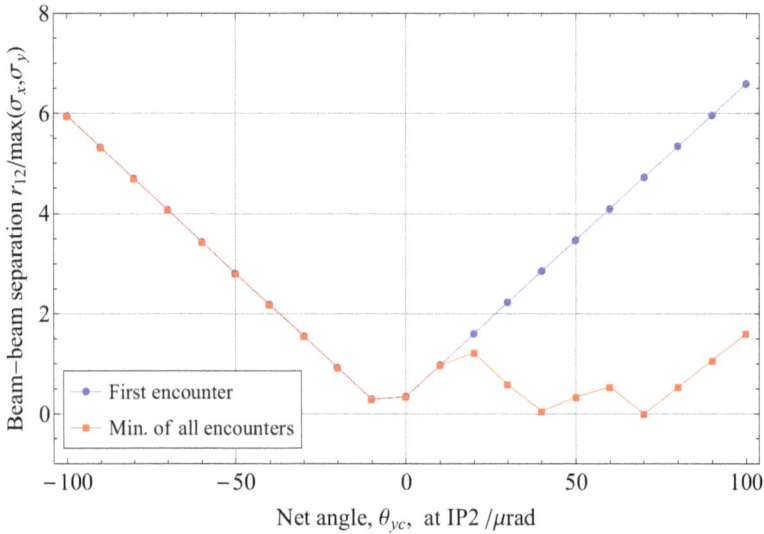

Fig. 1.   Beam–beam separation in IR2 as a function of the net half-crossing angle at the IP, θ_{yc} for $E_b = 7Z$ TeV with $\beta^* = 0.5$ m, the design Pb emittance of $\varepsilon_n = 1.5$ μm and a regular bunch spacing of 50 ns and with the ALICE spectrometer polarity such as to contribute a positive crossing angle at the interaction point. At $\theta_{yc} = +70$ μrad, the external angle is zero and parasitic head-on collisions occur. Separations are shown in units of the larger of the horizontal and vertical beam sizes, both at the first parasitic encounter (on either side of the IP) and at the minimum over all encounters excluding the IP itself.

2.3. Intensity, emittance and luminosity

Estimation of the future luminosity of Pb beams is complicated by the very large variation of bunch parameters [16], notably the emittances, ε_n, bunch populations, N_b, and bunch lengths, σ_z, due, principally, to intra-beam scattering (IBS) as bunch trains are built-up first in the SPS from batches injected from the PS. When these SPS batches are subsequently injected in the LHC, a similar pattern, on a larger scale, is imposed on the entire LHC bunch train as the SPS batches spend different times at the LHC injection energy. Injecting shorter trains in the SPS would allow its cycle length to be reduced so that the earliest injected bunches would suffer less. However the final LHC bunch train would contain more kicker gaps, reducing the total number of bunches and it would take longer to fill the LHC.

A model to provide realistic quantitative predictions for the Pb–Pb luminosity is in development. It works by first fitting data from the 2011 Pb–Pb run [16] to describe the intensity and emittance decay with time spent at injection in both the SPS and LHC. It then predicts the optimum number of PS injections per SPS cycle and can be used to compare different schemes for preparing batches in the PS. A further important ingredient is the minimum spacing of PS batches in the SPS which depends on the proposed upgrade of the injection kicker for ion beams in the SPS [15]. Here we shall assume a modest upgrade of the kicker which would allow a 100 ns spacing between the PS batches.

For present purposes, this model can be applied to one of the filling schemes [15] in which the PS delivers batches of four bunches, spaced at 50 ns to the SPS.[a]

On the basis of the injector performance in the 2013 run, it turns out the optimum SPS batches are built-up by injection of nine such PS batches, spaced at 100 ns. Finally 28 of these SPS batches are injected into each LHC ring. The model has to take into account the various cycle lengths and accumulation times in the filling scheme. The pattern of 1008 bunch intensities finally brought into collision in the LHC is shown in Fig. 2. Although this pattern includes major features such as the LHC injection kicker rise time and the abort gap in the LHC, the number of bunches may be slightly less in practice because of other constraints on

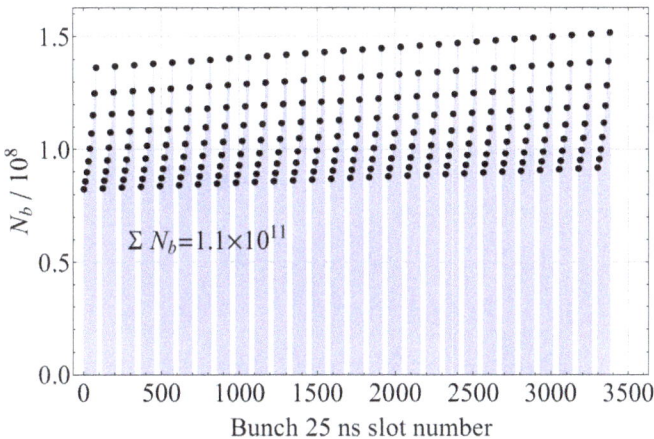

Fig. 2. Intensities of the individual Pb bunches at the start of collisions. The variation within each of the 28 batches injected from the SPS is due to IBS and other effects acting for different times at SPS injection. Each of the nine dots visible within each SPS batch represents four bunches injected simultaneously into the SPS from the PS. The abort gap appears at the right hand end of the plot.

---

[a] In more recent developments, a scheme providing two bunches spaced at 100 ns has been deemed the "Nominal" one for post LS2 operation. Moreover the manipulations required to produce the 50 ns spacing have not yet been demonstrated in the PS. On the basis of present injector performance the 100 ns scheme would yield higher single-bunch intensities and a similar final luminosity in the LHC. However we prefer to retain the 50 ns scheme here as it has greater potential for future improvement. The new "Nominal" scheme will be described in more detail elsewhere.

the filling scheme that are not yet included in the model. In addition the number of collisions occurring in a given experiment will vary according to where in the ring the leading bunch of each beam collides (up to now this has been IP1 and IP5).

Once in the LHC, the bunches suffer emittance increases and intensity losses from debunching, both effects arising from IBS, according to the time they spend at injection, the first train to be injected suffering the most. Further losses occur during the ramp and squeeze processes. These effects are also taken into account by the model, leading to a prediction for the final distribution of bunch populations as the beams are put into collision, N_b, as shown in Fig. 2. There is a corresponding distribution of emittances.

As the bunches collide pair-wise, the differences in their intensities and emittances produce a spectrum of initial luminosities which shows an even stronger variation along each SPS train and along the whole LHC train. Figure 3 shows an example of the luminosity distribution in an IP where the leading bunches collide.

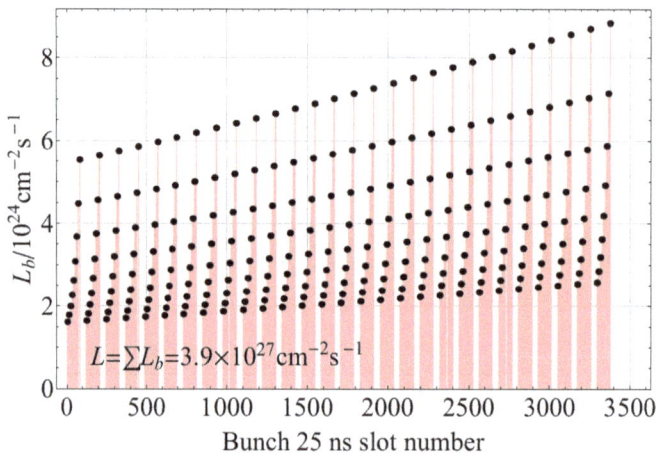

Fig. 3.   Luminosities of the individual Pb–Pb bunch pairs at the start of collisions.

3. Integrated Luminosity Projections

The next stage of the modelling is to look at the evolution over a fill using the Collider Time Evolution simulation program [16–18], a multi-particle simulation that includes the effects of IBS, luminosity burn-off, radiation damping, etc. Here again, of course, the spread in bunch parameters differentiates the rate of intensity and luminosity decay along the bunch train. To simplify the picture, Fig. 4 shows the luminosities of three bunch pairs, sampling each end, and the middle, of the luminosity spectrum. Naturally, the rate of luminosity decay is higher for the higher

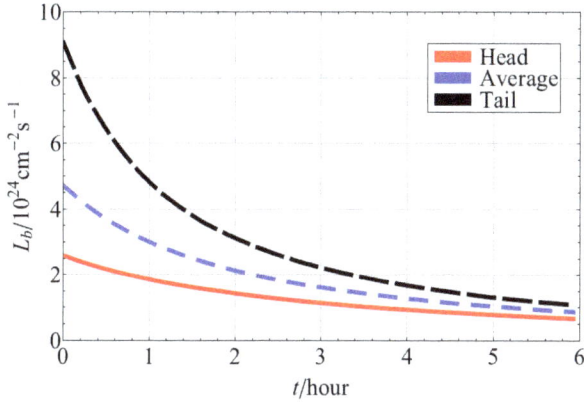

Fig. 4.   Evolution of the luminosity of three representative Pb bunches in the course of a fill.

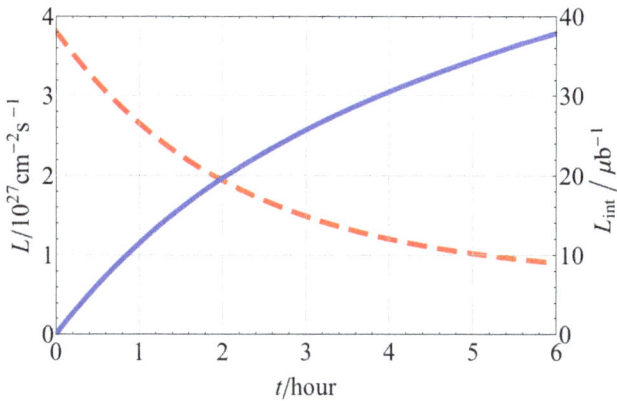

Fig. 5.   Evolution of the total luminosity, summed over all bunches (dashed line), and its integral (solid line) in the course of a fill.

luminosity bunch pairs so there is a tendency for all bunch pairs to equalize their luminosities after some time. This is nevertheless still a simplification since some bunches make fewer collisions than others.

Since each bunch pair luminosity is a smooth monotonic function, the model can fit them with similar functional forms. It then interpolates the fit parameters among the simulations run for representative bunches in order to estimate the luminosity evolution for the other bunch pairs with a good accuracy. Finally, the luminosities of all bunch pairs can be summed to predict the evolution of the total luminosity and the integrated luminosity, shown in Fig. 5.

For comparison, the most recent Pb–Pb run which took place in 2011 occupied about 24 days of calendar time and yielded 39 fills totalling just under 8 days of

Stable Beams time. The average during of the Stable Beams for physics was therefore 4.9 h. According to Fig. 5, such a fill would provide an integrated luminosity
of 34 μb^{-1}. So for the same run duration and operational efficiency as in 2011, the
total integrated luminosity would be 1.3 nb^{-1}.

A naive "Hübner factor" calculation, scaling the integrated luminosity obtained
in 2011 with the ratios of the peak luminosity then, 5×10^{26} $cm^{-2}s^{-1}$ to that shown
in Fig. 3 yields as similar value.

We emphasize here that these estimates are very conservative, assuming no
gains over the injector performance in 2011 or in operational efficiency. There
is every likelihood that the luminosity will be higher, particularly if the intensity
limits in LEIR can be overcome, the degradation of beam quality and intensity on
the SPS injection plateau can be reduced, or if schemes permitting the injection of
more high intensity bunches can be implemented.

3.1. *Mitigation of losses from collisions*

In the past decade, there have been extensive discussions of the potential limitation of peak Pb–Pb luminosity by the process of bound-free pair production
[1, 2, 4, 5, 19–22] in collisions. This creates a secondary beam of one-electron Pb
ions emerging from the interaction point and being lost in a well-defined spot in a
dispersion suppressor dipole magnet. At design luminosity, Eq. (1), the power in
the beam emerging from each side of the interaction points is some 25 W, potentially enough to induce quenches of the magnet. In 2003, when this problem first
came to light [3], it was too late for the obvious solution of installing collimators
or masks in the cryogenic sections of the dispersion suppressors to intercept these
beams. Although recent quench test measurements [7, 23] have indicated that the
quench limit is somewhat higher than estimated in the past, quenches remain likely
to occur with the upgraded Pb–Pb luminosity. For this reason, priority has been
given to the installation of collimators in the dispersion suppressors on either side
of the ALICE experiment as part of the High-Luminosity LHC upgrade [24].

Dispersion suppressor collimators may also be needed around ATLAS and
CMS also although there is some scope for mitigation of the effect using orbit
bumps [21], a method which was already tested in the 2011 run.

4. Performance with p–Pb Collisions

Although it had to work in a configuration that was almost unprecedented at previous colliders, the LHC performed remarkably well as a p–Pb collider for a single
fill in 2012 [25] and for a one month run in 2013 [10, 26–28].

Table 1. Indicative parameters of peak performance in the 2013 Pb–p production run and in future operation in Run 2 at higher energy; some numbers are averages because of the wide distribution of individual bunch parameters. Sets of four values correspond to the interaction points IP1(ATLAS), IP2(ALICE), IP5(CMS), IP8(LHCb).

Parameter	2013 production	Run 2
$E/(Z \text{ TeV})$	4	7
k_c	$(296, 288, 296, 39)$	$\simeq (400, 380, 400, 50)$
β^*/m	$(0.8, 0.8, 0.8, 2.0)$	$(0.5, 0.5, 0.5, 2.0)$
$\gamma\varepsilon_p/\mu\text{m}$	2	2
$\gamma\varepsilon_{Pb}/\mu\text{m}$	1.5	1.5
$N_{b,p}$	$1.8\text{--}5 \times 10^{10}$	1.6×10^{10}
$N_{b,Pb}$	14×10^7	14×10^7
$L/(10^{29}\text{cm}^{-2}\text{s}^{-1})$	$(0.98, 1.08, 0.98, 0.035/0.056)$	$\simeq 4\text{--}12 \times (1, 1, 1, 0.1)$

Operation in the first run was complicated by the need to supply collisions to the LHCb experiment, to several changes of operating conditions requested by the ALICE experiment, the intrinsic complexity of injection and ramping with unequal revolution frequencies and relocking RF frequencies at collision energy to collide "off-momentum".

Detailed plans for p–Pb runs in the HL-LHC era should take account of the potential increases in bunch intensity and bunch number in some detail. Injection schemes matching the proton bunch trains to the lead ones have to be worked out and the intrinsic beam-dynamical and instrumental (beam-position monitors, in particular) limitations of this mode of operation have to be further explored. For the moment we can only give some provisional estimates of the main parameters indicating some typical numbers of collisions at each IP, k_c, and a range of initial luminosities that could be accessed by varying the proton bunch intensity. Note that increasing this parameter will directly reduce the luminosity lifetime, leading to quite short fills.

5. Stochastic Cooling of Pb Beams

The RHIC collider has demonstrated the effectiveness of a bunched-beam stochastic cooling system against IBS [29] and a first study [30] has shown the potential of a similar installation to cool heavy ion beams in the LHC. In the HL-LHC, the benefits of the cooling lie in the reduction of colliding beam sizes at later stages of the fill. This maintains a high luminosity even when the bunch populations have been substantially eroded by the earlier luminosity burn-off. This is a much more efficient way to operate a collider since more of the particles stored in the beam are converted into collisions. A much smaller fraction are dumped at the end of the fill.

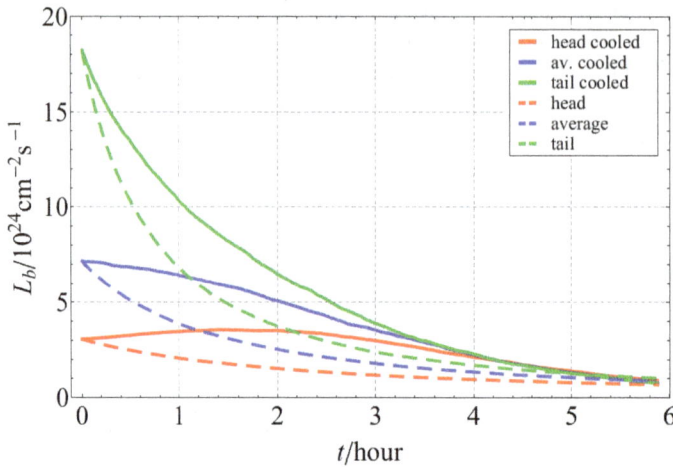

Fig. 6. Evolution of bunch-pair luminosity for three typical bunches in a train, with (solid curves) and without (dashed curves) the effect of stochastic cooling in all three planes. Bunch parameters are similar to the examples shown in the paper [30], except that luminosity burn-off is stronger with three experiments taking collisions. The bunch parameters are different from those shown in Fig. 4.

Studies are beginning to see whether the promise of a stochastic cooling system, at an apparently modest cost, can be realized in the LHC. Space for the system (roughly 20 m per beam for the kickers) must be found. They will be connected to the broadband pickups by fibre optics lines in the tunnel, avoiding chordal microwave links on the surface. As in RHIC, the kickers will have to come very close to the beam at physics energy so must open at injection where a larger aperture is needed. In the open position, the design must have a low enough high-order mode impedance to avoid overheating in the presence of the high-intensity proton beams.

6. Operational Aspects

Heavy-ion operation in the HL-LHC era will continue the advantageous pattern of a short run of about one month at the end of each year, generally preceding a short end-of-year shutdown. This allows the activation level of the accelerator complex due to high-intensity proton operation to decline before interventions. Each run so far has broken new ground and yielded important physics discoveries and the rhythm of data delivery and opportunities, changes of colliding species and other experimental conditions has been well-matched to the research programmes of the experimental and theoretical physics community.

Since each run is short, there is a high premium on being able to commission the new mode of operation very quickly. By the same token, the impact of any down-time can be significant.

7. Summary

High-luminosity heavy-ion operation of the LHC can start a few years earlier (in "Run 3") than the high-luminosity proton operation designated "HL-LHC" since it does not require the upgraded inner-triplet quadrupoles around ATLAS and CMS or the crab cavities. Conservative extrapolation from past performance indicates that it will be possible to accumulate an integrated Pb–Pb luminosity of $\simeq 1.3\ \text{nb}^{-1}$ in an annual one month run (cf. the initial goal of $1\ \text{nb}^{-1}$ for the entire first phase of LHC). This depends on upgrades to the Pb–ion injector chain allowing the implementation of new filling schemes.

Present expectations concerning quench levels of the magnets indicate that the peak luminosity will exceed the limits imposed by the BFPP losses and the installation of the dispersion suppressor collimators around experiments, as foreseen for ALICE, is necessary.

There are reasonable prospects for a further doubling of the integrated luminosity, particularly if intensity limitations in LEIR can be overcome, the intensity decay of Pb bunches in the SPS can be mitigated and/or a stochastic cooling system can be implemented at collision energy in the LHC.

It should be remembered that some annual runs will be devoted to p–Pb operation, where substantially higher luminosity can also be expected, and that there will also be requirements for p–p reference data at equivalent $\sqrt{s_{NN}}$ to the Pb–Pb runs.

References

[1] O. S. Brüning, P. Collier, P. Lebrun, S. Myers, R. Ostojic, J. Poole and P. Proudlock (editors), CERN-2004-003-V1 (2004).
[2] S. R. Klein, *Nucl. Instrum. Meth. A* **459**, 51 (2001).
[3] J. M. Jowett, in *Proceedings of the LHC Performance Workshop, Chamonix XII* (2003).
[4] J. Jowett *et al.*, in *Proceedings of the 2004 European Particle Accelerator Conference*, Lucerne, Switzerland (2004), p. 578.
[5] J. M. Jowett, J. B. Jeanneret and K. Schindl, in *Proceedings of the 2003 Particle Accelerator Conference*, Portland, Oregon (2003), p. 1682.
[6] R. Assmann, H.-H. Braun, A. Ferrari, J.-B. Jeanneret, J. Jowett and I. A. Pshenichnov, Collimation of heavy ion beams in LHC, in *Proceedings of the 2004 European Particle Accelerator Conference*, Lucerne, Switzerland (2004), p. 551.
[7] S. Redaelli, R. Assmann, G. Bellodi, K. Brodzinski, R. Bruce, F. Burkart, M. Cauchi, D. Deboy, B. Dehning, E. Holzer, J. Jowett, L. Lari, E. Nebot del Busto, M. Pojer, A. Priebe, A. Rossi, R. Schmidt, M. Sapinski, M. Schaumann, M. Solfaroli Camollocci, G. Valentino, R. Versteegen, J. Wenninger, D. Wollmann and Z. Zerlauth, in *Proceedings of the HB2012 Workshop*, Beijing, China (2012), p. 157.

[8] J. M. Jowett and C. Carli, in *Proceedings of the 2006 European Particle Accelerator Conference*, Edinburgh, UK (2006), p. 550.

[9] C. Salgado *et al.*, *J. Phys. G* **39**, 015010 (2012).

[10] J. Jowett, R. Alemany-Fernandez, P. Baudrenghien, D. Jacquet, M. Lamont, D. Manglunki, S. Redaelli, M. Sapinski, M. Schaumann, M. Solfaroli Camillocci, R. Tomás, J. Uythoven, D. Valuch, R. Versteegen and J. Wenninger, in *Proceedings of the 2013 International Particle Accelerator Conference*, Shanghai, China (May 2013), p. 49.

[11] S. Fartoukh and F. Zimmermann, *The HL-LHC Accelerator Physics Challenges*, in this book, pp. 45–96.

[12] R. Alemany-Fernandez, G. H. Hemelsoet, J. M. Jowett, M. Lamont, D. Manglunki, S. Redaelli, M. R. Versteegen and J. Wenninger, CERN-ATS-Note-2012-039 MD (2012).

[13] R. B. Appleby, J. M. Jowett and J. Uythoven, Moving the Recombination Chambers to Replace the Tertiary Collimators in IR2, CERN-ATS-Note-2011-014 MD (2011).

[14] K. Aamodt *et al.* (The ALICE Collaboration), *Journal of Instrumentation* **3**, S08002 (2008).

[15] D. Manglunki, *Challenges and Plans for the Ion Injectors*, in this book, pp. 311–319.

[16] M. Schaumann and J. M. Jowett, in *Proceedings of the 2013 International Particle Accelerator Conference*, Shanghai, China (May 2013), p. 1391.

[17] R. Bruce, J. Jowett, M. Blaskiewicz and W. Fischer, *Phys. Rev. ST Accel. Beams* **13**, 091001 (2010).

[18] R. Bruce, J. Jowett and T. Mertens, in *Proceedings of the 2011 International Particle Accelerator Conference*, San Sebastian, Spain (2011), p. 1840.

[19] H. Meier, Z. Halabuka, K. Hencken, D. Trautmann and G. Baur, *Phys. Rev. A* **63**, 032713 (2001).

[20] R. Bruce *et al.*, *Phys. Rev. Lett.* **99**, 144801 (2007).

[21] R. Bruce, S. Gilardoni, J. Jowett and D. Bocian, *Phys. Rev. ST Accel. Beams* **12**, 071002 (2009).

[22] J. M. Jowett, R. Alemany-Fernandez, R. Assmann, P. Baudrenghien, G. Bellodi, S. Hancock, M. Lamont, D. Manglunki, S. Redaelli, M. Sapinski, M. Schaumann, M. Solfaroli, R. Versteegen, J. Wenninger and D. Wollmann, in *Proceedings of the Chamonix 2012 Workshop on LHC Performance* (2012).

[23] M. Sapinski, T. Baer, M. Bednarek, G. Bellodi, C. Bracco, R. Bruce, B. Dehning, W. Hofle, A. Lechner, E. Nebot Del Busto, A. Priebe, S. Redaelli, B. Salvachua, R. Schmidt, D. Valuch, A. Verweij, J. Wenninger, D. Wollmann and M. Zerlauth, in *Proceedings of the 2013 International Particle Accelerator Conference*, Shanghai, China (May 2013), p. 3243.

[24] S. Redaelli *et al.*, *Cleaning Insertions and Collimation Challenges*, in this book, pp. 215–241.

[25] R. Alemany, M. Angoletta, P. Baudrenghien, R. Bruce, S. Hancock, D. Jacquet, J. M. Jowett, V. Kain, M. Kuhn, M. Lamont, D. Manglunki, S. Redaelli, B. Salvachua, M. Sapinski, M. Schaumann, M. Solfaroli, J. Uythoven, R. Versteegen and J. Wenninger, First proton-nucleus collisions in the LHC: the p-Pb pilot physics, CERN-ATS-Note-2012-094 MD (2012).

[26] R. Versteegen, R. Bruce, J. Jowett, M. McAteer, E. McLean, A. Langner, Y. Levinsen, T. Persson, S. Redaelli, B. Salvachua, P. Skowronski, M. Solfaroli Camillocci,

R. Tomás, G. Valentino, J. Wenninger and S. White, in *Proceedings of the 2013 International Particle Accelerator Conference*, Shanghai, China (May 2013), p. 1439.

[27] R. Alemany, P. Baudrenghien, D. Jacquet, J. M. Jowett, M. Lamont, D. Manglunki, S. Redaelli, M. Sapinski, M. Schaumann, D. Valuch, R. Versteegen and J. Wenninger, p-Pb Feasibility Test and Modifications of LHC Sequence and Interlocking, CERN-ATS-Note-2012-052 MD (2012).

[28] R. Versteegen, J. M. Jowett, A. Langner, Y. Levinsen, M. McAteer, E. McLean, T. Persson, P. Skowronski, M. Solfaroli, R. Tomas, J. Wenninger and S. White, Operating the LHC Off-Momentum for p-Pb Collisions, in *Proceedings of the 2013 International Particle Accelerator Conference*, Shanghai, China, CERN-ACC-2013-0180 (2013), pp. 1439–1441.

[29] M. Blaskiewicz, J. Brennan and K. Mernick, *Phys. Rev. Lett.* **105**, 094801 (2010).

[30] M. Schaumann, J. Jowett and M. Blaskiewicz, in *Proceedings of the COOL2013 Workshop*, Murren, Switzerland (Jun 2013), p. 76.

Chapter 22

Implications for Operations

G. Arduini, M. Lamont, T. Pieloni and G. Rumolo

CERN, BE Department, Genève 23, CH-1211, Switzerland

The HL-LHC will introduce a number of novel operational features and challenges including luminosity leveling. After a brief recap of the possible leveling techniques, the potential impact of the operational regime on overall efficiency is discussed. A breakdown of the operational cycle and the standard operational year, together with a discussion of fault time is presented. The potential performance is then explored and estimates of the required machine availability and efficiency for 250 fb^{-1} per year are given. Finally the e-cloud challenges, scrubbing runs requirements, and machine development potential are outlined.

1. Introduction

Come the commissioning and subsequent operation of the HL-LHC, the LHC itself will have been operational for over 10 years and a wealth of knowledge and experience will have been built up. The key operational procedures and tools will have been well established. The understanding of beam dynamics will be profound and refined by relevant measurement and correction techniques. Key beam related systems will have been thoroughly optimized and functionality sufficiently enhanced to deal with most challenges up to that point. Availability will have been optimized significantly across all systems. This collected experience will form the initial operational basis following the upgrade.

However the HL-LHC will pose significant additional challenges with the target integrated luminosity posing considerable demands on machine availability and operational efficiency. The planned beam characteristics will push beam dynamics to new limits. In the following the operational cycle is revisited in light of the key HL-LHC challenges. The expectations and issues relating to availability in the HL-LHC are outlined.

2. Leveling

As described in detail in preceding chapters, the planned bunch intensity, β^*, and compensation of the geometrical reduction factor lead to a potential bunch luminosity well above the acceptable maximum in terms of pile-up for the experiments. Thus an obligatory operational principle of the HL-LHC is luminosity leveling. The aim is to reduce a potential peak or virtual luminosity to a more manageable leveled value by some luminosity reduction technique. As the number of particles in the beam falls with time, appropriate adjustments keep the luminosity at the leveled value. The options are outlined below and a combination of some of them (e.g. dynamic change of of β^* and crab cavities) will be used in the HL-LHC. It has been realized recently that the longitudinal pile-up density is also another critical parameter for the experiments which must be taken into account during luminosity delivery.

2.1. Leveling options

2.1.1. Transverse offset

Transverse offsets of the beams at one interaction point have been used operationally for leveling in LHCb, and tested in 2011 with single bunches with HL-LHC like parameters. Offset leveling in one interaction point with two interaction points colliding head-on is certainly possible but concerns about coherent beam–beam instabilities and possible emittance growth rule this out as a universal solution.

2.1.2. Crab cavities

Crab cavities can be used to manipulate the beam overlap of the two beams in the luminous region, thereby reducing the effect of geometrical reduction factor. Details are given in Chapter 7.

The technique has the advantages of flexibility and IP independence. A possible operational scenario would be to start a fill with crab cavities effectively off and accept the full geometrical reduction factor. As the luminosity falls, the crab cavities could be used to an appropriate level to compensate the crossing angle, thus leveling the luminosity. The disadvantage is that the longitudinal vertex density of the fully uncompensated crossing angle exceeds the current performance expectations of the upgraded detectors. This implies that naive use of crab cavities would have to be supplemented with an additional technique. Alternatively the use of crab cavities in the separation plane can be used to tailor the longitudinal pile-up density and reduce its peak value while keeping the integral (i.e. events per crossing) constant [1].

Concerns about the use of crab cavities include phase or amplitude noise, synchronization errors that could lead to emittance blow-up and the fact that there is little experimental evidence that crab cavities can work effectively in proton machines.

2.1.3. *External crossing angle of the two beams*

If partial compensation of the long-range beam–beam interactions in the common vacuum system with the help of wires can be established as a viable technique, the crossing angle can be reduced and adjustment of the geometrical reduction factor can be used as leveling parameter. This method has similar limitations with respect to the longitudinal pile-up density as the leveling technique based on crab cavities.

2.1.4. *Dynamic change of β^**

Initial exploration of changing β^* during a fill have taken place with some success. This is a potentially interesting technique providing constant longitudinal vertex density and sufficient beam stability. The IPs can be treated independently, however any implementation must be operationally robust and must ensure that beams remain in collisions during the change in β^*.

2.2. *Leveling — formulation*

Relevant figures from the HL-LHC 25 ns baseline are shown in Table 1. Here k is defined as the ratio of the peak virtual luminosity to the leveled luminosity. The peak virtual luminosity is the peak luminosity that could be achieved if the geometric factor due to the crossing angle is fully compensated by crab cavities.

Table 1. Selected baseline HL-LHC parameters for 25 ns beam.

Protons per bunch	Number bunches	Total number protons	Peak luminosity without crab cavities $cm^{-2}s^{-1}$	Peak virtual luminosity $cm^{-2}s^{-1}$	Leveled luminosity $cm^{-2}s^{-1}$	k
2.2×10^{11}	2808	6.2×10^{14}	7.4×10^{34}	21.9×10^{34}	5×10^{34}	4.38

Considering luminosity burn-off and neglecting losses due to other mechanisms such as beam-gas and diffusion, the number of particles per beam as a function of time is given by Eq. (1):

$$N(t) = N_0 - n_{ip}L_{lev}\sigma_{vis}t \tag{1}$$

where N_0 is the initial number of particles, n_{ip} is the number of interaction points, L_{lev} is the leveled luminosity, and σ_{vis} the visible cross-section. Equation (1) can be re-expressed in terms of an effective lifetime τ_{eff}.

$$N(t) = N_0 \left(1 - \frac{t}{\tau_{\mathrm{eff}}} \right) \tag{2}$$

where

$$\tau_{\mathrm{eff}} = \frac{N_0}{n_{\mathrm{ip}} \sigma_{\mathrm{vis}} L_{\mathrm{lev}}}. \tag{3}$$

Assuming no additional losses due to diffusion/collimation during the fill and no transverse or longitudinal emittance blow-up, the number of particles at the end of the leveling period is:

$$N_e = \frac{1}{\sqrt{k}} N_0. \tag{4}$$

The leveled luminosity L_{lev} may be delivered for the leveling time T_{lev} before the luminosity starts to drop naturally below L_{lev}. T_{lev} is given by:

$$T_{\mathrm{lev}} = \left(1 - \frac{1}{\sqrt{k}} \right) \tau_{\mathrm{eff}}. \tag{5}$$

In an ideal world, T_{lev} is trivially calculated and one might naively imagine to let a fill run for T_{lev}. For the numbers shown in Table 1 T_{lev} is 10.6 hours. However, additional losses, primarily to the collimation system, will undoubtedly be present as a consequence of a number of emittance blow-up mechanisms (for example: beam–beam, noise, IBS, elastic scattering at the IP). These mechanisms will also contribute to an additional luminosity reduction because of the resulting emittance increase. An estimate of the virtual luminosity lifetime is possible and realistic numbers give a virtual luminosity lifetime of around 5 hours and a leveled time of 6 to 7 hours.

Once the luminosity falls below L_{lev} one could imagine keeping the fill for a certain amount of time (t_{decay}) before dumping. The contribution to the integrated luminosity from this part of the fill, assuming a constant luminosity lifetime (τ_L), is shown in Eq. (6):

$$L_{\mathrm{decay}} = L_{\mathrm{lev}} \tau_L \left(1 - e^{-\frac{t_{\mathrm{decay}}}{\tau_L}} \right). \tag{6}$$

Optimization of fill length given the turnaround time and average fill length will be possible. Further consideration of the possibilities is given below when estimating potential integrated luminosity performance.

3. General Operations Considerations

3.1. *Nominal cycle*

The nominal operation cycle provides the framework underpinning luminosity production. As of 2012 the nominal operational cycle is well-established for 50 ns and bunch population exceeding nominal. A brief outline of the phases of the nominal cycle follows.

(1) The working point of the various magnetic elements is reset by an appropriate magnetic cycle ("Precycle").
(2) The machine settings are verified with the injection of a limited number of bunches having a reduced population (pilot) or nominal population (intermediate). This is the so-called "Set-up" phase.
(3) The two LHC rings are progressively filled with trains of bunches transferred from the SPS to the LHC at a momentum of 0.450 TeV/*c* ("Injection").
(4) Once the machine is filled with the maximum number of bunches possible the beams are accelerated to top momentum (up to 7 TeV/*c*) ("Ramp").
(5) The beam size at the interaction point is minimized with the aim of maximizing the luminosity once the beams are brought in collision. This is obtained by varying the current in the quadrupole magnet circuits in the two rings according to pre-calculated functions ("Squeeze").
(6) During the whole process of injection, ramp and squeeze the two counter-rotating beams are separated to avoid collisions and the detectors are in a safe state to avoid damage resulting from losses. At the end of the squeeze the beams are brought in collision at the interaction points ("Adjust").
(7) Once all the above procedures are completed and no abnormal conditions are detected (excessive losses or orbit excursions...) the conditions are met for safely switching on the detectors and the start of data taking ("Stable Beams").

The differences between the key operational parameters before and after the start of the HL-LHC era are shown in Table 2.

Table 2. Comparison of key Run I and HL-LHC performance parameters.

Parameter	Value end Run 1	HL-LHC
Bunch population 25 ns	1.2×10^{11}	2.2×10^{11}
Normalized emittance [mm mrad]	2.5	2.5
β^* [cm]	60	15

The principal challenges during the cycle involve the injection, and transmission through the cycle of high bunch and high beam current while preserving small emittances. Given the experience of LHC operations thus far, the following potential issues may be identified.

- High total beam intensity brings with it possible issues with electron cloud, UFOs [2], beam induced heating, cryogenic heat load.
- High bunch population implies very good parameter control and the need to properly control beam loss through the cycle. Beam instabilities will have to be fully under control using a variety of measures.
- Low β^* implies strong non-linearities from the inner triplets and associated parameter control.

The following outline gives the results of folding these issues into the nominal operational cycle.

- **Precycle**: Following extensive experience during Run 1, it is known that the magnetic state of machine can be re-established before every fill by rigorous application of either: a combined ramp-down/precycle of the main circuits and a "de-Gauss" of the corrector circuits; or, following an access, by a full cycle of all circuits.

  The ramp-down/precycle stage represents the adopted operational strategy for coming down from high energy: the main bends, quadrupoles, independently powered quadrupoles, inner triplets, etc. are ramped down to their pre-injection levels, while the other circuits are put through their normal "de-Gauss" cycle which puts them on the right hysteresis branch at injection.
- **Set-up of machine at 450 GeV**: Here the correction of the basic parameters (tune, chromaticity, orbit, RF phase and synchro errors) and the sequence of pre-injection checks can be assumed to have been mastered.
- **Injection**: The injection process should be very well optimized by the time of the upgrade. However high bunch intensities will have to be anticipated and all necessary mechanisms to minimize loss and maximize reproducibility should have been deployed. This process will certainly have to work like clockwork to ensure machine safety and high operational efficiency.

  The persistent current decay at 450 GeV has a powering history dependence which is well described by the magnet model [3]. The correction of the effects of the decay are well established and correction mechanisms are in place. Any new magnets will, of course, have to be appropriately measured and characterized and integrated into the magnetic model.
- **Ramp**: The ramp should hold no surprises. The characteristics of snap-back and correction of associated parameter swings can be taken as given. The 10 A/s

ramp rate of main dipole power converters will still hold. Transverse feedback and tune feedback and their healthy cohabitation should be anticipated. Orbit feedback will be mandatory. Controlled longitudinal blow-up will certainly be required.

- **Squeeze**: The squeeze mechanics including feedbacks should and will have to have been fully mastered. Issues of beam instabilities should have been resolved, these might include squeezing with colliding beams — again this should have been made operational before the HL-LHC era. The possibility of combining the part of the squeeze process in the ramp is being considered to reduce the time required for the squeeze process.
- **Collisions**: There is some flexibility in sequencing the order in which the beams are collided in the different interaction points. The main issue here will be to avoid the loss of Landau damping during the process. Experience and understanding gained in future runs will be essential for validating the models presently being developed for explaining the observed instabilities.
- **Stable Beams**: With a leveled luminosity, the luminosity lifetime is, of course, infinite. There will, however, be a virtual luminosity lifetime which will fold in emittance growth, losses to diffusion, and luminosity burn and will give an operational leveling time somewhat below the ideal. In general, this phase is expected to be stable with overly sufficient Landau damping from beam–beam and small movements in beam overlap thanks to good orbit stability. Transverse and longitudinal emittance growth from intra beam scattering and noise sources should be estimated.

Given the higher than ever stored beam energy of the HL-LHC era, the nominal cycle must be fully mastered for effective, safe operation. In the performance estimates that follow this is assumed.

3.2. *Turnaround*

The turnaround time is defined as time taken to go from Stable Beam mode back to Stable Beam mode in the absence of significant interruptions due to fault diagnosis and resolution. A breakdown of the foreseen ideal HL-LHC turnaround time is shown in Table 3.

From Table 3, one can see that realistically a three hour minimum turnaround time may be assumed. The main components are the ramp-down from top energy, the injection of beam from the SPS, the ramp to high energy and the squeeze. The ramp-down, the ramp and the squeeze duration are given by the current rate limitations of the power converters. Of note is the 10 A/s limit up and down for the main bends; and the need to respect the natural decay constants of the main quadrupoles,

Table 3. Breakdown of turnaround with estimated minimum times shown.

Phase	Time [minutes]
Ramp down/precycle	60
Pre-injection checks and preparation	15
Checks with set-up beam	15
Nominal injection sequence	20
Ramp preparation	5
Ramp	25
Squeeze	30
Adjust/collisions	10
Total	180

the individually powered quadrupoles and the triplets during the ramp-down and the squeeze. These quadrupoles are powered by single quadrant power converters and take a considerable time to come down. A faster precycle via upgrades to the power converters may be anticipated. Two quadrant power converters for the inner triplets, for example, would remove them as a ramp-down bottle neck.

In practice, the turnaround has to contend with a number of issues which could involve lengthier beam based set-up and optimization, or fault resolution. Typical beam based optimization might include: the need to re-steer the transfer lines, occasional energy matching between the SPS and LHC and the need for the SPS to adjust scraping during the injection process. Injector and LHC tuning and optimization are accounted for in the average turnaround time at present.

3.3. Availability and faults

Availability is defined as the overall percentage of the scheduled machine time left to execute the planned physics program after removing the total time dedicated to fault resolution.

Faults cover an enormous range from a simple front-end reboot to the loss of a cold compressor with corresponding loss of time to operations from 10 minutes to potentially days. Typical concerns are:

- Injector downtime.
- Hardware faults (broken PFNs, broken switches, faulty power supplies, broken computer components ...). Access is often required. Preparation and recovery from access comes out of availability.
- Software and control system problems.

- Cryogenics availability, quenches and quench recovery.

As will be seen below high availability will be key to getting anywhere near the stated HL-LHC target integrated luminosity [4] This will require far-sighted, targeted, consolidation in the years leading up to HL-LHC operation.

Operations can also be affected by a number of other factors.

- The challenges of high intensity and high energy, such as instabilities which can provoke losses and in the limit a beam dump.
- Premature dumps not related to faults which reduced the average fill length — for example: UFOs; vacuum spikes.

These factors reduce overall operation efficiency and during Run 1 were taken in account in either the turnaround time (fills lost in ramp and squeeze) or the average fill length (premature dumps in Stable Beams).

3.4. *Operational year*

The longer term operational model appears to be settling into a series of long years of operation interspersed with long shut-downs of order of a year or more. The long shut-downs are foreseen for essential plant maintenance, experiment upgrades and so forth. It is estimated that required length of a standard shut-down in the HL-LHC era will alternate between 16 and 20 months. The approximate breakdown of a generic long year is:

- 13-week Christmas technical stop including 2-week hardware commissioning (HWC) (this would count 3 weeks at the end of a year and 10 weeks at the start of the following year);
- around 160-day proton–proton operation;
- three technical stops of 5-day duration during the year;
- a 4-week ion run;
- and time for special physics and machine development.

A more detailed breakdown is shown in Table 4.

The longer period for scrubbing will be required after long shut-downs during which a significant fraction of the machine will be warmed-up and vented to air.

3.5. *Integrated luminosity*

A standard heuristic estimate for integrated luminosity takes the product of the scheduled physics time (minus MD, scheduled stops, etc.) and the peak luminosity and multiplies the result by the so-called Hübner factor to give an estimated integrated luminosity for said period. The Hübner factor is an approximation that

Table 4. Potential breakdown of a standard HL-LHC year.

Activity	Time assigned [days]
Christmas technical stop including HWC	91
Commissioning with beam	21
Machine development	22
Scrubbing	7 (to 14)
Technical stops	15
Technical stop recovery	6
Proton physics running including intensity ramp-up	160
Special physics runs	8
Ion run setup	4
Ion physics run	24
Contingency	7

implicitly takes into account luminosity lifetime, turnaround time, unplanned interventions, etc. During the LEP era it was typically around 0.2. During the first years of LHC operation the value of Hübner factor climbed with experience and targeted improvements in system availability and reached 0.18 in 2012. However luminosity leveling implies that this approach is no longer directly applicable.

Before considering any integrated luminosity estimates, the relevant terms are recalled.

- The scheduled proton physics time (SPT) is the total time assigned to high luminosity production. This is typically expressed in days per year. It does not include initial beam commissioning, special physics runs, ions, MD, technical stops etc. It does include the intensity ramp-up following initial commissioning.
- Machine availability (A) is the fraction of time realised for physics production after time lost to faults and fault recovery has been subtracted. The actual time available for physics production is thus: machine availability × scheduled physics time.
- Physics efficiency (PE) is the fraction of the scheduled physics time spent in Stable Beams.
- The turnaround time (T_{around}) is defined as time taken to go from Stable Beam mode back to Stable Beam mode in the absence of significant interruptions due to fault diagnosis and resolution. An average turnaround time maybe assumed based on explicit consideration of the operational phases, and experience. Recall that unsuccessful fill attempts that do not make it into Stable Beams are absorbed into the turnaround time.
- An average fill length (T_{fill}) may be assumed based on experience. It should be noted that an average fill length is based on a distribution of a certain number

of fills with given lengths that can span 0 to, say, 20 hours, and the ensemble can have a large number of low length fills, and longer fills that span lower luminosity delivery rates. At the HL-LHC, longer fills ($t_{fill} > T_{lev}$) will span the exponential decay region detailed above and this must be taken into account. The distribution of fill lengths motivates the use of a Monte Carlo [5] or an analytic approach [6].

Both the average fill length and fill length distribution will play an important role in the overall exploitation of the LHC. They will also be key factors in any estimates of future performance.

As the first step in an integrated luminosity estimate, the total time spent in Stable Beams for a given period is calculated.

- Reduced the scheduled physics time SPT by the availability factor.
- Assume an average turn around and an average fill length in the time that is left. Reduce the available time by the number of fills times the turnaround time to get the total time spent in Stable Beams (defined above as the Physics Efficiency (PE)).

The PE may thus be expressed as:

$$PE = \frac{(A \times SPT - N_f \times T_{around})}{SPT} \tag{7}$$

where N_f the number of fills given by:

$$N_f = \frac{A \times SPT}{T_{fill} + T_{around}}. \tag{8}$$

Thus:

$$PE = A \times \left(1 - \frac{T_{around}}{T_{fill} + T_{around}}\right). \tag{9}$$

Given the PE, the next question is how much luminosity might one hope to produce in said time. At this juncture consideration of the fill length distribution becomes of interest. It has already been established that for a fill of length t_{fill}:

- If $t_{fill} \leq T_{lev}$ the delivered luminosity is simply $t_{fill} \times L_{level}$.
- If $t_{fill} > T_{lev}$ the delivered luminosity is given by:

$$L_{int} = T_{lev} \times L_{level} + L_{lev}\tau_L(1 - e^{-\frac{t_{decay}}{\tau_L}}). \tag{10}$$

The fill length distribution is a non-trivial consideration. For example, the 2012 fill distribution showed a lot of short, prematurely dumped fills compensated by a number of long fills to give an average of around 6.0 hours. The cost of the short fills is a corresponding number of extra turnarounds which fold directly into lost time for physics. The longer fills naturally include less productive latter stages.

A non-leveled adjustment factor (NLAF) may be introduced. This factor takes into account luminosity production in the exponential decay regime after leveling has been exhausted. The NLAF will depend on a number of factors including the leveling time, the fill length distribution, and the luminosity lifetime during the exponential phase.

The NLAF for a single fill may expressed in terms of these factors (see Eq. (11)):

$$\text{NLAF} = \frac{T_{\text{lev}} + \tau_L \left(1 - e^{\frac{(t_{\text{fill}} - T_{\text{lev}})}{\tau_L}}\right)}{t_{\text{fill}}}. \tag{11}$$

The aggregate NLAF will depend on the actual fill distribution (as well as the actual distribution of luminosity lifetime and leveling time). An estimate can be extracted from past experience.

A leveled Hübner factor (LHF) may be defined:

$$\text{LHF} = A \times \left(1 - \frac{T_{\text{around}}}{T_{\text{fill}} + T_{\text{around}}}\right) \times \text{NLAF}. \tag{12}$$

The LHF multiplied by product of the scheduled physics time and the leveled luminosity will give an estimate of integrated luminosity for an extended period.

The LHF defined here illustrates the factors that will contribute to the overall HL-LHC operational performance, namely: availability, turnaround time, mean fill length and, via the NLAF, the fill length distribution.

3.6. *Integrated luminosity estimates*

An estimate of the potential integrated luminosity per year may be made using: the above formulation, the 2012 data outlined in Table 5 and estimates of a realistic leveling time and luminosity lifetime.

- 2012 saw an average fill length of 6.0 hours and an average turnaround time of 5.5 hours.
- The HL-LHC virtual luminosity lifetime during leveling is estimated from the luminosity burn-off, an estimate of single beam losses to other processes, and an estimate of emittance blow-up. It is difficult at this stage to claim accurate predictions for lifetime estimates in the very high luminosity regime. Luminosity burn-off is given, reasonable assumptions on losses due to other processes and emittance growth give a leveled time of around 6 hours, to be compared with luminosity only leveled time of 10.6 hours calculated above.

Table 5.   Overall operational performance 2012.

Scheduled physics time	201 days
Availability	71%
Physics efficiency	$\approx 37\%$
Average fill length	6.0 hours
Average turnaround	5.5 hours
Mean luminosity delivery rate	12.97 pb^{-1}/hour
Peak luminosity delivery rate	≈ 25 pb^{-1}/hour

- Taking leveled time of 6 hours, a luminosity lifetime of 5 hours in the exponential decay phase, mapped on to 2012 fill length distribution one gets a NLAF of 0.88.

Given this, and bearing in mind the provisos outlined above, a leveled Hübner factor of around 0.33 is obtained. With a leveled luminosity $= 5 \times 10^{34}$ cm^{-2}s^{-1} and a scheduled physics time per year of 160 days this gives approximately 225 fb^{-1}. A leveled Hübner factor of 0.36 is required to integrate 250 fb^{-1} per year.

If one directly maps the 2012 fill time distribution on to a fill scenario with the same numbers for T_{lev} and luminosity lifetime, assumes a 13-hour fill length cut-off, and naively scales to 160 days (this implies the same availability and average turnaround time), one obtains a total integrated luminosity of around 221 fb^{-1} for a standard HL-LHC year [4].

A Monte Carlo approach [5] which also extends the 2012 figures to the full HL-LHC (and assumes the average turnaround time is increased from 5.5 to 6.2 hours) gets a figure of 213 fb^{-1}. The team also simulates the impact on the integrated luminosity of SEUs, UFOs, quenches and gives a range of 180 to 220 fb^{-1} in simulations that attempt to take these factors into account.

The details of the calculations are unimportant but what is clear is that given 2012's availability and turnaround, and reliable operation with leveling, approximately 85% of the HL-LHC annual target could be achieved. This is encouraging but clearly the already good availability must be maintained and improved if the ambitious goals of the HL-LHC are to be reached. Operational experience at or near design energy will provide valuable further input.

4. Scrubbing

Electron cloud build-up resulting from beam induced multipacting (see Chapter 4, Section 1.2.4) is one of the major limitations for the operation of the LHC with beams with close bunch spacing and in particular with 25 ns spacing. Electron

clouds induce unwanted pressure rise, heat loads on the beam screens of the superconducting magnets and beam instabilities leading to emittance blow-up and to
a reduction of the luminosity. Multipacting is occurring when the values of the
so-called Secondary Electron Yield (SEY) of the beam screen surface are larger
than a given threshold value SEY_{th} depending on bunch spacing and to a lesser extent on bunch population (above the nominal one). Typical values of the SEY for
uncoated surfaces exposed to air as those of the beam screens after venting to air
can exceed 2.3–2.5. Operation with 25 ns beams below the multipacting threshold
at 7 TeV implies reducing the SEY below 1.4 in the dipoles and below 1.2 in the
main quadrupoles. These estimations are based on the present knowledge of the
SEY dependence on electron energy and in particular of the probability of elastic
reflection at low energy obtained from the benchmarking with experimental data
in the SPS and LHC. Operation with 50 ns beams is more tolerant with threshold
SEY of 2.1 in the dipoles and 1.6 in the main quadrupoles.

Laboratory measurements have shown that a progressive reduction of the SEY
can be achieved by electron bombardment with an approximate dependence of the
secondary electron yield on the logarithm of the integrated dose of electrons [7, 8].
In practice this can be achieved in an accelerator environment by operating the
machine in multipacting regime (scrubbing). In the LHC this has been done successfully for allowing operation with 50 ns beams up to 4 TeV in 2011–2012.
Operation at 450 GeV with 50 and 25 ns beams in multipacting regime has been
demonstrated and in 2012 the machine could be operated with 25 ns beams, gradually increasing the number of bunches up to 2748 within the limits imposed by the
maximum heat load on the beam screens acceptable by the cryogenic system and
by the vacuum levels reached in some critical high voltage components like the
injection kickers. The studies conducted at the end of 2012 [9] in the LHC have
shown a significant decrease of the rate of reduction of the SEY as a function of
the dose for SEY below 1.45 in the main dipoles and quadrupoles, however so far
the operation in an electron cloud free environment with 25 ns beams has not been
demonstrated for a large number of bunches.

It is expected that a scrubbing run is needed after every long shut-down when
significant fractions of the machine will be warmed up and possibly vented, while
reduced scrubbing runs will be required for the operation with 25 ns beams after
short stops where the machine sectors will not be vented and kept at low temperature. Even in this case, signs of de-conditioning have been observed. As an
example, during the Christmas stop 2011–2012 the SEY has increased from 1.55
to 1.65 in the main dipoles and a similar increase was again found at the beginning of the scrubbing run in December 2012 [9]. From the present experience,
the benchmarking with the simulations and the existing models/measurements for
the evolution of the SEY as a function of the electron dose [10] the following sce-

narios can be envisaged. After a long shut-down and after the machine set-up at low intensity, a dedicated scrubbing run with increasingly longer trains of bunches spaced by 50 ns and (later) by 25 ns will be required for vacuum conditioning and for lowering the SEY in the arcs to a value below the threshold for electron cloud build up for 50 ns beams. It is expected that this mode of operation would require approximately 7–9 days of scheduled time, in which the second half of the period would see the injection and accumulation of trains of bunches spaced by 25 ns at 450 GeV. This period includes the setting-up of the injection of bunch trains and the progressive intensity ramp-up which will be driven by vacuum (in particular in critical areas like at the injection kickers) and maximum heat load and heat load transient on the beam screens. After this period, physics at 7 TeV with trains of bunches spaced by 50 ns will be possible allowing for a gradual increase of the number of bunches brought in physics according to machine protection considerations. Once this phase is completed, operation with 25 ns beams can be envisaged only after an additional scrubbing period at 450 GeV with trains of bunches spaced by 25 ns up to the maximum number of 2780 bunches. This period is expected to take approximately 5 days. After that it will be possible to accelerate trains of a few hundred bunches to 7 TeV with reduced electron cloud effects. According to the 2012 experience, the scrubbing process described above will however not be sufficient to suppress multipacting with 25 ns beams and a period with degraded beam characteristics has to be expected during the intensity ramp-up in physics. The length of this period is presently being estimated through a careful analysis of the data collected in 2011–2012 [11]. Scrubbing runs will be required also to reduce the occurrence of beam dumps due to UFOs.

5. Machine Development

A robust programme of machine studies will be required in preparation and in the initial phase of the operation of the HL-LHC. A non-exhaustive list of studies is presented below:

- Validation of the scrubbing run scenarios for 25 ns operation by extending the measurement of the evolution of the SEY in the warm and cold section of the LHC as a function of the electron dose.
- Investigation of scenarios to enhance the electron dose rate for more efficient machine conditioning.
- Validation of the dynamic aperture models and triplet error compensation for the present LHC machine and later for the HL-LHC layout.
- Validation of the leveling schemes to control the pile-up density at the experiments: β^* leveling, bunch flattening, etc.

- Operation of the machine (including leveling and luminosity scans) in the presence of large crossing angles, beam–beam tune spreads larger than 0.03[a] and noise sources (power converters, transverse damper, crab cavities, etc.) and benchmarking with the weak-strong beam–beam simulation codes (including noise effects).
- Study of the long range effects in HL-LHC regime (e.g. pacman orbits).
- Halo measurement and control in collision in view of the operation with crab cavities.
- Operation of the crab cavities and their impact on transverse emittance blow-up with separated beams and in collision.

Approximately 20–22 days per year are expected to be devoted to the machine studies at least during the operation preceding the HL-LHC upgrade and during the first years of operation after the upgrade.

References

[1] S. Fartoukh, Pile up management at the High Luminosity LHC and introduction to the crab-kissing concept, CERN-ACC-2014-0076 (2014).

[2] T. Bär, Very Fast Losses of the Circulating LHC Beam, their Mitigation and Machine Protection, CERN-THESIS-2013-233.

[3] E. Todesco, N. Aquilina, M. Giovannozzi, M. Lamont, F. Schmidt, R. Steinhagen, M. Strzelczyk and R. Tomas, The Magnetic Model of the LHC during the 3.5 TeV run, in *Proc. of the 2012 International Particle Accelerator Conference* 2194-6 (2012), also in *CERN ATS 2012-195 (2012)*.

[4] M. Lamont, How to reach the required availability in the HL-LHC era, in *Review of the LHC and Injectors Upgrade Plans*, https://indico.cern.ch/event/260492/.

[5] A. Apollonio, M. Jonker, R. Schmidt, B. Todd, S. Wagner, D. Wollmann and M. Zerlauth, HL-LHC: Integrated Luminosity and Availability, in *Proc. of the 2013 International Particle Accelerator Conference* 2968 (0213).

[6] J. Wenninger, Simple models for the integrated luminosity, in *CERN-ATS-Note-2013-033 PERF*.

[7] N. Hilleret *et al.*, Electron Cloud and Beam Scrubbing in the LHC, LHC Project Report 290.

[8] V. Baglin *et al.*, Measurements at EPA of vacuum and electron-cloud related effects, in *Proceedings of the LHC Workshop – Chamonix XI*, ed. J. Poole, CERN SL/2001-003(DI), pp. 141–143.

[9] G. Iadarola, G. Arduini, H. Bartosik and G. Rumolo, Electron Cloud and Scrubbing in 2012 in the LHC, in *Proceedings of the LHC Beam Operation Workshop – Evian 2012*, CERN-ATS-2013-045, pp. 119–127.

---

[a]As those expected for the HL-LHC when operating with collisions at three Interaction Points and with crab cavities compensating completely the crossing angle at the high luminosity experiments.

[10] References in *Proceedings of ECLOUD'12*, eds. R. Cimino, G. Rumolo and F. Zimmermann, CERN-2013-002, INFN-12-26/LNF, EuCARD-CON-2013-001.

[11] G. Rumolo, G. Arduini, H. Bartosik and G. Iadarola, Electron cloud scrubbing run and strategy for 2015, presented at the 7th meeting of the CERN Machine Advisory Committee, 14–15 March 2013, CERN, http://indico.cern.ch/event/235072/contribution/1/material/slides/0.pptx

Index

www.ingramcontent.com/pod-product-compliance
Lightning Source LLC
Chambersburg PA
CBHW081041220326
41598CB00038B/6951